Urban Sustainability

Editor-in-Chief

Ali Cheshmehzangi, Architecture & Built Environment, University of Nottingham Ningbo China, Ningbo, Zhejiang, China

The Urban Sustainability Book Series is a valuable resource for sustainability and urban-related education and research. It offers an inter-disciplinary platform covering all four areas of practice, policy, education, research, and their nexus. The publications in this series are related to critical areas of sustainability, urban studies, planning, and urban geography.

This book series aims to put together cutting-edge research findings linked to the overarching field of urban sustainability. The scope and nature of the topic are broad and interdisciplinary and bring together various associated disciplines from sustainable development, environmental sciences, urbanism, etc. With many advanced research findings in the field, there is a need to put together various discussions and contributions on specific sustainability fields, covering a good range of topics on sustainable development, sustainable urbanism, and urban sustainability. Despite the broad range of issues, we note the importance of practical and policy-oriented directions, extending the literature and directions and pathways towards achieving urban sustainability.

The series will appeal to urbanists, geographers, planners, engineers, architects, governmental authorities, policymakers, researchers of all levels, and to all of those interested in a wide-ranging overview of urban sustainability and its associated fields. The series includes monographs and edited volumes, covering a range of topics under the urban sustainability topic, which can also be used for teaching materials.

Xianchun Zhang

Regional Integration and Governance Reshuffling in the Making of China's City-Regions

Springer

Xianchun Zhang
School of Public Affairs
Zhejiang University
Hangzhou, Zhejiang, China

The Hong Kong Institute of Asia-Pacific
Studies
The Chinese University of Hong Kong
Hong Kong, China

ISSN 2731-6483 ISSN 2731-6491 (electronic)
Urban Sustainability
ISBN 978-981-99-2791-3 ISBN 978-981-99-2792-0 (eBook)
https://doi.org/10.1007/978-981-99-2792-0

This Springer imprint is published by the registered company Springer Nature Singapore Pte Ltd.
The registered company address is: 152 Beach Road, #21-01/04 Gateway East, Singapore 189721, Singapore

"With an incredibly in-depth study of industrial, infrastructure and regional development in the Pearl River Delta, this excellent book provides insightful observations on state-centred city-region building. Beyond rescaling, Xianchun Zhang reveals how a wide range of state and non-state actors have collaborated to achieve the strategic objective of coping with the crises and challenges in China's development. As a vivid illustration of state entrepreneurialism, the inter-city collaboration of industrial parks is driven by both entrepreneurial endeavours and multi-scalar geopolitics. This remarkably detailed book is an essential reference on Chinese regional development and governance."

—Fulong Wu, *Bartlett Professor of Planning, University College London*

"This is a significant book on recent economic and institutional integrations and the underlying governance reshuffling process in China's city-regions. The Shenzhen-Dongguan-Huizhou sub-region in southern China is chosen for detailed case study in terms of regional governance and integrations. The book is an important reference for scholars and professionals in the fields of urban and regional planning, management and human geography."

—Jianfa Shen, *Department of Geography and Resource Management, The Chinese University of Hong Kong*

"This book contributes significantly to the urban study literature by offering a meticulous and comprehensive analysis of China's city-regionalism strategy. The book's rigorous analytical framework and case studies provide valuable insights for scholars, policymakers, and practitioners interested in understanding the dynamics of regional governance, economic integration, and institutional restructuring in China and beyond."

—Zhigang Li, *School of Urban Design, Wuhan University*

Acknowledgements

This book would not have been completed without the valuable assistance and support of many people. First, I would like to extend my sincere gratitude to my supervisor, Prof. Rebecca L. H. Chiu, for her constructive advice, inspiring comments, and mentorship during the 4 years of my Ph.D. study in the University of Hong Kong. Being selected as one of her students is an honor. Without her consistent and enlightening instructions, this book will not have realized its present form. Apart from her insightful comments and guidance for my book, she has also offered numerous suggestions to improve the research methodology and writing skills and several opportunities to expand my personal boundaries in terms of academic research. I would also like to thank the Department of Urban Planning and Design and the University of Hong Kong for providing me with financial support, an academic research and working environment, and valuable resources for my Ph.D. study. I also would like to express my heartfelt gratitude to the Philip K. H. Wong Foundation for its generous support for my research studies in HKU.

I would like to express my deepest gratitude to Dr. Xiaofan Luan of the School of Urban Design in Wuhan University, for helping me conduct field trips and data collection for my research. I also want to show my appreciation to Dr. Heng Chao, Mr. Jiacheng Li, Ms. Jiarong Nie, and Mrs. Hua Gong for their assistance in collecting the data and organizing interviews. Without them, I would not have completed this research. I am also extremely grateful for the support offered by Mr. Haoyu Chen, Mr. Yuchao Chen, Ms. Wenyu Chen, and Ms. Zijing Li in data collection. Special thanks to Dr. Zhuoran Shan, Dr. Jingxin Nie, Mr. Chengjie Lu, and Ms. Guohua Hu from the School of Architecture and Urban Planning in Huazhong University of Science and Technology for generously supporting the field trips that I have to take during the course of my research. I am also grateful to the interviewees from planning, transportation, and housing departments of local governments, Management Committee of Shenzhen-Shanwei Special Cooperation Zone, and Guangdong Provincial Railway Construction Investment Group. In addition, I am extending my gratitude to Dr. Li Tu from the College of Architecture and Art in Hefei University of Technology, for her insightful suggestions regarding data collection.

Furthermore, I am indebted to Dr. Roger C. K. Chan, Prof. Jianfa Shen, Prof. Shenjing He, Dr. Mandy Lau, and Dr. Weifeng Li for their generous support and for providing inspiring comments regarding my research. I am also thoroughly grateful to Dr. Yi Sun from the Department of Building and Real Estate in Hong Kong Polytechnic University, for his perceptive recommendations and constant encouragement. In addition, I would like to express my gratitude to all my colleagues and friends in the University of Hong Kong, who have provided me with help, care, advice, and encouragement when times get tough during the 4 years of my Ph.D. study. Thank you, Xi Wang, Chang Liu, Zifeng Chen, Xiaoyan Mu, Chen Shi, Shouqiang Chen, Xu Wang, Dunxu Wu, Peng Lin, Kun Wang, Bo Wang, Afsana Haque, Linchuan Yang, Yiguan Ma, Jie Li, Changchang Zhou, Yishiqing Li, Mengdi Guo, Rong Cai, Tianzuo Wen, Jixiang Liu, Fang Bian, and Yiling Luo, for all the precious times we spent together. I am especially grateful to Dr. Jing Song, Mrs. Yueli Xu, and Dr. Tianyao Zhang, for their continuous encouragement and recommendations.

Last but not least, I am beholden to my parents, for always understanding me and giving me unconditional love. They are always there to accompany me, take care of me, and encourage me when I feel frustrated. Their unconditional support and infinite patience are powerful motivational forces that drive me to move ahead. Finally, I would like to dedicate this book to my family members, teachers, and friends.

About This Book

Under China's reorientation toward city-regionalism as the main economic-spatial strategy, some regions that are affected by global economic restructuring and ridden with crises of economic development deadlock are officially designated as city-regions governed by integration-oriented institutions. This government-induced city-regionalism receives critical attention in the political economy literature. Nonetheless, the theoretical framework that documents economical and institutional integrations is still ambiguous, which calls into question the political construction process of regional governance in enabling effective economic integration. Therefore, it is imperative to thoroughly investigate the extent of economic and institutional integrations and the underlying governance reshuffling process. In this study, the Shenzhen-Dongguan-Huizhou sub-region (SDH) in southern China is anchored since it has been acting as a testbed for the nation's "one-step-ahead" policy for a long time.

Specifically, this study systematically assesses the extent of the economic integration of SDH by critically gauging and investigating market, industrial, and infrastructure integrations. It has been found that industrial integration and infrastructure equality have became gradually weakened, constituting to the underlying obstacles of city-regionalism. To counteract this, the market gradually became more integrated, and infrastructure connectivity rather than equality was emphasized. Shenzhen, as a regional core city, was well integrated with its adjacent cities: Dongguan and Huizhou, in both the market and infrastructure dimensions, but industrial isomorphism was high, signaling keen competition among the three cities. Regarding integration between the regional core city and outlying cities (i.e., Shanwei and Heyuan), market and industrial integration was high, while the infrastructure connectivity and equality were weak.

This study formulated a multi-level governance-based analytical framework to assess inter-scalar, inter-city, and state-market-civil society relations in order to understand the institutional integration of SDH. It unveiled that the institutional integration of SDH remained a government-led integration process. Also, an inter-scalar relation examination formed that intensifying central administrative control was entangled with flourishing bottom-up institutional innovations. The lack of regional authority led to the provincial government's dominance in constructing the sub-regional fabric. Regarding inter-city relation, it was a voluntary process of economy-first and growth-oriented city grouping, stabilized by formal institutions and centered on regional core city. In addition, institutional integration remained a government-led integrating process, and the market and social actors with limited influence stayed at the edge.

Furthermore, this study investigated the reshuffling process of the governance structure in response to industrial and infrastructure integrations and the interconnected economic and institutional integrations. It was found that the government-led integration process reshuffled the governance structure toward a collaborative mode, particularly facilitating the cross-boundary industrial collaboration. This process was characterized by the political mobilization and deregulation initiated by provincial government, and the highly integrated inter-city cooperation based on complementation of aims and relative advantages across different cities. To propel regional infrastructure project, the government-led integration process has restructured inter-scalar interaction which is ridden with collaboration, but currently resistance and compromise. These two divergent trajectories of regional governance restructuring show that China's city-regionalism is essentially a state-orchestrated and institution-based process in which the state's role varies case by case, rather than being a mechanical and two-tracked "upscaling and downscaling" process.

Contents

Abbreviations

CCGP	Construction Committee of Guangdong Province
CDGP	Construction Department of Guangdong Province
EU	European Union
FDI	Foreign direct investment
MLG	Multi-level governance
NDRC	National Development and Reform Commission
PRD	Pearl River Delta
SC	State Council
SDH	Shenzhen-Dongguan-Huizhou sub-region
SGSIR	Sui-Guan-Shen Inter-city Railway
SSSCZ	Shenzhen-Shanwei Special Cooperation Zone
TCSM	Transport Commission of Shenzhen Municipality
TDDM	Transport Department of Dongguan Municipality
TDHM	Transport Department of Huizhou Municipality
YRD	Yangtze River Delta

List of Figures

List of Tables

Chapter 1
Introduction: Background and Research Significance

1.1 Introduction

In the latter half of the twentieth century, the shift of production mode from Fordism to post-Fordism in the advanced capitalist countries transformed energy dependence and mass production to dependence on information and technology, which triggered the finding that the space of place is superseded by the space of flow (Castells, 1989, 1996). A high-speed economic flow facilitated the rise of a global cities, such as London, Hong Kong, and New York, whose economy was decoupled from the national economy and further transcended across borders of nation-state to create a global city network (Brenner, 1999; Sassen, 1991, 2006). Multinational corporations accommodated in a global city gave rise to the redistribution of industrial chains from the urban core to a more extensive geographic site. The restructuring of regional industrial divisions, which was beneficial to growth-oriented governments and corporations, reconstituted linkages among urban cores, semi-urban areas, and outskirts to foster city-regions, such as the Greater Dublin, the Central Belgium, and the Greater South East. The hollowing-out of nation-states in advanced capitalist countries triggered nation-states to rescale themselves through transferring power upwards to super-national organizations and downwards to sub-national organizations. This transfer of power was a response to the restructuring of the global economy and global–local relations, manifesting as an emerging governance structure—multi-level governance (MLG) (Brenner, 1999).

Regional development in socialist China after the market reform in 1978 was concurrently remolded by globalization, decentralization, and marketization processes (Wei, 2001; Wu, 2002). City-regionalism was the prevailing sociopolitical strategy to proactively propel regional development. This strategy has been documented and ensured by substantial national agendas and spatial plans in the past three decades. The spatial reorientation of China from city to city-region presupposes that the city-region increasingly served as the new geographical site to accelerate capital accumulation and achieve socioeconomic prosperity. Institutional designs that aimed to foster effective cross-boundary partnerships were formulated from

X. Zhang, *Regional Integration and Governance Reshuffling in the Making of China's City-Regions*, Urban Sustainability, https://doi.org/10.1007/978-981-99-2792-0_1

central to local government to ensure that the city-region could operate as a regional economic unit; these integration-oriented designs were driven by a series of political and economic incentives, such as the enhancement of national and urban strategic status and competitiveness, political promotion for local officials, and demand for urban industrial development (Li et al., 2014). However, strong intervention by the central government and an entrepreneurial local government constantly triggered the emergence of weakly or partially integrated regions; in these regions, areas that faced severe dilemmas of cooperation and underlying crises are designated as city-regions in national and local agendas (Luo & Shen, 2008; Shen, 2018; Wu, 2015). These high-profile government-planned city-regions seriously call into question the extent of economic and institutional integration and the internal reshuffling process of regional governance in response to the rapidly changing socioeconomic environment at the local scale (Li & Wu, 2013; Luo & Shen, 2008, 2009).

Pearl River Delta (PRD) is a highly populated area in China, comprising nine cities administratively. PRD has strategically undertaken city-regionalism for more than two decades and has been taken out of recent national efforts on regional development. The segmental development of a city-region in PRD occurred in three sub-regions, namely, the Shenzhen–Dongguan–Huizhou sub-region (SDH), the Guangzhou–Foshan–Zhaoqing sub-region, and the Zhuhai–Zhongshan–Jiangmen sub-region. These sub-regions were reinforced by a series of integration-oriented institutional arrangements that originated mainly from different levels of government. This strategy was an ingenious solution to territorial fragmentation in PRD, but vicious inter-city competition and uncoordinated regional development inevitably occurred within and beyond the three sub-regions. These sub-regions confronted fairly disparate local politico–economic environments, but the regional integration of SDH was the most complicated and haphazard. The rise of Shenzhen's economy elicited an overflow of socioeconomic activities, such as the decision of Huawei to relocate its supply chain from Shenzhen to Dongguan. Another example is the escalating crisis of land shortage that propelled Shenzhen to prioritize SDH as the overreaching geographical scaffolding for breaking the development deadlock. However, Dongguan and Huizhou encountered an insurmountable hurdle in promoting an industrial upgrade toward high-end sectors and boycotting the low-end industry coming from Shenzhen. Existing integration-oriented institutions attempted to enlarge SDH spatially by incorporating the outlying cities (i.e., Shanwei and Heyuan) to address economic-spatial inequality. Thus, SDH was the region of greatest concern and the institutional testbed for regional development, but its actual situation calls into question the effectiveness of the segmental integration strategy of PRD, i.e. by way of a three sub-region structure with reference to city-regionalism.

This chapter describes the overall research context in Western countries and marketized socialist China and sets out the theme and significance of this book. This chapter consists of eight sections. Section 1.2 portrays the international background of the rescaling of city and state in response to the restructuring of the global economy, setting out the politic-economic nature of the city-region, and the rise of the new governance mode. Section 1.3 probes the regional development process and the role of government to shed light on city-regionalism in marketized socialist

China. Section 1.4 introduces the contradiction within SDH for industrial development driven by divergent local political and economic contexts and the deadlock in economic development. Section 1.5 discusses the research aims and research questions of this book. Section 1.6 presents the research significance and scope and the structure of the book followed by a section of conclusion.

1.2 International Background: Globalization and Rescaling of City and State

Technological progress in the early twentieth century facilitated the second revolution of science and technology. The Fordist mode, which was the predominant production mode, was adopted in Western countries with a particular emphasis on customization and mass production (Brenner, 1999). The urban economy served as a significant engine for the growth of the national economy. Hence, Fordist cities, such as Detroit, Birmingham, and Pittsburgh, increasingly played a guiding role in advancing capitalist countries economically, politically, and socially. These cities were characterized by efficiency, the use of a production line, and large inputs of resources and labor, which heavily depended on mechanization and the resource endowment of localities.

The global economic crisis at the beginning of the 1970s transformed the economic mode from Fordism to post-Fordism, and thereby Fordist mass-production became outdated. High technology, knowledge, capital, and information superseded the basic and traditional factors of production (i.e., the input of labor, resources, and the production line). High-speed capital and information flow gave rise to the disassembly of bundles between the national economy and the urban economy. Some emerging cities accordingly served as new economic powerhouses in the global urban system and were redefined as global cities or post-Fordist cities; examples of these cities were London, Paris, New York, Tokyo, and Hong Kong (Ancien, 2011). This detached process was renamed as "capital deterritorialization" (Appadurai, 1996); the increasingly tight connection among cities across the borders of nation-states involved the economy plus political and social systems and has restructured global–local relations. Additionally, the globalizing process was characterized by high-speed flows of capital and information and diminishing nation-states; the powerful effect of agglomeration and the diffusion of economic activities propelled the integration of global cities between vast suburban and rural areas, thereby forming global city-regions (Kantor & Nelles, 2015). Nonetheless, deterritorialization was entangled with the re-territorialization process, which is marked by the "geographically fixed configuration of the territorial organization" (Brenner, 1998, p. 462), subsequently restructuring the global urban system and spatial form of a city-region under economic globalization.

Nation-states up-scaled themselves in the wake of socioeconomic post-Fordism to transfer some powers upward to supra-national organizations, such as the North

American Free Trade Agreement, the Association of Southeast Asian Nations, and the European Union (EU). This institutional reconfiguration began to gain increasing attention. At the supra-national scale, transnational organizations formulated their own regulations, laws, policies, and governance structure to tackle cross-border issues. For example, member states under the organizational framework of the EU needed to adopt the principle of subsidiarity and territorial cohesion to insert a quasi-democratic set of policies (Pemberton & Morphet, 2014). At the sub-national scale, several regional core cities strengthened the linkage with their suburban areas and outskirts, which facilitated the formation of city-regions, and were empowered with more authorities and obligations from the central state. The restructuring of the state space promoted the rise of these city-regions at the EU scale and that of MLG; this process was regarded as the strategy of neo-liberalism in the realm of industrial restructuring and crisis-management (Brenner, 1999). The deterritorialization process and emergence of flow space eroded the relationship between national and urban economies (Taylor, 1995). Spatial plans, regional regulations, and investments reinforced the status of these city-regions in the national and global city systems. However, the downscaling of the state led to a trade-off between the central state and the city-regional authority (Brenner, 1999).

Besides vertical state rescaling (i.e., up-scaling and down-scaling), horizontal statization and destatization in state and non-state domains emerged, forming more intricate state–market–civil society relations (Li et al., 2014). Destatization refers to the power transfer from state domains to different types of non-state actors and their partnerships to reshuffle state-centric governance (Xu et al., 2009). Statization, started in America from the 1900s to the 1930s, was featured as the centralization process through the consolidated approach to restructuring administrative space. For example, the area of New York has increased from 44 to 299 km^2 since 1988 by consolidating surrounding municipalities (Yi, 2010). Los Angeles acquired a large territory by combining eight cities (Yi, 2010). After the consolidation process, the new metropolitan government repaired territorial fragmentation and enhanced administrative efficiency. However, statization was more apparent and exhibited nuanced manifestation in European countries, such as France, Britain, and Germany. The new metropolitan governments in Europe rose in the 1960s. After World War II, the government intervened in the national economy and implemented pump-priming policies, such as raising taxes, spending cuts, and investment in major infrastructure projects, to inhibit economic recession and address market failure (Ferris et al., 2012). This trend, which was renamed "Keynesianism", further strengthened the power of the territorial state in terms of state and non-state relations. However, the deepening economic crisis after the 1970s generated the bankruptcy of the welfare state policy and neo-liberalism to get rid of Keynesianism. Neo-liberalism stressed the importance of a free market. The government must adjust to the economy and create conditions for the market to engage in economic growth in accordance with an entrepreneurial urban regime. Under the context of neo-liberalism, city-regionalism became the prevailing ideological approach that provided an integration-first institutional framework in terms of power configuration among different stakeholders to foster partnerships at variegated geographical scales. Collaborative governance as a mainstream governance paradigm under city-regionalism asserted that different stakeholders should occupy various types of resources; thus, actors who pursued their goals should obtain optimized choices by fostering partnerships (Ansell & Gash, 2008). In the territorial praxis, various formal and informal institutional organizations, such as the Smart Growth Network, Urban Community, Regional Councils, and Voluntary Inter-communal Syndicates were established in America and Europe to tackle urban and regional public issues.

In summary, the transition from Keynesianism-Fordism to Neo-liberalism post-Fordism transformed the global economy from the mass production mode to information-, capital-, and knowledge-led production. Traditional industrial cities that heavily depended on local resources, the input of labor, and highly efficient production were powerless during the transformation of the production mode; this transformation gave way to global financial cities, which were referred to as agglomerated cities with headquarters economies. Subsequently, the geographical arena for capital accumulation extended to a wider arena and the politics of distribution extensively occurred at the city-regional scale; this strategy was adopted to overcome the crisis of city-based accumulation regimes, such as vicious inter-city competition and inefficient provision of public service (Wu & Zhang, 2010). The underlying formation of global city-regions was still driven by state-centric governance, such as those in the Manaus Metropolitan Region and Yangtze River Delta (YRD) (Zhang et al., 2008); Neo-liberalism was widespread in the formation process of city-regions (Klink & Denaldi, 2012). The restructuring of economic production mode under globalization enabled city-regions and well-rounded regional entities to operate and profoundly altered territorial governance and the role of nation-states (Newman, 2000). The rescaling of the state facilitated deep changes to city-regional governance, one of which was the power reconfiguration across hierarchical governments and state–market–civil society relations. On the one hand, the city integrated into the world urban hierarchy and the territorial state was regulated by supra-national organizations. On the other hand, tremendous restructuring was observed in urban governance, which was manifested as the political decentralization of administrative power from a central state downward to region-based authorities (Brenner, 1999, 2004). In addition, statization and de-statization processes remolded sociopolitical forces across state and non-state sectors. The restructuring of the global economy reconstituted and intensified the connections among core cities, including their peripheral areas and outskirts. These core cities further promoted their integration as a new locus for socioeconomic activities. In the wake of the functional integration process of city-regions, the nation-state rescaled itself to build more effective institutional arrangements by reshaping the structure of city-regional governance in response to the post-Fordist economic order at the city-regional scale as a result of the second technological revolution in the early twenty-first century.

1.3 China's Regional Development: Market Economy and Governmental Intervention

After the economic reform and opening up in 1978, the urban economy in socialist China was transformed from a planned economy to a market-oriented one (Wu, 2002). This socioeconomic transformation suggests that regional development was tightly connected with the macro-economic landscape of China (Xu & Yeh, 2005). Deng Xiaoping introduced the market economic mode into China and carried out a host of opening strategies that facilitated regional development to address the crises exhibited by the planned economic mode before 1978. These crises included poor efficiency in resource allocation, low income of citizens, and structural stiffening of the market and government. The regional development led to the formation of a policy-driven opening-up pattern in the late 1980s. The central government designated a series of special economic zones in 1980 (for Shenzhen, Zhuhai, and Xiamen), 1981 (for Shantou), and 1988 (for Hainan) respectively. These areas were granted substantial concessionary policies, such as reductions of enterprise taxes, preferential land-use terms, and simplification of administrative procedures, to enhance territorial attractiveness for foreign investments and multinational corporations. In May 1984, the State Council (SC) proposed the opening of coastal port cities, such as Dalian, Qinhuangdao, Tianjin, Yantai, Qingdao, and Lianyungang. Some urban economic patterns exhibited unanticipated vitalities, such as those of Wenzhou, Suzhou, Tianjin, and Quanzhou. Therefore, an urban economy prevailed throughout the country. However, the prosperity of these different urban economies implied various pathways to boost economic growth, which was partly characterized by strong governmental intervention driven by the mechanism of market choice (Zeng et al., 2015). Integrated socioeconomic connections between cities then appeared in a number of coastal areas. These densely populated regions with impressive economic competitiveness acted as the engine of the national economy, such as the Circum–Bohai–Sea Economic Zone, YRD, and PRD.

From 1990 to 1998, the central government boosted the socioeconomic status of three city-regions through promulgating a series of national concessionary strategies and policies. These measures contained infrastructure investment projects and tax incentives to enhance economic growth in these regions. In the early twenty-first century, PRD, YRD, and Beijing–Tianjin–Hebei city-region were the major contributors to the national economy. In 2007, statistics show that the GDP and population of the three city-regions accounted for 49.9% of the national GDP and 25.5% of the national population, respectively.[1] The integration of city-regions was the precondition for the promulgation of the national policy and strategy at this stage. However, almost all of these concessionary policies and strategies led to a fundamental transformation in socio-economic development. These city-regions in the urban system moved from being contributors to the national economy toward the important nodes

[1] People's Daily Overseas Edition. (2008, November 27). PRD, YRD, and Circum-Bohai-Sea economic zones lead the national economic growth. Retrieved from http://202.116.197.15/cadal/Fulltext/4780_20131021_115114_19.pdf.

in the global city network. From 1999 to 2005, the four-plate strategies, namely, the West Development Strategy, Revitalizing Northeast China, Facilitating the Rise of Central China, and Encouraging the Eastern Region to Lead the Development of the Whole Country, were proposed by the central government continuously. These strategies were bundled with concessionary policies in infrastructure development, such as railway construction, the West–East Electricity Transmission Project, and the West–East Gas Pipeline Project. Moreover, the central government invested in ecological environment and education to foster corresponding regions, thereby enabling them to operate as new loci and competitive interconnected areas in accommodating socioeconomic activities.

Several emerging cross-boundary issues required a high degree of inter-jurisdictional partnership at the city-region scale in China, which ran against the socioeconomic trend of regionalization (Xu & Yeh, 2013). Local governments fostered cross-boundary partnerships and created new city-regions to strive for national concessionary policies. Typical examples include the Wuhan and Changsha–Zhuzhou–Xiangtan urban agglomerations, which are referred to as the Two-Type Society Trial Area (*Liangxing shehui shifanqu*, 两型社会示范区) in the national agendas. Given the driving effect of PRD, YRD, and the Beijing–Tianjin–Hebei city-region because of their overt success, and the demands of territorial states to foster economic growth, city-regional development was reoriented by different levels of government as a significant geographical scaffolding for capital circulation, encouraged by central and local governments. The spatial plans of city-regions were formulated by different levels of government. Parts of these city-regions were designated across provincial-level administrative units, such as urban agglomeration in the Middle Reaches of the Yangtze River and Ji-Lu-Yu. In doing so, China has created 32 city-regions, which are grouped into three categories: developed, developing, and potential urban agglomeration.[2] Unlike city-regionalism in the Western context, the experience of socialist China revealed a distinctive trend of economic and institutional integration, a political narrative among different levels of government, and a re-articulation of regional development. For political consideration and local demand for development, several city-regions without tight socioeconomic linkage were articulated in government policies. Some urban clusters were not fostered with interconnected economy and cooperative willingness, but these areas could resurface and be reinforced by political factors or national concessionary policies, such as urban agglomerations in the middle reaches of the Yangtze River (Li & Cai, 2015; Li & Sun, 2013). Central and local governmental interventions played a crucial role in the integration process of the city-region by offering a series of incentives, such as regional infrastructure investment, industrial funds, and favored institutional design (Li et al., 2014). The concept of "city-regionalism" was diffused through China's reorientation of the locus of economic development from cities to city-regions, giving rise to the restructuring of central- and inter-local relations.

[2] Sina News. (2013, December 19). Building 32 urban agglomerations in 2030. Retrieved from http://news.dichan.sina.com.cn/2013/12/19/989132.html.

In recent national agendas and policies, the multilevel government has focused on the establishment of integration-oriented institutional architecture to undertake the strategy of city-regionalism, forming a governmental agenda approach referred to as a regional coordinated mechanism. The spatial plans of the city-region, inter-jurisdictional partnerships, regional institutions, and restructuring of administrative space were extensively utilized as fragmentation-repairing institutional tools to establish and achieve city-regional governance and integration in China (Wu, 2016; Zhang et al., 2015). These institutional designs were employed in city-regions, such as at the Mayor's Joint Conference on the Coordination of Urban Economy in YRD, the Outline Plan for the Reform and Development of the PRD (2008–2020) (hereafter the *Outline Plan*), and the Urban Coordinated Development Meeting in the Beijing–Tianjin–Hebei city-region. A number of institutional innovations went hand in hand with regional infrastructure projects and cross-boundary economic cooperation zones. An example is the Jiangyin–Jingjiang Industrial Transfer Park, fostering a new cross-boundary mode of governance in industrial cooperation. However, in terms of city-regional integration, the effectiveness of these integration-first institutional arrangements to operate as integrating entities in propelling functional integration needs to be examined contextually under specific city-regionalism.

In summary, the new era after 1978 can be partially viewed as an era where Neoliberalism was practised territorially in socialist China, marked by a grand transformation from a centrally-planned economy to a market-oriented economy. Unlike the development pathways in Western countries, the past intensive governmental intervention in China significantly remolded the city-region integration process in the post-reform era. The introduction of a market economy stimulated the activity of local governments and corporations, but the central government still played a dominant role in territorial economic growth, especially after the reform of the tax-sharing system in 1994. From the vertical perspective, the central government remained the massive financial and political controller of local development through tax-sharing, local cadre evaluation systems, and project approval apparatuses. The local government merely generated revenue from land acquisition, land auction, and levying of corporate tax due to the reform of the tax-sharing system. An equally crucial aspect of horizontal governance was the fact that local governments were inclined to cooperate with business elites and corporations in encouraging local economic growth. This phenomenon was conceptualized as the creation of local corporatist states (Oi, 1992). In addition, the GDP-based performance evaluation mechanism for local officials accelerated the inter-jurisdictional competition to attract mobile capital and reinforced the administrative hurdle for regional cooperation (Liu, 2004; Liu & Chen, 2003).

However, regional development policy at the national level provided forceful incentives to enable local governments to seek city-regionalism by conducting city-regional integration. Under such circumstances, partly-integrated areas operated as a city-regional entity through constructing trans-territorial infrastructure and formulating cross-boundary institutional design. Given the politico-economic logic of the city-region, probing the formation process of city-regions answers two major questions related to urban and regional politics in China. Can a region, which is partly integrated and driven through the initiatives of making a city-region, operate as a politico-economic entity to drive the role of governmental intervention at horizontal and vertical levels? How do existing institutional arrangements overcome the administrative hurdle to enable the functional integration of a city-region? Merely a few studies were conducted to unveil the underlying reshuffling process of regional governance to enable the economic integration of city-region. In addition, the extent to which government initiatives to promote a city-region can be translated into concrete institutional arrangements across government hierarchies to induce inter-city cooperation, and infiltrate the everyday practices of market and civil society remains questionable. These issues need to be meticulously and quickly analyzed and evaluated. The extent of economic and institutional integrations of city-regions in China and how the existing regional governance structure reshuffles itself to enable city-regions to operate as economic entities should be examined in different local milieus theoretically and empirically.

1.4 Local Context: New Situations Emerging as a Result of the City-Regionalism of SDH

In recent years, environmental crises as the by-products of economic development triggered inter-jurisdictional cooperation to cope with the issues of socio-environmental sustainability. A notable crisis was the haze problem in the Beijing–Tianjin–Hebei city-region, which was a result of gas emissions from motor vehicles and industries. The cross-boundary pollution forced the three provincial and municipal governments and related regional institutions to build mutually beneficial inter-jurisdictional partnerships, especially in treating polluting industries and establishing regional ecological security systems. In March 2015, three provincial and municipal governments jointly issued the "Medium- to Long-Term Planning on Air Pollution Control of Beijing-Tianjin-Hebei city-region and Its Neighboring Area" (*Jingjinji ji zhoubian diqu daqi wuran ji fangzhi zhongchangqi guihua*, 京津冀及周边地区大气污染防治中长期规划).[3] The Committee of Experts for the Control of Air Pollution in the Beijing-Tianjin-Hebei city-region was subsequently established, and the mechanism for joint law enforcement in controlling air pollution was to be reinforced through the enforcement of new environmental protection law.[4] This development indicates that effective city-regional institutional integration will play a contributive role in the control of air pollution in the Beijing–Tianjin–Hebei

[3] Xinhuanet. (2015, March 21). The mayor of Beijing (Wang, Anshun): "Beijing, Tianjin, and Hebei will cooperate with each other to make medium to long term of plan to control air pollution." Retrieved from http://www.gov.cn/xinwen/2015-03/21/content_2837010.htm.

[4] NetEase News. (2015, April 26). Regional cooperation promotes the control of air pollution in Beijing-Tianjin-Hebei city-region. Retrieved from http://money.163.com/15/0426/14/AO4PMAV800253B0H.html.

city-region. In addition to the environmental crisis, industrial and infrastructure integrations spread from urban to city-regional scope. Inter-city metro and railway were built at a larger territory within a host of city-regional spatial plans to enforce the "One Hour Commuting Circle" and integrate regional resources, facilitate industrial integration, and contain the commuting behavior of citizens. The state-planned institutional arrangements of cross-boundary cooperation contributed to the implementation of large-scale infrastructure projects (Li et al., 2014; Xu & Yeh, 2013). Diverse economic situations and industrial demands in larger territories, such as Beijing, Shanghai, Shenzhen, Guangzhou, and Tianjin reconstituted and strengthened the linkage between the urban core and hinterlands. This situation was elaborated through the spillover effect of industry in metropolises, the industrial transfer process at city-region scale, and the construction of cross-boundary economic cooperation zones (Luo & Shen, 2006). Institutional integration of city-regions is empirically significant to guarantee effective and benign functional integration and echoes with crucial regional matters.

To address new challenges in the development of city-regions, and foster national competitiveness, the central state set forth three new national strategies related to regional development, including "Coordination Development of the Beijing–Tianjin–Hebei city-region," "Yangtze River Economic Belt," and "Belt and Road Initiatives." These three strategies covered the vast majority of provinces and city-regions in China. The politico-economic statuses of YRD and the Beijing–Tianjin–Hebei city-region were reinforced under these national agendas. However, compared with the Beijing–Tianjin–Hebei city-region and YRD, the role of PRD was downplayed by recent national strategies. The "Belt and Road Initiative" defined Guangzhou and Shenzhen as key cities but marginalized the role of PRD. Unlike the Beijing–Tianjin–Hebei city-region and YRD, PRD was merely a marginal part of the "Belt and Road Initiative". Thus, the integration process of PRD, which dated back to the 1980s, was taken out of the top-down national effort of regional development. Such a move of the central government seemed to signal that PRD obtained fruitful achievements in regional integration and development through its bottom-up institutional arrangements, and national integration-oriented policies were thereby unnecessary to continuously forge PRD integration. However, despite the reorientation of the focal point of the central government, the actual extent of PRD regional integration still needs to be further investigated.

Regional problems did emerge in PRD and severely challenged the institutional arrangements. To fulfill the leading role of the core cities in the region, achieve integration of regional resources, and obtain balanced and coordinated development, PRD implemented a decentralization and segmental strategy through defining three sub-regions in the city-region. The institutional design of the Guangdong Provincial Government and related municipal governments in 2008 was due to the government's attempt to construct these three sub-regions as three city-regions. City-regions strove to become integrated institutions and economies by formulating spatial planning, investing in regional infrastructure projects, and constructing industrial transfer parks. Some of these artificial sub-regions that emerged under prevalent city-regionalism were partly integrated and they even witnessed paradoxical development incentives. Conflicting local demands intensified after the economic crisis in 2008. SDH, which is one of the three sub-regions, was the most-complicated region: it was ridden with rock-ribbed contradictions for regional development between cities.

After 2008, the urban economy of Shenzhen experienced rapid growth compared with other metropolises in China. The GDP of Shenzhen remained in fourth place among all cities of China until 2016; it increased from 778.6 million Yuan in 2008 to 1949 billion Yuan in 2016 (Fig. 1.1). In terms of per capita GDP, Shenzhen was also developing better than other main metropolises in China, such as Beijing, Guangzhou, Shanghai, Suzhou, and Chongqing, reaching 183.1 thousand Yuan in 2017 (Fig. 1.2). The total and average size of the economy of Shenzhen demonstrated significant achievements in recent years *vis-à-vis* other metropolises. The industrial structure saw a steady upgrading process. The ratios of primary, secondary, and tertiary industries respectively, were 0.1:48.9:51 in 2008 respectively. In 2014, the percentages were upgraded to 0.1:42.7:57.3. Six emerging strategic industries within the industrial upgrading process, namely, the bio-industry, Internet, new energy industry, information technology, advanced materials production, and creative industry, showed great potential with increases to 6.4%, 15.5%, 9.7%, 14%, 7.1%, and 15.6% of GDP, respectively, in 2014 compared with the data in 2013.

Given the powerful development momentum, Shenzhen had to restore the dilemma between rapid economic growth and insufficient land supply. The permanent resident population of Shenzhen in 2014 was 10.78 million, whereas those of Beijing, Shanghai, and Guangzhou were 25.52, 24.26, and 13.08 million, respectively, or 2, 2.25, and 1.2 times the population of Shenzhen. However, the areas of

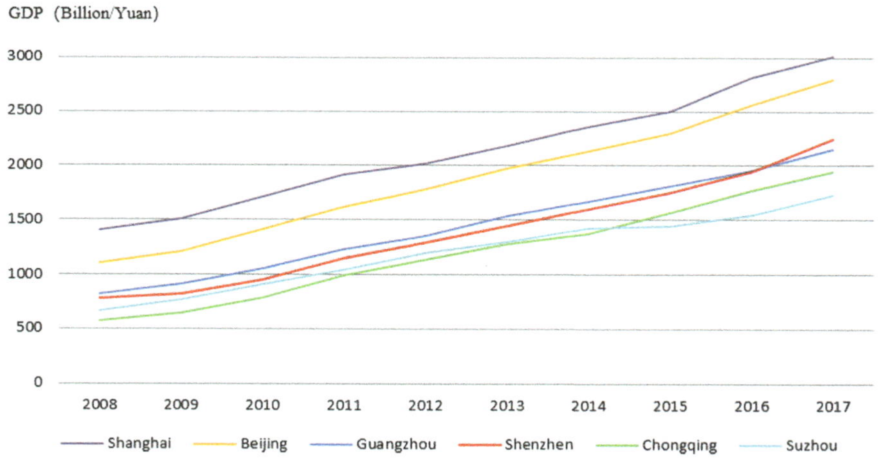

Fig. 1.1 GDP of Shanghai, Beijing, Guangzhou, Shenzhen, Chongqing and Suzhou from 2008 to 2017. *Source* Economy and society developed statistical bulletin of Shanghai, Beijing, Guangzhou, Shenzhen, Chongqing and Suzhou (2008–2017)

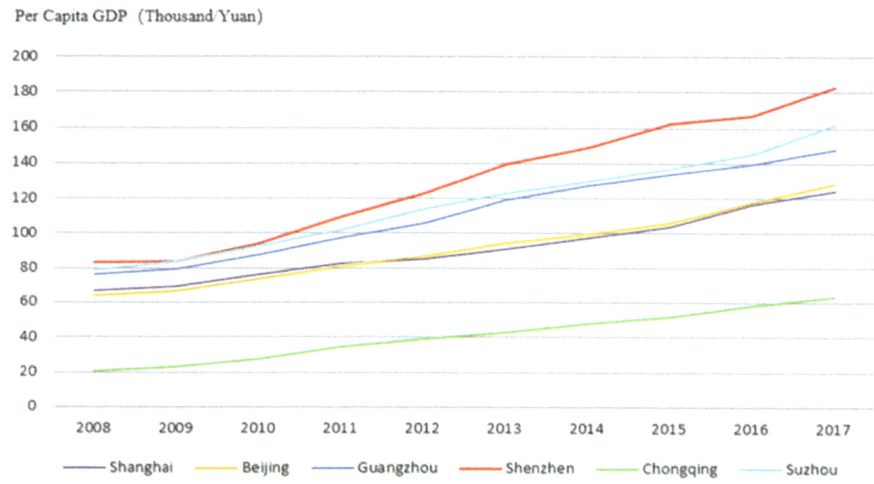

Fig. 1.2 Per capita GDP of Shanghai, Beijing, Guangzhou, Shenzhen, Chongqing and Suzhou from 2008 to 2017. *Source* Economy and society developed statistical bulletin of Shanghai, Beijing, Guangzhou, Shenzhen, Chongqing and Suzhou (2008–2017)

Beijing, Shanghai, and Guangzhou were roughly 8.2, 3.2, and 3.7 times of Shenzhen.[5] Thus, Shenzhen confronted a more serious deadlock in development. Moreover, Shenzhen has only 30 km^2 incremental construction land quotas from 2016 to 2020.[6] In other words, Shenzhen can expand an average of 5 km^2 of construction land every year, which is the primary constraint to the strategy of building global cities. Most of the incremental construction lands will be utilized to build infrastructure and social facilities in the future. The land used to accommodate industrial activities is arranged in merely 2 km^2 every year.[7] To achieve additional urban space, Shenzhen accelerated its urban redevelopment and renewal process. At the same time, it also attempted to seek lands and restructure the industrial distribution at SDH's scale by conducting industrial and infrastructure integration. Under this situation, the spillover of the low-end industry in Shenzhen was inevitable, and this intention was resisted by other municipalities in SDH. For example, the "Pin(*Pindi*)-Qing(*Qingxi*)-Xin(*Xinyu*) Industrial Cooperation Demonstration Park" was built to redistribute the industrial chain among Shenzhen, Dongguan, and Huizhou. Due to the contradictory initiatives of the local development request, Shenzhen transferred several low-end

[5] Sohu Finance. (2015, June 5). Exploration the reason why the price of Shenzhen housing estate experienced a rapid growth: The shortage of land and the suppression of demand. Retrieved from http://business.sohu.com/20150605/n414473939.shtml.

[6] Tubatu News. (2016, January 24). Shenzhen merely has 30 km^2 incremental construction land quota from this year. Retrieved from http://news.to8to.com/article/116843.html.

[7] Sina News. (2014, December 4). The incremental construction land of Shenzhen only has 30 km^2 in the next five to six years. Retrieved from http://news.sina.com.cn/o/2014-12-04/150731245731.shtml.

industries into the park, which was deeply inconsistent with the industrial transformation and upgrading of Dongguan. Therefore, this industrial transfer park remained stagnant under the unfettered inter-jurisdictional competition.[8]

The global financial crisis in 2008 largely destroyed the manufacturing industry in China, especially for PRD. Manufacturing revenues and jobs experienced a rapid decline. This deindustrialization process exacerbated the manufacturing of SDH, especially that of Dongguan, which was traditionally known as "the workshop of the world". Leading industries included textiles, furniture manufacturing, paper-making, and food. Before 2008, the average GDP of Dongguan even reached the standard of moderately developed countries. However, the severe financial crisis and subsequent economic depression resulted in a high increase of labor and material costs and the loss of advantages of cheap labor and materials, leading to the failure of many businesses in Dongguan.[9]

According to the Fig. 1.3, the growth rate of the GDP of Dongguan was smoother than that before 2008, which had witnessed rapid growth. In addition, the growth rate of the secondary industries dropped dramatically and even declined by −3.7% in 2009. Although the growth rate of the secondary industry experienced a rapid increase from 2009 to 2010, the spurt was lower than that before 2008. The growth rate of tertiary industries from 2007 to 2008 remained at a high level, but the figure was lower than that before 2008. This figure reveals that the growth rate of GDP of manufacturing in Dongguan experienced a dramatic decrease after 2008 and that of the tertiary industries significantly dropped as well. The two occurrences triggered the smooth growth of GDP after 2008 compared with the rapid growth ratio before 2008. While the GDP of Dongguan has seen steady growth from 2002 to 2017, the growth rates of secondary and tertiary industries have witnessed some fluctuations after 2006 and remained at a relatively low level compared with the rates before 2006. Unlike Shenzhen, Dongguan confronted a more serious dilemma in industrial transformation and upgrades. On the one hand, Dongguan lost the advantages of traditional manufacturing. On the other hand, the radiating and exemplary role of its tertiary industry was not good enough. Based on this status quo, Dongguan aspirated to develop a tertiary industry and cooperate with Shenzhen to achieve industrial transformation and upgrades. This desire can be investigated in the *Outline Plan*, the "Main Functional Area Planning," and the "SDH Coordinated Development Plan (2012–2020)." Compared with the strong appeal of Shenzhen, this aspiration resulted in severe obstruction of SDH integration with reference to industrial collaboration.

The situation of Huizhou is different from Dongguan and its GDP experienced a smooth increase from 52.7 billion Yuan in 2002 to 383 billion Yuan in 2017. The growth rate of the secondary industries experienced several fluctuations from 2002 to 2017 and growth from 11.5% in 2008 to 23.7% in 2010. Rates after 2010

---

[8] Jinyangnet. (2013, April 26). The projects are stagnant, which is the difficulty of "Pin-Qing-Xin industrial transfer park." Retrieved from http://myjjb.ycwb.com/2013-04/26/content_4430354.html.

[9] NetEase. (2015, July 25). The dilemma of bosses in Dongguan: the happy life stopped after the economic crisis in 2008. Retrieved from http://money.163.com/15/0725/05/AVBIMAL500252603.html.

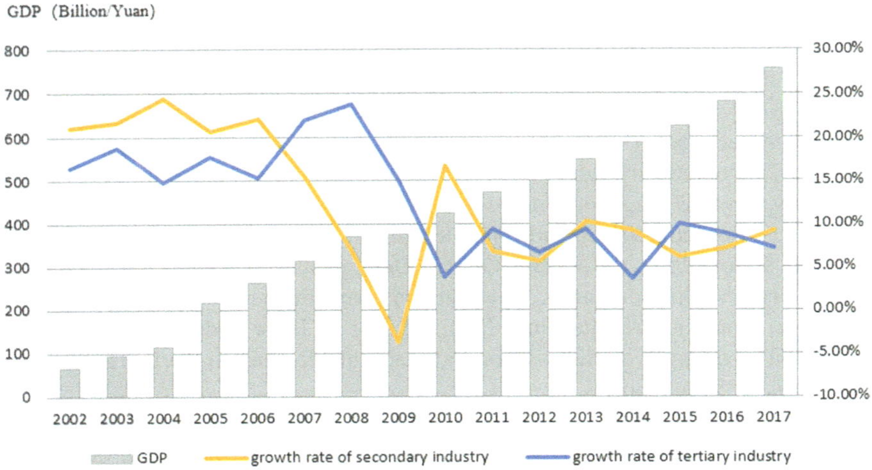

Fig. 1.3 GDP, the growth rate of secondary and tertiary industry of Dongguan from 2002 to 2017. *Source* Dongguan Economy and society developed statistical bulletin (2002–2017)

experienced a steady decrease from 23.7% in 2010 to 7.9% in 2017. Hence, the manufacturing of Huizhou was less affected by the financial crisis than Dongguan was. However, the growth rate of tertiary industries witnessed a sustained decline from 21.7% in 2004 to 7.6% in 2017 (Fig. 1.4). On the one hand, the Huizhou Municipal Government attempted to upgrade its manufacturing from low-end to high-end by building a headquarters economy and attracting high-tech enterprises from Shenzhen and Dongguan.[10] On the other hand, the government facilitated the upgrading of tertiary industry in terms of technology level and diversity of genres as well.

In summary, Shenzhen transferred its low-end manufacturing to SDH, but Dongguan and Huizhou were keen to upgrade their tertiary and manufacturing industries from low- to high-end. This developmental contradiction impeded the integration of SDH, especially in the industrial sector.

To fulfill the driving role of the three sub-regions in PRD, the Guangdong Provincial Government formulated the document "Significant Tasks of Further Promoting the Development of Eastern, Northern and Western Areas of Guangdong Province" and proposed the strategy of "Revitalization and Development of the Eastern, Western and Northern Guangdong" in 2014.[11] Shanwei and Heyuan were spatially incorporated into SDH because of this policy. However, Shanwei and Heyuan stayed in

[10] DeanZeng. (2015, January 5). Current situation: The development of manufacturing in Huizhou have advantages and disadvantages. Retrieved from http://hz.southcn.com/content/2015-01/05/content_115565965.htm.

[11] The Guangdong Provincial Government. (2015, April 16). Significant tasks of further promoting the development of eastern, northern and western areas of Guangdong province. Retrieved from http://zwgk.gd.gov.cn/006939748/201504/t20150430_578853.html.

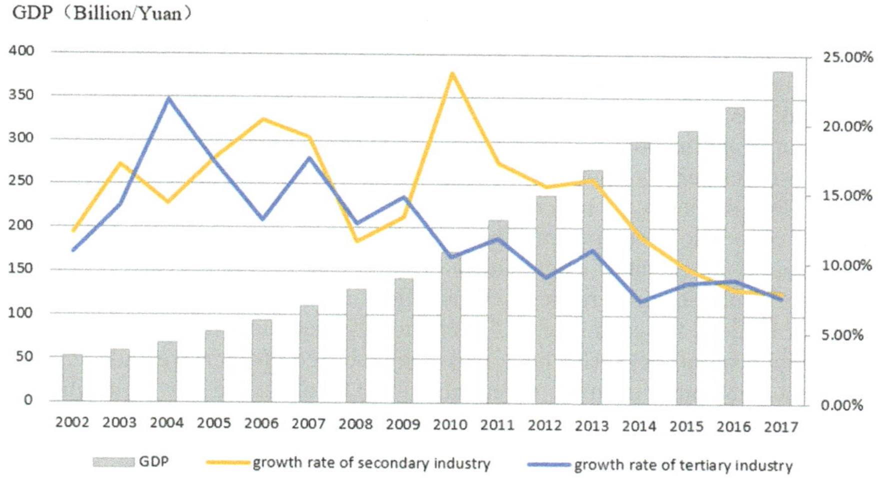

Fig. 1.4 GDP, the growth rate of secondary and tertiary industry of Huizhou from 2002 to 2017. *Source* Huizhou Economy and society developed statistical bulletin (2002–2017)

different urban development stages *vis-a-vis* Shenzhen, Dongguan, and Huizhou. The GDP of the two cities in 2014 were 71.6 billion Yuan and 75.8 billion Yuan, respectively, whereas that of Shenzhen, Dongguan, and Huizhou reached even 1600, 588, and 300 billion Yuan, respectively (Figs. 1.1, 1.3, and 1.4). From the perspective of industrial structure, the percentages of primary, secondary, and tertiary industries of Shanwei and Heyuan in 2014 reached 15.9:47.4:36.7 and 11.9:50.7:37.4, respectively. These two cities were keen to increase their GDP and achieve economic growth through industrialization. Under this situation, Shanwei and Heyuan had better possibilities of cooperating with Shenzhen in terms of industrial transfer than Dongguan and Huizhou did.

The deep contradiction largely depended on the institutional design to coordinate the diverse interests among different stakeholders. The institutional arrangement of SDH involved multiple aspects, such as joint conferences, integration of planning systems, and building cross-boundary economic cooperation zone. A series of questions should be articulated in analyzing SDH integration. These questions are: How does Shenzhen cooperate with other cities in SDH to cope with the dilemma in industrial integration? Could Dongguan and Huizhou attract high-end industries from Shenzhen to achieve its industrial upgrading? Are Shanwei and Heyuan the best choices for Shenzhen for industrial transfer? How do Shanwei and Heyuan respond to the call for SDH's integration and how can they be involved in the integration process? These issues are situated in China's pursuit of city-regionalism, spearheaded by different levels of government. Thus, there is a need to understand the interplay of dynamics between the regional economy and governance. Five municipal governments were involved in the SDH integration process, which included building an inter-city transit system, construction of a cross-boundary industrial transfer park, and establishing city-regional joint conferences. The development demands of five cities involve the following questions: Does institutional design effectively promote SDH's integration as a successful functional entity under the city-regionalism of PRD? This idea depends largely on the assessment of the extent of economic and institutional integrations of SDH. In other words, whether SDH fosters a successful city-regional coordination mechanism provides a case to fathom city-regional development under China's reorientation to the city-region as a new locus of economic development. Therefore, this book uses SDH as a case study to investigate the extent of economic and institutional integrations and reshuffling processes of regional governance to enable functional integration and to introduce recommendations for economic and institutional integration under the strategy of city-regionalism of PRD. This book may facilitate a comprehensive understanding of the economic and institutional interplay of SDH under the city-regionalism of PRD, thereby helping to rethink city-regionalism plans in socialist China.

1.5 Research Aim and Research Questions

1.5.1 Research Aim

As discussed above, city-regionalism is manifested in the implementation of spatial-economic strategies, which are constantly driven by globalization and invariably involve regional politics (Li & Wu, 2018; Scott, 2001). The central and local governments in China are keen to develop city-regions into new loci of economic development. Institutional re-configuration is inevitable in the process of city-regional integration. Given the diverse economic and spatial characteristics among the cities of variegated geographical scales, some cities may only be weakly or partly in a city-regional entity, which nonetheless can still be considered as a regional economic unit.

Given the above context, this book investigates the extent to which initiatives from governments to promote city-region development can evolve into concrete institutional arrangements across governmental hierarchies, inter-city cooperation in specific development projects, and market activities to establish a city-region, operating as a regional economic unit. Institutional arrangements will be interpreted with the concept of MLG adopted from the EU context and their extent in integrating the economic development of a city-region is evaluated, using case studies in the SDH sub-region. The conventional theory in demonstrating China's city-regionalism, notably the theory of state rescaling, will be problematized to challenge its two-track "upward" and "downward" framework to construe the changing governance reshuffling in which the institutions are reconfigured case-by-case. In addition, integration-oriented policy and institutional design are formulated in accordance with the territorial praxis of SDH to enable successful city-regionalism in China.

The case study involves, first, the assessment of the economic integration of SDH in the dimensions of market, industrial, and infrastructure integration. Second, institutional integration in SDH is examined through the lens of inter-scalar, inter-city, and state-market-civil society relations by employing an organization-embraced approach of institutions and drawing on an adopted MLG paradigm. Finally, this book takes the Shenzhen–Shanwei Special Cooperation Zone (SSSCZ) and Sui(*Guangzhou*)–Guan(*Dongguan*)–Shen(*Shenzhen*) Inter-city Railway (SGSIR) as empirical cases within SDH to examine more deeply the intrinsic mechanism of regional governance reshuffling in response to the economic integration of SDH, such as industrial and infrastructure integration. This approach reveals a specific reconfiguration process in regional governance in the regional integrating process. Four research questions are formulated accordingly.

1.5.2 Research Questions

(1) ***To what extent does SDH operate as an economically integrated entity as reflected in price similarity, industrial structure similarity, and infrastructure development?***

Economic activities and their linkages in Western countries and socialist China underlie city-regional integration in theoretical and practical domains. Thus, the primary task of city-regionalism is to strengthen and open up an economic connection between and among cities to create a regional economic unit. Repacking the production elements is therefore involved. Economic connection should be assessed first in this book. The question is answered by investigating economic integration along three dimensions: (a) market integration, by evaluating relative price variance; (b) industrial integration, by examining industrial structure measured by the degree of industrial structure similarity with the modified Krugman's index; and (c) infrastructure integration, assessed by the coefficient of variation, degree of infrastructure development similarity, and infrastructure connectivity index.

(2) ***To what extent does SDH operate as an institutionally integrated entity by redefining the governance structure across government hierarchies, between cities, and between the state and non-state domains?***

The institutional integration of city-region is assessed by building an MLG-based analytical framework, which is adopted and developed from the territorial practice of MLG in the EU context. To address this question, this book investigates the governance structure of the city-region across the government hierarchies, between cities, and between the state and non-state domains. Apart from secondary data collected through desktop sources, social media data on inter-city cooperation from the search engine and governmental official websites are tapped to provide evidence for inter-city and state-market-civil society relations.

(3) ***How does the governance structure reshuffle itself to enable or encumber the functioning of SDH as an integrated economic entity under China's reorientation strategy to develop city-region as a locus of economic development?***

After assessing the economic and institutional integrations of SDH, this book unfolds the mechanism by which governance structure reshuffles itself to facilitate socioeconomic integration to enable SDH to develop into a locus of economic development. Given the wide scope of economic integration and city-regionalism of PRD and SDH, this book necessarily focuses on the restructuring of regional governance in the industrial and infrastructure integrating process of two specific empirical cases: SSSCZ and SGSIR. This book examines the reshuffling process of the governance structure in these two cases by examining the inter-scalar and inter-city relations, and assesses how the restructured governance contributes to the performance of these two projects.

(4) ***How does the territorial praxis of SDH contribute to the theoretical debates on China's city-regionalism?***

By tackling the above four questions, the rationale of city-regionalism of China and the city-regional politics among different actors for controlling space is systemically outlined by a representative case of SDH. According to the major empirical findings, this book will explore how these findings contribute to city-regionalism of China, and it will add to the theoretical debates of state rescaling theory in illustrating regional governance, inter-scalar, and inter-city relations. In addition, this book will formulate institutional and policy recommendations to bring forward regional development and regional integration in SDH specifically and in China generally.

1.6 Research Significance and Research Scope

1.6.1 Research Significance

This book adds to the existing literature on China's territorial development by critically examining the development process of a city-region in economic and institutional domains. The study probes the successful mechanism of city-region governance. The findings will inform future regional policy formulation on developing the city-regions as new loci for China's economic development and territorial restructuring. This study is anchored in SDH in PRD, which has for a long time been acting as a testbed for the nation's "one-step-ahead" policies. This book provides state-of-the-art insights into South China's latest territorial restructuring and its socioeconomic, spatial, and institutional implications. This book holds both theoretical and empirical value.

This book conducts comprehensive examinations on the economic and institutional integration of SDH. Based on the recent economic development and institutional design of SDH under the restructuring of the global economy, this book offers substantial contributions to the understanding of the long-term endeavor of implementing city-regionalism in PRD. This approach is applicable in the context of the underlying crisis and recent regional practices in SDH triggered by globalization and political decentralization and marketization. Two empirical cases that illustrate the governance restructuring process broaden conventional literature on the subject by exploring the multi-dimensional and multi-actor interactions.

This book investigates the nature of economic integration by looking into market, industrial, and infrastructure integration and by establishing a quantitative analytical framework to assess each of the dimensions to provide a multifaceted and context-sensitive grasp of economic integration. This book systematically presents the relations between institutional integration and MLG through drawing on the territorial practice of MLG in the EU milieu and building an MLG-based analytical framework, encapsulating the inter-scalar, inter-city, and state–market–civil society relations. Finally, the intrinsic mechanism between the economy and the institution of city-regions is unfolded through two case studies to enhance theories on the formation and recognition of city-regions.

1.6.2 Research Scope

The case study is an interplay between institutional configuration and economic integration, which is applied to respond to how and why questions. The case study approach is used. The two state-led projects in SDH are examined. SDH is unique in city-regional integration aspects for two reasons. First, the new situation concerning industrial integration involves the rise of serious industrial isomorphism and building a problem-solving cooperation mechanism across different stakeholders. Second, the integration of SDH has been strategically encouraged by the Guangdong Provincial Government for more than two decades. Given the host of institutional designs, the applicability of these governmental endeavors on the city-regional integrating process is heavily associated with the local context and emerging situations. The exploration of SSSCZ aims to identify how the reshuffling process of governance structure facilitates the economic integration of SDH in industrial collaboration. An examination of SGSIR aims to answer how the restructuring of regional governance responds to the socio-economic integration of SDH in infrastructure construction.

Given the long-developed city-regionalism in PRD, emerging conflicts and struggles in industrial development led to the possibility of incurring partial or weak SDH integration. On the one hand, the central government viewed PRD as a well-integrated region that could self-sustain and self-elevate its economic and institutional integrations, and PRD was increasingly taken out of the national strategy of making city-region and related concessionary policies. On the other hand, PRD, especially the SDH region, was deeply affected by the restructuring of the global economy and financial turbulence, and the local demand for industrial transformation and upgrading left irreconcilable conflicts in institutional design, industrial collaboration, and infrastructure development. Hence, this book selects SDH which is ridden with struggles and tensions, to investigate the extent of economic and institutional integrations and the reshuffling process of governance structure in order to achieve the prevalent city-regionalism in China. SSSCZ and SGSIR are selected to elaborate the reshuffling process of governance structure in response to the socioeconomic integration of SDH under city-regionalism. These regional projects served as the laboratory for investigating the internal mechanism of city-regionalism of SDH and how it contributes to the performance of regional integration.

1.7 Structure of the Book

This book has eight chapters that respond to the research questions. The first four chapters elaborate the research background and questions, conceptual and empirical reviews, and research methodology. The next three chapters answer the relevant research questions. The conclusions and institutional and policy recommendations are presented in the final chapter according to major findings.

The first chapter provides an introduction of the international background on globalization, the rescaling of the state and rise of city-regions, China's regional development, and the new situation that emerged in SDH integration under the restructuring of the global economy. The research question and significance of the book are discussed in the final section. Chapter 2 establishes the theoretical foundation and seeks the conceptual position for this book. Existing literature on the contextualized theory of city-regional integration, which specifically refers to economic and institutional integration, and conventional literature on governance structure reshuffling, are reviewed and discussed. The theoretical gap is presented in the final section. Chapter 3 reviews city-regional development from the macro (national), meso (PRD), and micro (SDH) scales. The emerging situation is positioned into the discourse of SDH integration by examining the territorial praxis of SDH in institutional arrangement, industrial, and infrastructure integrations. The empirical significance is presented in the final section. Chapter 4 illustrates the research methodology, which encompasses the analytical framework, data collection, and analysis methods.

The next three chapters undertake the empirical study of SDH to answer the first three research questions. Chapter 5 comprises two parts, which assess the economic and institutional integration of SDH. The assessment of the economic integration of SDH is quantitatively conducted from three perspectives, namely: (1) market integration, (2) industrial integration, and (3) infrastructure integration. The assessment of institutional integration of SDH is qualitatively conducted as an MLG-based analytical framework through three lenses, namely: (1) inter-scalar, (2) inter-city, and (3) state-market-civil society relations. Chapter 6 focuses on how the governance structure reshuffles itself to enable industrial integration of SDH by deciphering the empirical case of SSSCZ. The intrinsic mechanism is primarily investigated in inter-scalar, inter-city relations. Chapter 7 focuses on how the governance structure reshuffles itself in response to the infrastructure integration of SDH by elaborating the empirical case of SGSIR. The intrinsic mechanism is primarily investigated in inter-scalar relations.

Chapter 8 summarizes the major findings, theoretical contributions, and conclusions of this book. It also further formulates the institutional and policy implications in facilitating city-regional development in China's reorientation to city-regions as the locus of economic development to address the last research question.

References

Ancien, D. (2011). Global city theory and the new urban politics twenty years on the case for a geohistorical materialist approach to the (new) urban politics of global cities. *Urban Studies, 48*(12), 2473–2493.

Ansell, C., & Gash, A. (2008). Collaborative governance in theory and practice. *Journal of Public Administration Research and Theory, 18*(4), 543–571. https://doi.org/10.1093/jopart/mum032

Appadurai, A. (1996). *Modernity at large: Cultural dimensions of globalization*. University of California Press.

Brenner, N. (1998). Between fixity and motion: Accumulation, territorial organization and the historical geography of spatial scales. *Environment and Planning D: Society and Space, 16*(4), 459–481. https://doi.org/10.1068/d160459

Brenner, N. (1999). Globalisation as reterritorialisation: The re-scaling of urban governance in the European Union. *Urban Studies, 36*(3), 431–451. https://doi.org/10.1080/0042098993466

Brenner, N. (2004). Urban governance and the production of new state spaces in Western Europe, 1960–2000. *Review of International Political Economy, 11*(3), 447–488.

Castells, M. (1989). *The informational city: Information technology, economic restructuring, and the urban-regional process* (p. 24). Basil Blackwell Press.

Castells, M. (1996). *The rise of the network society* (Vol. 1). Basil Blackwell Press.

Ferris, J. S., Winer, S. L., & Grofman, B. (2012). Do departures from democratic accountability compromise the stability of public finances? Keynesianism, central banking, and minority governments in the Canadian system of party government, 1867–2009. *Constitutional Political Economy, 23*(3), 213–243. https://doi.org/10.1007/s10602-012-9123-6

Kantor, P., & Nelles, J. (2015). Global city region governance and multicentered development: A North American perspective. *Environment and Planning C: Government and Policy, 33*(3), 475–495. https://doi.org/10.1068/c13133

Klink, J., & Denaldi, R. (2012). Metropolitan fragmentation and neo-localism in the periphery revisiting the case of Curitiba. *Urban Studies, 49*(3), 543–561. https://doi.org/10.1177/0042098011408936

Li, L., & Cai, L. (2015). The time and space evolution characteristics of economic connection of the urban agglomerations in the middle reaches of Yangtze River. *Urban Problems, 34*(7), 62–70 (in Chinese). https://doi.org/10.13239/j.bjsshkxy.cswt.150709

Li, X., & Sun, B. (2013). Regional integration of the Yangtze River Middle reaches urban agglomerations: Measuring and comparison. *Resources and Environment in the Yangtze Basin, 22*(8), 996–1003 (in Chinese).

Li, Y., & Wu, F. (2013). The emergence of centrally initiated regional plan in China: A case study of Yangtze River Delta Regional Plan. *Habitat International*, 39, 137–147. https://doi.org/10.1016/j.habitatint.2012.11.002

Li, Y., & Wu, F. (2018). Understanding city regionalism in China: Regional cooperation in the Yangtze River Delta. *Regional Studies*, *52*(3), 313–324. https://doi.org/10.1080/00343404.2017.1307953

Li, Z., Xu, J., & Yeh, A. G. (2014). State rescaling and the making of city regions in the Pearl River Delta, China.*Environment and Planning C: Government and Policy*, *32*(1), 129–143. https://doi.org/10.1068/c11328

Liu, J. (2004). The analysis of "administrative area economy" in China's economic transformation period. *Theoretical Front*, (10), 20–22 (in Chinese).

Liu, J., & Chen, Z. (2003). Thinking of revolution of administrative system in Yangtze River Delta. *Exploration and Free Views*, *18*(6), 12–14 (in Chinese)

Luo, X., & Shen, J. (2006). Urban growth between cities: The case of Jiangyin economic development zone in Jingjiang. *Acta Geographica Sinica*, *61*(4), 435–445 (in Chinese)

Luo, X., & Shen, J. (2008). Why city region planning does not work well in China: The case of Suzhou–Wuxi–Changzhou. *Cities*, *25*(4), 207–217. https://doi.org/10.1016/j.cities.2008.04.003

Luo, X., & Shen, J. (2009). A study on inter-city cooperation in the Yangtze river delta region, China. *Habitat International*, *33*(1), 52–62. https://doi.org/10.1016/j.habitatint.2008.04.002

Newman, P. (2000). Changing patterns of regional governance in the EU. *Urban Studies*, *37*(5/6), 895. https://doi.org/10.1080/00420980050011145

Oi, J. C. (1992). Fiscal reform and the economic foundations of local state corporatism in China. *World Politics*, *45*(1), 99–126. https://doi.org/10.2307/2010520

Pemberton, S., & Morphet, J. (2014). The rescaling of economic governance: Insights into the transitional territories of England. *Urban Studies*, *51*(11), 2354–2370. https://doi.org/10.1177/0042098013493484

Sassen, S. (1991). *The global city: New York, London, Tokyo, Princeton*. Princeton University Press.

Sassen, S. (2006). *Cities in a world economy* (3rd ed). Pine Forge Press.

Scott, A. J. (2001). Globalization and the rise of city regions. *European Planning Studies*, *9*(7), 813–826. https://doi.org/10.1080/09654310120079788

Shen, J. (2018). *Urbanization, regional development and governance in China*. Taylor & Francis.

Taylor, P. J. (1995). World cities and territorial states: The rise and fall of their mutuality. In P. Knox, & P. Taylor (Eds.), *World cities in a world-system* (pp. 48–62). Cambridge University Press.

Wei, Y. D. (2001). Decentralization, marketization, and globalization: The triple processes underlying regional development in China. *Asian Geographer*, *20*(1–2), 7–23. https://doi.org/10.1080/10225706.2001.9684073

Wu, F. (2002). China's changing urban governance in the transition towards a more market-oriented economy. *Urban Studies*, *39*(7), 1071–1093. https://doi.org/10.1080/00420980220135491

Wu, F. (2015). *Planning for growth: Urban and regional planning in China*. Taylor & Francis.

Wu, F. (2016). China's emergent city-region governance: A new form of state spatial selectivity through state-orchestrated rescaling. *International Journal of Urban and Regional Research*, *40*(6), 1134–1151. https://doi.org/10.1111/1468-2427.12437

Wu, F., & Zhang, F. (2010). China's emerging city region governance: Towards a research framework. *Progress in Planning*, *73*(1), 60–63.

Xu, J., & Yeh, A. G. (2005). City repositioning and competitiveness building in regional development: New development strategies in Guangzhou, China. *International Journal of Urban and Regional Research*, *29*(2), 283–308. https://doi.org/10.1111/j.1468-2427.2005.00585.x

Xu, J., & Yeh, A. G. (2013). Interjurisdictional cooperation through bargaining: The case of the Guangzhou–Zhuhai railway in the Pearl River Delta, China. *The China Quarterly*, 213, 130–151. https://doi.org/10.1017/S0305741013000283

Xu, J., Yeh, A. G., & Wu, F. (2009). Land commodification: New land development and politics in China since the late 1990s. *International Journal of Urban and Regional Research*, *33*(4), 890–913. https://doi.org/10.1111/j.1468-2427.2009.00892.x

Yi, C. (2010). Centralization and decentralization: The practical analysis on American metropolitan governing process. *Urban Development Studies*, *17*(7), 73–79 (in Chinese)

Zeng, G., Shang, Y., & Si, Y. (2015). The convergent evolution of China's regional economic development models. *Geographical Research, 34*(11), 2005–2020 (in Chinese).
Zhang, J., Luo, X., & Yin, J. (2008). Polycentric mega-city regions and multi-level governance of the Yangtze River Delta. *Urban Planning International, 23*(1), 65–69 (in Chinese)
Zhang, X., Zhao,Y., Shan, Z., Chen, Y., & Hong, S. (2015). City-regional governance from the perspective of comparison: Conception clarification, theory development and research progress. *Economic Geography, 35*(7), 6–13 (in Chinese). https://doi.org/10.1016/j.landusepol.2019.104170

Chapter 2
Conceptual Background: The City-Region and Its Economic and Institutional Integration

2.1 Introduction

This chapter aims to lay a theoretical foundation and build analytical perspectives on city-regionalism by reviewing extant literature on (1) economic integration, (2) institutional integration, and (3) governance structure reshuffling. By doing so, the chapter shows how city-regionalism in China can be analyzed. Economy and institution are the primary factors that drive and facilitate city-regionalism. Thus, this chapter lays a theoretical foundation and analytical framework for a broad understanding of the economic and institutional integration of the city-region. This chapter also explores how the literature interprets, conceptualizes, and theorizes the pathways in which governance structure is reshuffled in response to socioeconomic reconstruction.

Accordingly, this chapter comprises five sections. The first section presents two underlying and important concepts, i.e., the city-region and city-regional integration. The second section reviews the concept and assessment approaches of the city-region's economic integration. The methodologies on the economic integration of city-regionalism are categorized in two directions, i.e., the long-distance urban system and the proximity relation. The third section focuses on the concept of the city-region and the assessment of its institutional integration. Three key integrative approaches are identified by employing an organization-based approach to re-ascertain institutions and by drawing the territorial praxis of MLG in EU integration. These approaches shed light on the function of the analytical framework in assessing the institutional integration of the city-region. The fourth section is a thorough literature review that probes the reshuffling of the governance structure. The case-specific paradigm decodes city-regionalism by focusing on the cross-boundary economic cooperation zone and a large-scale infrastructure project. The fifth section presents theoretical gaps.

X. Zhang, *Regional Integration and Governance Reshuffling in the Making of China's City-Regions*, Urban Sustainability, https://doi.org/10.1007/978-981-99-2792-0_2

2.2 City-Region and City-Regional Integration

In the past decades, globalization and the transformation of the accumulation regime from Fordist-Keynesianism to post-Fordist Neoliberalism have triggered tremendous restructuring of territorial organizations worldwide (Brenner, 2004; Harvey, 1989; Yeh et al., 2015). The restructured accumulation regime has reconstituted territorial space based on new economic, institutional, and social orders. The restructuring of territorial space between urban centers and hinterlands has given rise to city-regions. Emerging city-regions are characterized by the agglomeration of economic activity (Krätke, 2007), which is an effect of the post-Fordist economic sector (Scott, 2001). The new international division of labor has complicated economic geography. It has not only formed a global city system but also turned city-regions into crucial territories for capital accumulation (Derudder, 2006), as well as locomotives for the economy of a nation-state (Scott & Storper, 2007).

2.2.1 Concept of the City-Region and City-Regional Integration

2.2.1.1 City-Region

The term "city-region" is an intricate and rich one, varying in meaning in accordance with variegated scales, such as urban agglomeration (Gu et al., 2007; Partridge & Rickman, 2008; Tabuchi, 1998), a global city-region (Scott, 2001; Pain, 2008), a metropolitan area (Masek et al., 2000; Salet et al., 2003; Yuan et al., 2005), a cosmopolitan metropolis (Germain, 2011; Keith, 2005), a megacity-region (Hoyler et al., 2008; Taylor et al., 2008), and a megalopolis (Gottmann, 1957; Lang & Knox, 2009; Mori, 1997). According to Vogel (2010), scale largely limits the definitions of relative concepts, such as city, metropolis, city-region, and region. A global city is governed by a metropolitan government administratively, whereas a city-region has spatially expanded beyond the municipal boundary to encompass additional administrative units into its territory.

In the early stage, the city-region was an academic term in the discourse of continuous change and adaptation (Wise, 1965). A city-region is a context-sensitive form that overrides the jurisdiction of the local government. Several academic terms are used to refer to city-regionalism or regionalization. Such terms were created in response to the relationship between an urban core and its surrounding territories. When several cities continue to expand beyond their administrative boundaries, they may form a metropolitan district (Simmonds & Hack, 2000), megalopolis (Gottmann, 1957; Mori, 1997), and urban field (Friedmann & Miller, 1965). In such cities, Bennett (1985) identified three criteria to define a city-region: (1) more than 45% out-commuting and more than 30% commuting to the core; (2) enclosed by a public transport isochrone of one hour; and (3) with less than 46% agricultural employment (p. 70). Friedmann (1986) defines city and city-region in the context of globalization and regionalization:

> A city in these terms is a spatially integrated economic and social system at a given location or metropolitan region. For administrative purposes the region may be divided into small units which underlie, as a political or administrative space, the economic space of region. (p. 70)

In the book *The World City*, Hall (1984) construes the basic characteristics of a world city, defining it as a metropolis that has intensive political, economic, and cultural control over the world or the majority of countries. Sassen (1991) puts forward the concept of "global city" to represent a new social space based on the early work of Hall and the "*The World City Hypothesis*" proposed by Friedmann in 1986. In a global city, social space is a scattered spatial structure in terms of production activities, but a highly centralized spatial structure in terms of management and control activities under the driving force of globalization (Sassen, 1991). Yeung and Olds (2001) set forth three categories of global cities: hyper-global cities, emerging global cities, and global city-states. This division redefines global cities as "some other cities that strive to become global cities" (Yeung & Olds, 2001, p. 20). The focus shifted to the relationship between a global city and a hinterland region. Scholars have recognized the importance of inner links between cities in the urban system and developed a new concept, a "global city-region" (Hall, 2002; Hall & Pain, 2006; Scott, 2001). Hall and Pain (2006) define global city-region as follows:

> More than simply center cores; they are viewed as more complex urban regions, encompassing several cities, networked in a polycentric structure. (p. 14)

Scholars define the features of a city-region from various perspectives to distinguish it from other complementary concepts. According to Vermeijden (2001), a city-region should have a hierarchical spatial structure, where the urban core is surrounded by satellite towns. Davoudi (2003) highlights that a city-region is socially, economically, and culturally dominated by the core city. The comprehensive grips of a city-region are constructed by Parr (2005), who identifies two basic elements of city-region and their mechanisms: the core city (the C zone) and the territory surrounding it (the S zone). The C zone refers to a city or a regional metropolis with vibrant economic activities, an urbanized area, and a population. The S zone, which contains urban and rural populations, is excluded from but easily affected by the C zone. This preliminary modelling deduction is amplified by the mechanisms within a city-region, namely, labor-market flows and capital movements, among others. This preliminary investigation proffers a profound understanding of the structural elements and formation of a city-region. However, a systematic definition that clarifies the interaction and relationship between C and S zones is rare. Jonas and Ward (2007) identify the basic functions of a city-region, those involving economic (trade) and political (regulatory) functions. According to Jonas and Ward (2007), a city-region is "the product of a particular set of economic, cultural, environmental and political projects, each with their own logics" (p. 176). Scott (2011) captures a common theme in recent literature on city-regions and embeds a social dimension into the concept:

> City-regions, then, are functioning more and more as assertive systemic units of the global economic and political system [......] In a nutshell, they are powerful and increasingly self-affirming configurations of social, economic, and political activity within a multifaceted, multitiered system of global emergence. (p. 862)

Based on this definition, Rodríguez-Pose's (2008) pioneering work establishes a comprehensive concept. This concept not only integrates spatial dimensions into the city-region but also argues that the nature of the city-region is affected by the economic and social development among the urban core, semi-urban, areas and rural hinterland (Rodríguez-Pose, 2008). This explanation specifies the territories of C and S zones, dividing them into the urban core, semi-urban areas, and rural hinterland, *vis-à-vis* the preceding definitions. Rodríguez-Pose (2008) further elucidates that the relationship between the urban core and hinterland varies in accordance with specific contexts, embodying "a combination of economic, housing market, travel-to-work, marketing, or retail catchment factors" (p. 1027). Moreover, the urban core often features a polycentric structure that also forms the polycentric geographical structure at the city-regional scale (Parr, 2003; Rodríguez-Pose, 2008). The hinterlands of different city-regions even geographically overlap with one another (Davoudi, 2003; Rodríguez-Pose, 2008).

Based on the foregoing comprehensive elaborations as summarized in Table 2.1, the different geographical-spatial forms are also integrated into the concept of a city-region. Such an integration proves that a metropolitan area, urban agglomeration, and megacity-region are synonymous with a city-region. Compared with these academic terms, the city-region as a broadly-accepted concept can more flexibly depict different forms of spatial structure as well as the relationship between the urban core and hinterland. In sum, the city-region is an emerging spatial form characterized by tight social, economic, and political connections among the urban core, semi-urban areas, and rural hinterland.

2.2.1.2 City-Regional Integration

City-regional integration varies depending on the different spatial meanings of city-region from the transnational to the local level. A majority of conventional research focuses on cross-border integration across different nation-states, particularly European integration (Perkmann, 2007a; Savini, 2012; Sohn et al., 2009). At the trans-national scale, a city-region is conceptually referred to as a "bi-national city" (Ehlers & Buursink, 2000; Ehlers et al., 2001) or borderland (Brunet-Jailly, 2005; Szczepański, 1999). National borders and territories have generated barriers, distances, and control functions in integration (Anderson, 1996). According to Sohn et al. (2009), the role of a state border is "always likely to put the brakes on exchanges and cooperation, given in particular the cultural, institutional, and regulatory difference that it instigates" (p. 937). In cross-border integration, the differences in culture, institution, and regulation brought by the different nation-state systems are overt. The inherent linkage brought by globalization and the transformation of the global economic mode helps overcome the economic, institutional, and social stumbling

Table 2.1 Definitions of city-region

Scholar	Year	Definition	Core
Bennett	1985	(1) "more than 45% out-commuting and more than 30% commuting to the core"; (2) "enclosed by a public transport isochrone of one hour"; (3) "with less than 46% agricultural employment" (p. 70)	Commuting flow; public transportation system; less agricultural employment
Vermeijden	2001	It has a hierarchical spatial structure, which means satellite town surrounds urban core	Hierarchical spatial structure
Davoudi	2003	It is socially, economically, and culturally dominated by the core city	Social, economic, and cultural dimensions
Jonas and Ward	2007	It has dual economic (trade) and political (regulatory) functions	Economic and political functions of city-region
Rodríguez-Pose	2008	The essence of city-region is the interactions of economic and social development among urban core, semi-urban areas, and rural hinterland	Urban core, semi-urban and rural hinterland
Rodríguez-Pose	2008	The tie includes "a combination of economic, housing market, travel-to-work, marketing, or retail catchment factors"	–
Scott	2011	"[...] In a nutshell, they are powerful and increasingly self-affirming configurations of social, economic, and political activity within a multifaceted, multi-tiered system of global emergence"	Social, economic, and political activity
Conclusion		City-region is an emerging spatial form, characterized by the tight social, economic, and political connections among urban core, semi-urban, and rural hinterland	

Source Author

stocks among state borders. Moreover, empirical studies shed light on cross-border integration. By employing the "Growth Triangle" in the Singapore–Indonesia border zone, Grundy-Warr et al. (1999) identify genuine transnational institutions and a harmonious regulatory system as the two core structural factors in defining a well-integrated borderland. However, both factors are based on merely politics and institutions. Previously, Oscar (1994) defined the integrated borderland as occurring when "neighboring nations eliminate all major political differences between them as well as existing barriers to trade and human movement across their mutual boundary (p. 9)". Social cohesion, inter-region human movement, and economic trade are crucial elements in eliminating stumbling stocks in borderlands.

In city-regions within the nation-state, the nature of boundaries between different administrative units changes over time in conjunction with the development of the political, cultural, and economic systems. Previous literature on city-regional integration focuses on city-regional integration at the sub-national level. As the term "border" is synonymous with "national border," if used within the nation-state, it is replaced with "boundary" (*bianjie*). Thus, cross-boundary integration stands for the integration of a city-region within the territory of a nation-state (Gass et al., 2009; Lidström, 2013; Loo, 2010; Yang & Li, 2013).

The economic, political, and cultural discourses of cross-border and cross-boundary areas differ. However, consensus on the city-regional integration is reached with particular reference to regional development and governance. The city-region is regarded as the basic geographical site for social production and distribution (Jonas & Ward, 2007; Parr, 2005). It is tightly associated with the network and group that share common citizenships (Hveem, 2000; Lewis & Wigen, 1997). Commuting modes and the range of economic activities have become the major standards in delimiting the scope of a city-region (Coombes, 2014; Sultana & Weber, 2014). The functional urban region serves as the main criterion for measuring the scope of a city-region, especially in terms of economic function. Hence, vast literature replaces city-regional integration with city-regional economic integration (Breslin & Higgott, 2000; Mansfield & Milner, 1999). Tinbergen (1954) asserts that city-regional integration involves negative and positive integration. Negative integration refers to the disappearance of a physical border or boundary, whereas positive integration means how regulation and institutions modify any failure of the free market and reinforce unity. Fang and Zhang (2014) use the term "integration of urbanization area" to refer to the integration of a city-region, further providing a comprehensive illustration. An integrated city-region comprises two or more adjacent cities that are easily accessible, functional, well-organized, and in possession of a cultural identity (Fang & Zhang, 2014). Based on the recognition of urban differences and the importance of complementarities among urban functions, integration seeks approaches and outcomes of coordinated development in terms of economy, society, culture, ecology, space, and governance among adjacent cities. The operating mode is considered a high degree of agglomeration and diffusion of elements; eventually, the integrated city-region forms a highly collaborative network mode (Fang & Zhang, 2014). Furthermore, Li and Yin (2012) define city-regional integration as

> [……] the process of forming regional economic union among two or more than two countries or regions. In this process, different governments aim to seek the common development by assigning economic cooperation agreements, forming the final picture which is the free flow of goods and production within the city-region. (p. 91)

In line with Scott's definition of a city-region, the political and social lenses on the city-region attract substantial attention among scholars. Sohn et al. (2009) emphasize the importance of city-regional integration from a comprehensive lens as follows:

> [......] it emphasizes the process more than the form and it positions the analysis in a systemic perspective where the relationships between the system elements prevail over their attributes. (p. 925)

Extant reviews refer to city-regional integration as an integration with a final comprehensive aim. It forms a systemic and well-organized economic system that fosters effective institutional arrangements to strengthen the cross-boundary connection. Social cohesion, which is attained by forming city-regional citizenships, can be achieved in integration.

2.2.2 *Content of City-Regional Integration*

Environment, demography, sociology and culture, technology, economy, and consumption are the crucial components of a city-region (Hall, 2009). Harrison (2012) views a city-region as the nurturing and anchor nodes of dense political, social, and economic activities in the era of globalization. Economy is treated as a crucial factor that articulates city-regional integration; thus, scholars pay attention to economic integration (Yang, 2005; Cheung, 2015). However, a city-region cannot escape the existing territorial mosaic of regional political-administrative units (Harrison, 2010). The political-administrative dimension remains central to the city-region and regional governance. As a mature cross-border community, city-region should share a common value and identity, which are central to social integration. The city-region is affected largely by economic and institutional integrations.

Contemporary research on the integration of a city-region focuses on economic, institutional, and social domains (Andrew & Doloreux, 2012; Feiock, 2009; Nicholls, 2005; Scanlon & Adlam, 2011). At least three dimensions are distinguished: economic, institutional, and social integration (Shen, 2014). In addition, Sohn et al. (2009) classify city-regional integration into functional and institutional integration. Functional integration refers to the "form and intensity of socioeconomic interaction," whereas institutional integration is "the form and intensity of interactions between actors who are potentially willing to cooperate" (Sohn et al., 2009, p. 925). These two classifications share the same benchmark to conceptualize the content of city-regional integration. Functional integration can be categorized into economic and social integrations.

City-region space is divided into two categories: functional and institutional spaces (Coombes, 2014). The functional integration of a city-region is defined as "the form and intensity of socioeconomic interactions observed from one part of the border to the other" (Sohn et al., 2009, p. 925). Socioeconomic interaction incorporates various aspects, such as the market (Cheung & Phillimore, 2014; Nepal &

Jamasb, 2012), the industrial sector (Al-Mohannadi & Linke, 2016; Zayati et al., 2012), infrastructure cooperation (Tolone et al., 2004; Xu & Yeh, 2013), and social integration (Andrew & Doloreux, 2012). Correspondingly, institutional integration refers to "the rule of the game in society, or more formally, are the humanly devised constraints that shape human interaction" (North, 1990, p. 3). In addition, the interactions between institutional and functional integration have been the subject of various research (Li et al., 2014; Sohn et al., 2009; Xu & Chen, 2014). However, their relationship remains ambiguous and debatable. Sohn et al. (2009) argue, "there does not necessarily have to be a reciprocal link between the intensity of the socioeconomic interactions and extent of the cooperation instigated by the territorial institutions" (p. 936). Other scholars use empirical cases to prove the inevitable effect of city-regional governance on functional integration (Feiock, 2009; Gass et al., 2009; Salet et al., 2003). Therefore, the mechanism of city-regional integration must be traced back to ways that economic, institutional, and social factors interact with one another. Moreover, substantial empirical evidence must be derived from case studies in different countries and cultural and political systems, especially in socialist China.

2.3 Economic Integration of City-Region

2.3.1 Concept of Economic Integration of City-Region

International trade research considers economic integration a crucial component of city-regional integration. Extant literature focuses on economic integration at the international scale under globalization. Tinbergen (1954) argues that economic integration aims to remove international trade barriers between goods and services, achieving high collaborative efficiency to optimize the international economic structure. Balassa (1961) identifies economic integration as a process and condition. As a process, economic integration eliminates related impediments among countries. As a condition, economic integration in its final state, is the disappearance of the different treatments among integrative countries. Streeten (1964) notes that economic integration should be considered not only a process and tool but also an objective to achieve peaceful development and co-prosperity. Pinder (1969) argues that economic integration is the free movement of commodities, and the factors of production within the integrated area and the nation-state government eliminate the related stumbling blocks via policy coordination. International economic integration takes various forms in the real world, such as customs, free trade areas, a common market, and economic and monetary unions (Robson, 2002). In addition, Robson (2002) notes that the theory must expand far beyond market integration. Bryant and Javalgi (2016) enrich the essence of economic integration and elaborate such theoretical underpinnings as Foreign Direct Investment (FDI) and international trade theories. The theoretical deduction defines global economic integration as a multi-faceted construction process that includes foreign direct investment and international trade.

Given the evidence supporting economic integration at the international scale, Robson (2002) defines economic integration as follows:

> Economic integration is basically concerned with the promotion of the efficiency in the resource use on a regional basis. Necessary conditions for its fullest attainment include: the elimination of all barriers to the free movement of goods and factors of production within the integrated area; and of discrimination on the basis of nationality amongst the members of the group in that respect. In addition, where resources are allocated by the price mechanism, measures will be required to ensure that the market provides the right signals. Institutions will also be required to give effect to the integrating force of the market. (p. 2)

Economic integration is an integrative process in which members of nation-state in the integrated region attempt to eliminate the barriers of international trade between goods and factors of production. The absence of barriers attracts additional foreign direct investments. Moreover, the differential treatment within an integrated region will finally disappear. The evolution of the term "economic integration" sheds light on its assessment at the international scale. Economic integration is measured not only via the final state, such as the degree of market integration and price of international goods, but also through the economic factors within the integrated process.

Economic integration at the city-regional scale does not receive much attention in the literature. However, several pioneering works shed light on the understanding and explanation of how an economy works in the formation of a city-region. Lv and Zhang (2010) note that economic integration can be used to examine both the international and the city-regional scales. Moreover, economic integration bonds municipalities, thereby forming a regional economic unit. Li and Yin (2012) argue that the economic integration of a city-region is an economic entity in which different jurisdictions assign agreements on economic cooperation. They remove barriers and achieve the free movement of goods and factors of production to promote economic co-development (Li & Yin, 2012). Subsequent to these foundational works, the economic integration of a city-region can be best understood as a process that eliminates the barriers between the production of goods and factors of production across territorial entities. A nation-state is not necessarily the basic member of economic integration in a city-region. Often, the jurisdictions and territorial authorities within the integrative city-region are the primary stakeholders in the process of regional integration and economic co-development.

2.3.2 *Assessment of Economic Integration of City-Region*

A vast literature has assessed economic integration across variegated geographical scales. Conventional studies on economic integration articulate two main types of spatial relations. The first, the long-distance relation, also known as a network relation, denotes the urban system at the supra-regional level. The second is the proximity relation, which refers to the city-regional (regional or sub-regional) scale (Dematteis, 1997).

(1) Economic integration of long-distance urban systems

The relation for a long-distance urban system is conceptualized using Friedmann's World City Theory and Sassen's Global City Theory. World City Theory was proposed by Friedmann and Wolf in 1982. This pioneering theory introduced the terms "world city" and "global network of cities", which vividly perform as the "command centers" and the "new international division of labor" (Friedmann & Wolff, 1982). Furthermore, Friedmann (1986) proposes the following concept:

> Key cities throughout the world are used by global capital as 'basing points' in the spatial organization and articulation of production and markets. The resulting linkages make it possible to arrange world cities into a complex spatial hierarchy. (p. 71)

In addition, Friedmann defines a world city as "the organizing nodes of a global economy system" (Friedmann, 1995, p. 25, cited in Alderson & Beckfield, 2004). According to this hypothesis, global capital is embodied by the internationalization of production, in the wake of "new international division of labor" (Derudder, 2006; Friedmann, 1986). The key feature of a world city is the city-regional location, which possesses a high extent of spatially integrated economic and social systems. Additionally, the divisions between small economic units are employed for administrative purposes (Friedman, 1986). The specific form of a world city resembles a polycentric city-region, such as Randstad Holland, the Ruhr Area, and Kansai Region (Friedmann, 1995), rather than a sole urban core (Derudder, 2006). Connected with the concept of the city-region (see Sect. 2.2.1), the production of economic-geographic inequality of a tri-polar system for a city-region, comprising an urban core, semi-urban areas, and rural hinterlands, is the basic structure of the World City Network (Derudder, 2006).

Sassen's Global City Theory has embedded urban economy into the discourse of the global economy, identifying the transnational urban network that reconfigures the global core-periphery layout (Sassen, 1991; Derrudder, 2006). Global City Theory enriches and perfects the World City Theory by employing the dimension of attraction of advanced producer services instead of multinational enterprises. Global City Theory stems from World City Theory—the former utilizes the advanced producer service to examine how the headquarter economies are shaped by local knowledge-rich and technology-enabled environments (Derrudder, 2006). Shifting from the command power in the World City Theory, Global City Theory has emphasized the functional centrality of a city through investigating the role of advanced producer services in the global production and reproduction processes. Sassen (1995) notes that a global city is the command and power operation center of an agglomeration economy; it is also the center of production activities due to its global control capability and internationally-connected infrastructure. Producer service firms are viewed as the major catalyst in creating global office networks (Derrudder, 2006). Thus, the conventional core-periphery spatial structure, manifesting as the North–South divides, is reconstituted by the transnational urban system, which is "a new geography of centrality and marginality" (Derrudder, 2006, p. 2033). In Global City Theory, an urban center is more deeply connected with other urban centers within the Global City Network than with its surrounding areas or hinterlands (Sassen, 1991). Given the spatial unit and specific form of agglomeration of advanced producer service, Global City Theory takes the traditional central business district (CBD) or a grid of intense business activities as its basis of territory instead of a metropolitan region or city-region (Derrudder, 2006).

As elaborated above, the differences between World City and Global City Theories are summarized in Table 2.2.

Apart from the characteristics of a world city and a global city, scholars also focus on the World and Global City Networks, which are based on the "Space of Flows" (Neal, 2011; Taylor, 2001). There are distinct theoretical interpretations of these networks. However, methods and databases are convergent without obvious distinctions when assessing global or world city networks. For example, within the structure of a node-link, Taylor (2001) divides a world city network into three levels: (1) cities as nodes; (2) the world economy as the supra-nodal network level; and (3) the sub-nodal level, which is formed by advanced producer service firms. The strategy in selecting office locations at the global scale triggers the formation of an interlocking network. Hence, a world city network focuses on advanced producer services (Taylor, 2001).

Based on this recognition, Taylor (2001) further constructs four types of inter-city relational matrices: elemental, proportional, distance, and asymmetric. Such matrices describe the economic relations among world cities from the dimension of advanced producer services. In a different dimension, Alderson and Beckfield (2004) examine the power and position of the world city in the world city system using the data of branch locations of the top 500 multinational corporations. They measure power and prestige via several indictors, such as *indegree, outdegree*, and *closeness*. According to Alderson and Beckfield (2004), *indegree* is an indicator that is used to outline the prestige and influence of city A in the network through calculating "the number of ties it receives" (p. 824); *outdegree* refers to the connectivity of city A with other cities, in other words, how many cities connected with city A; *closeness* has been

Table 2.2 Comparison between World City and Global City Theories

	World City	Global City
Representative	Friedmann	Sassen
Function	Power	Advanced servicing
Key agents	Multinational corporations	Producer service firms
Structure	Reproduces tri-polar spatial inequality in the capitalist world-system	New geography of centrality and marginality, reconstituting existing core-periphery pattern
Territorial unit	Metropolitan region	Traditional CBD or a grid of intense business activity

Source Derrudder (2006)

employed to measure the distance of city A to other cities within the World City Network. Moreover, the authors employ block modelling to conceptualize the role and position of a world city in the world city system. Generally, they agree with Friedmann's definition of a world city, with multinational corporations as the key agents. These works focus on the position of a world city, referring to the power and influence in the World City Network, such as the measurement of *indegree, outdegree,* and *closeness*. However, Neal (2011) notes that centrality is completely distinct from power. He further categorizes the world city into three types (i.e., quintessential, hub, and gateway world cities) using the criteria of centrality and power.

Apart from corporate organization data, a few scholars adopt infrastructure data to outline the characteristics of the world and global city networks. Moss and Townsend (2000) measure the inter-metropolitan links using transcontinental data transport services (i.e., internet backbone bandwidth) from the dimension of the development of advanced telecommunications services. Smith and Timberlake (2002) use the travel data of passengers to measure the connection of the world city network. Derudder et al. (2008) demonstrate that air passenger flow data are the proper information to measure the global city network. Short et al. (1996) list five criteria to assess the position of a world city or global city, including "major financial centers, corporation headquarters, telecommunications nodes, transportation nodes, site of global spectacles" (p. 698).

To summarize, the economic integration of a long-distance urban system has not yet generated a universally acknowledged paradigm. The functions of a world city have a weak focus on advanced producer services but a strong focus on power in contrast with the function of the global city. Therefore, the basic territorial scope of a world city is the metropolitan area (core city and surrounding areas of production) and that of a global city is the traditional CBD or a district of intense business activity, but its linkage and influence are of international scale. Although these concepts are apparently different, the research on the global and world city networks are methodologically integrated. For example, Taylor (2001) employs the criterion of advanced producer services to categorize the levels of a World City Network. The data for the assessment manifest as two categories: (1) business contract data on corporate organization, which refers to producer service firms and multinational enterprises, and (2) data on infrastructure between cities, which refers to passenger and freight flow.

(2) Economic integration of city-regions

Given the availability of data on inter-firm relations and imperfect methodologies, a smattering of literature embarks on empirical studies to measure the economic integration of city-regions. However, few studies on inter-firms relations and imperfect methodologies uncover the economic dynamics of city-regions through network analysis. Van Oort et al. (2010) divide urban economic complementarities into spatial and functional integration and further use inter-firm data to measure the spatial structure of Randstad in the Netherlands. The data in relation to inter-firm relations includes information on the customers and suppliers of these firms with reference to their daily business (i.e., physical goods, services, and business information). The data involves more than 2,000 firms. The method used to measure spatial integration is the gravity model, and the method for functional integration is location quotients. The major findings challenge the existing views on the spatial structure of Randstad, which is commonly perceived as a typical polycentric city-region. Conceptually speaking, a relatively high hierarchy exists in the spatial independence in Randstad economically. Yeh et al. (2015) employ intra- and inter-firm data in producer services to measure urban connectivity in PRD. The results show that PRD is hierarchically organized and spatially localized from the dimensions of intra- and inter-firm linkage on city connectivity. Previously, Lüthi et al. (2010) used the same type of data to identify the spatial structure of the megacity-region of Munich through the value chain approach. The results reveal that Munich shows not only the characteristic of being hierarchically organized but also a strong high-grade localized system of the value chain. Hall and Pain (2006) conduct a well-conceived empirical study in which they analyze the polycentric spatial structure of eight metropolitan areas on the basis of various intra- and inter-city flows such as commuting flow, capital flow, communication data flow, and passenger flow. The eight metropolitan areas include South East England, the Randstad, Central Belgium, Rhine Ruhr, the Rhine-Main Region of Germany, the EMR of Northern Switzerland, the Paris Region, and Great Dublin. This fuller dissection integrates quantitative and qualitative analyses to measure the polycentric structure of these metropolitan areas. Specifically, they use the data on inter- and intra-city flows to measure the polycentric structure quantitatively, and interview a large number of experts and take field trips to qualitatively make up the missing information in the quantitative analyses.

In sum, the existing literature on the connectivity of city-region using inter- or intra-firm data forms a powerful, mature, and useful method for exploring the city-region connection, and the method is mature and exercisable. However, for the economic integration of city-region, such data are difficult to collect, as they present a huge workload (e.g., by using surveys) and are inconsistent in statistical caliber. In addition, the selected data to measure inter-city linkage are based on the flow within the sector of advanced producer services, which largely simplifies the local context of the study area. For example, in the context of rapid urbanization and industrialization in China, advanced producer services may play a less prominent role in the economic and industrial system *vis-à-vis* the manufacturing industry. Thus, the representativeness and rationality of one-fold flow data are weak in the context of China with reference to the changing local socioeconomic environment of different city-regions.

Aside from the lens of intra- or inter-firm relations, some scholars select other criteria to assess the economic integration of a city-region. In light of the definition of economic integration at the international and city-region scales, Shen (2014) uses FDI and international trade as indictors to examine the degree of economic integration between Hong Kong and Shenzhen. Other scholars focus on the dimension of the market in the economic integration of a city-region. Braun (2008) sets up an integrating approach to clarify a city's marketing, emphasizes how governance structure enables the formation of city marketing, and embeds city marketing into a wider policy framework. Li and Yin (2012) identify two crucial parts of the economic integration of a city-region, specifically referring to market integration and integration of production factors. The integration of production factors is further refined into the agglomeration of industry and local geographical specialization. Based on this recognition, Liu (2015) enriches the study of regional integration by using the integration of Nanchang–Jiujiang as an example of widening the scope of investigation by involving market integration, industrial integration, infrastructure integration, and public service integration. According to the existing literature, the conventional method to evaluate market integration is by the price method (Goldberg & Verboven, 2005; Ravallion, 1986). For example, Ravallion (1986) employs a mode of spatial price differentials to measure market integration through selecting the monthly rice price. Regarding industrial integration, how it actually operates depends on the local context, such as the stage of industrial development, the existing structure of the industry and the primary challenges that the city-region confronts (Chen, 1998; Frank, 1925). In addition, the infrastructure and public service integration are also under-discussed; their actual content and approach to integration still need to be investigated in future studies. Thus, no generally acknowledged paradigm exists to measure the economic integration of a city-region, and the actual contents and methodologies of such a paradigm should be context-sensitive and embedded in the local economic environment.

2.3.3 Research Gap

In sum, as discussed above, the economic integration of a long-distance urban system is extensively studied through the network-based analysis paradigm. World City Theory and Global City Theory provide solid theoretical and methodological foundations for the measurement of economic linkages and depiction of the role of an individual city and inter-city relations within the world city system. The network analyses where inter-firm or intra-firm data are used in assessing city-regional economic integration in China have two deficiencies. The first one is data availability. The inter-firm or intra-firm data are difficult to gather, and historical data across different time periods are even more challenging, which poses an obstacle to demonstrating the formation and evolution of economic integration. Second, this method reveals merely a one-fold dimension or a single type of industrial linkages in the formation of city-region.

However, there is scant literature systemically deciphering the economic integration of a city-region and providing the generally acknowledged paradigm and methodology to measure the economic integration of a city-region. Unlike the world city system which selects the fixed standards and dimensions to define a world city, the cities in the city-regional scale normally stay at the different economic statuses, and selecting agreed standards to measure the inter-city relations is difficult. Also, the local socioeconomic environment of city-region changes case-by-case, so selecting the standards to define and depict all of the economic linkages in a city-region is difficult (Hall & Pain, 2006). For example, Shen (2014) selects the FDI and international trade to measure the economic relations between Shenzhen and Hong Kong which are specialized by a political border. However, such indictors are difficult to apply in the city-regions in inland areas and some disputed areas where the official definitions of city-region are disputable. This is the reason why Hall and Pain (2006) employ a variety of intra- and inter-city flows to outline primary metropolitan areas in the EU. Compared with the network-based analysis, which is prevalent in global city or world city research, the assessment of economic integration of a city-region is in dire need of comprehensive investigation. The local socioeconomic environment needs to be fully rearticulated, research methods and indictors need to be properly selected and formulated and problem-oriented domains of economic integration within the local socioeconomic environment are to be carefully crafted. In other words, the economic integration of a city-region needs to follow a case-by-case paradigm and identify the local socioeconomic environment of every concrete city-region. In addition, economic integration should be treated as a process, which is central to the government's endeavor, to eliminate the barriers to the free flow of goods and factors of production. Thus, historical evolution needs to be investigated further in assessing the economic integration of a city-region.

2.4 Institutional Integration of City-Region

A city-region is an agent of wealth creation, and its dual functions reconstitute the relation of trade, which binds the urban core and wider hinterlands, including economic (trade) and political (regulatory) functions (Jonas & Ward, 2007). Jonas and Ward (2007) utilize "political territories" to refer to the characteristic of developing "regulatory and decision-making capacities" autonomously (p. 171). Aside from the conventional research in socioeconomic integration within a city-region, currently, a growing literature probes city-regional integration through the political–institutional lens. The literature maintains the belief that a city-region is constructed politically (Brenner, 2004; Ward & Jonas, 2004).

2.4.1 Changing Concept of an Institution: Deploying the Organization-Embraced Approach

The concept of an "institution" stems from social anthropology, consolidating language, morality, law, business, hierarchy, and technology through a culture-based perspective. Under this recognition, the institution is regarded as "what appeared in this perspective as an institution was what had also been characterized in the legal tradition as 'institute' or institution'" (Brie & Stölting, 2012, p. 21). An institution is also bounded with sociological research, focusing on "the element of expectations regarding one's own activities and that of others by means of which stability and regularity are constituted" (Brie & Stölting, 2012, p. 21). North (1990) provides an insightful view to emphasize the feature of societal rule, by describing an institution as "the rule[s] of the game in a society or, more formally, the humanly devised constraints that shape human interaction" (p. 3).

According to the institution-as-rule framework, an institution can be segmented into formal and informal institutions. A formal institution encompasses law, regulation, state, and global economic structure, whereas an informal institution incorporates culture and local customs (Hudalah et al., 2014). Most prior research focuses on observable and formal institutions (Hudalah et al., 2014). To distinguish the difference between formal institution and informal institutions, Brie and Stölting (2012) note five differences between them, involving the difference between social and group norms, the difference between individual action and formal institutional condition, the difference between role conformity and deviance, the difference between the public and non-public spheres, and the difference between a reference to common social welfare and group interests.

However, such an "institutions-as-rules" framework is criticized by Greif (2006) for the neglect of organization and private order and treating motivation as merely exogenous. Institution is reinterpreted as "equilibrium phenomena in which they constitute the structure that influences behavior, while the behavioral responses of agents to this structure reproduce the institution" (Greif, 2006, p. 14). More importantly, Greif (2006) holistically conceptualizes an institution as "a system of rules, beliefs, norms, and organizations" (p. 35) and emphasizes the organization as either a component of an institution or an institution due to its motivating function for behaviors. This "organization-embraced" approach sheds light on the conventional wisdom for the mutually determinant relations between an organization and an institution (Greif, 2000, 2006).

Resonating with such a monolithic approach, political economists identify territorial organizations as useful tools to re-ascertain the institution-building of a city-region (Swyngedouw, 1992) as the "geographically fixed configurations of the territorial organization" (Brenner, 1998, p. 462). A city and a territorial state serve as two specific forms of territorial organization where the geographical scales in the circulation of capital are deployed. In particular, a territorial state is clarified as the institutional mediator for reducing economic-geographic inequality (Brenner, 1998, 1999). According to Mongelli et al. (2005), the main manifestation of an organization at the international scale is the government of countries. Similarly, in Peng's (2000) definition at the regional scale, institutional integration is "integration through regional integrative organization and comprehensive legal agreements" (p. 177). However, the specific forms of a regional integrative organization and comprehensive legal agreement remain ambiguous, perplexing, and amorphous conceptually. For example, in Peng's definition, what the specific forms of the regional integrative organization remain unclear. Is it an independent, semi-independent, or loose union across member cities in a city-region? Thus, institutional research has not yet found a universally acknowledged paradigm, and the existing literature on institutional integration is necessarily inherits a context-sensitive lens through exploring specific institution-building processes in every concrete city-regions. Correspondingly, working from the institutional practices of the EU in facilitating city-regionalism and fostering regional affinity, Salet and Thornley (2007) emphases that triumphant metropolitan strategic policies have occurred, owing to full-fledged institutional preconditions that triggered regional connectivity. In this respect, institutional building in the EU actually represents a successful example of an integration-oriented and synergetic regionalization process that offers various pathways across territorial organizations.

2.4.2 *Rationale of MLG and Institutional Praxis in the EU Integration Context*

Globalization and the global economic restructuring from Fordism to post-Fordism are the driving forces in reshaping the integrative institution of city-regions, exerting economic and political impacts across variegated geographical scales (Brenner, 2004; Delwaide, 2011; Nicholls, 2005; Perkmann, 2007a). Global economic restructuring splits national and territorial economies, where the "space of flow" overrides "space of place" (Castells, 1989, 1996). This economic detachment thereby erodes the role of the nation-states and rebuilds a worldwide urban hierarchy (Sassen, 1991, 2006). The city-based accumulation regime is evocatively depicted as follows: "cities are therefore no longer to be conceived as the sub-national components of self-enclosed, auto-centric and nationally scaled regimes of accumulation" (Brenner, 1999, p. 437). In addition, the concomitant geographic reconfiguration deemed equally crucial is the economic connection between the urban core and its hinterlands that is reinforced and fueled by the agglomeration effect of post-Fordist economic sectors, which transcends the urban border, enabling cross-border interactions (Scott, 2001).

At the global scale, the world urban hierarchy is formed to reconstitute a new core-periphery layout, and inter-city competitions are intensified by proactive engagement with international circulatory capital. At the national scale, the national urban system is integrated into a global or supra-regional city system, and the economy of a world city is disjointed from the national sphere. At the urban-regional scale, the exopolis appears, in which the restructuring of urban forms incurs polycentric city-regions and new industrial spaces (Brenner, 1999). In particular, urban space is restructured as the emerging metropolitan areas and megaregions, which are increasingly being canonized as chiefly geographical sites for reshuffling socioeconomic regulation and wealth creation, such as Greater London's and Randstad's polycentric regions.

The territory of nation-states has long been assumed to be eroded under the globalizing regime (Ruggie, 1993; Strange, 1996). However, there is growing literature demonstrating that the reconfiguration of the state's regulatory capacity is entangled with urban rescaling to remold governance at local, sub-regional, national, and supra-national levels (Brenner, 1999; Scott, 2001). The EU, which consists of selected nation states, is undeniably represented as the typical trans-territorial region affected by urban and state rescaling (Brenner, 2004; Nicholls, 2005; Raunio, 2011). The nation-states rescaled themselves in response to the post-1970s wave of globalization (Brenner, 1999). With the "hollowing-out" of national power and the decisively weakened political capacity from Fordist Keynesianism to post-Fordist Neoliberalism, nation-states transfer powers upward to supra-national organizations to ensure socioeconomic regulation for governing cross-border affairs and European development strategies (Breathnach, 2010; Jessop, 1996). Moreover, in light of the dis-embedding of the local-territorial economy from the national economic system, national power is decentralized to localities through deregulation and privatization (Breathnach, 2010; Brenner, 1999; Swyngedouw, 2000). Generally speaking, under the harsh climate of globalization, the hollowing-out of the nation-state, and dramatic spatial restructuring, the territory of statehood has been correspondingly reconstructed through stratifying the power of nation-states in accordance with the multiple tiers of super-national, national, regional, and urban scales. This crisis-induced power reconfiguration is thereby conceptualized as a state re-territorialization process, particularly manifesting as the creation of place-specific locational advantages by the local statehood in attracting international circulatory capital (Brenner, 1999).

Regarding the governance structure, the politico-economic evolution of the EU can be seen as what Marks (1993, p. 392) terms "multilevel governance," marked by dispersive decision-making powers across variegated geographical scales and jurisdictions to depart from the centralized state bureaucracy (Liesbet & Gary, 2003; Marks, 1993). According to Yang (2005), MLG in the EU is a governance mode in which "sub-regional, regional, national, and supranational authorities interact with each other in two ways: first, across different levels of government (vertical dimension); and second, with relevant actors within the same level (horizontal dimension)" (p. 2149). This organization-based concept of MLG enriches the scope of the territorial organization from the multiple levels of government to non-governmental sectors, such as corporations, non-governmental organizations (NGOs), and the community.

Connected to the socioeconomic evolution of the EU, MLG exerts an escalating influence on the trajectory and direction for the multiplication of institutional integration, endangering three prevailing pathways among territorial organizations within the vertical and horizontal state apparatus and between public and private sectors. First, various forms of region-based authority are widely established, representing the interests of a city-region. The prevalent form is the regional government, taking responsibility for the provision of region-wide public service (Salet & Thornley, 2007). The territorial meso-government varies across the regions in the EU to tie up with the individual state regimes (Keating, 1998). It is also typical that other independent or semi-independent regional authorities, such as the Committee of the Regions, and Regional Office, serve as principal channels of subnational representation (Hooghe & Marks, 1996). The specific forms of regional authority are contextual and differ enormously in keeping with the politico-cultural traditions of nation-states to enhance efficiency and accountability (Hooghe & Marks, 1996; Moore, 2008). Second, the construction of the inter-jurisdictional network is also conceived out of the spatial reconfiguration of the territorial state in the cross-border regions (CBRs) to enable MLG (Blatter, 2004). Aside from regional institutions, "a horizontal border-crossing network of local authorities" is a solid platform to integrate territorial organization in CBRs and manifests as a voluntary process of city grouping, crystalized by a formalized agreement (Perkmann, 2007b). The example of EUREGIO CBRs suggests that the formalized inter-jurisdictional agreement with a technocratic nature, essentially constituting problem-solving modes of cooperation, functions multi-objectively (e.g., for economic development, social matters, and daily issues) beyond the pure bargaining and is instituted as the pivotal level for MLG (Perkmann, 2007b). Likewise, in the case of the Autonomous Community of Basque Country, the cross-border cooperative body is legally grounded by a formalized agreement across nation-states (García-Álvarez & Trillo-Santamaría, 2013). Finally, building the partnership across public and private sectors as the major socio-political force and prevalent practice to deal with regional affairs has spatially integrated the territorial organizations in response to the harsh climate of globalization. Alongside the privatization of public infrastructure in the later 1980s, the full engagement of quasi- and non-state actors broadly occurred in multiple fields of public services, such as water supply (Chong et al., 2006), transportation infrastructure (Kwak et al., 2009), and healthcare facility (McKee et al., 2006). These institutional realignments entail the structural consolidation of multiple local authorities and private actors in close conjunction with rescaled urban space.

2.4.3 Institutional Integration in China's City-Regionalism: A Critical Review

MLG demonstrates a changing spatial division for socioeconomic regulation under post-Fordist Neoliberalism. The now voluminous literature affirms that the rescaling of territorial organizations is frequently seen as the production of MLG at a city-regional scale through restructuring the hierarchical state apparatus, intergovernmental relations, and social rules across multiple actors (Brenner, 2004; Jessop, 2002). Under these theoretical deductions, the socioeconomic evolution and regulatory activities are scaled and geographically conjoined with how territorial organizations respond to capital movement under post-Fordist Neoliberalism through the spatial infrastructure.

Against this backdrop, growing political economy theses probe the socioeconomic evolution in post-reform China, notably *vis-à-vis* the territorial-political processes of the city-region, and the relations across territorial organizations are re-articulated within the case-specific paradigm (Li et al., 2014; Xu & Yeh, 2013; Zhang et al., 2021a). Although institutional integration receives substantial attention from policymakers and researchers, at least two theoretical gaps emerge among the political economy theses. The first gap concerns the conceptual framework of MLG or state rescaling. State rescaling is a theory that portrays the changing geographies of statehood under modern capitalism, marked by dynamic and periodic movement of state space (Brenner, 1999, 2009). Specifically, the power configuration of the hollowing-out of a nation-state is dynamically abstracted to a self-adjusting historical motion that moves upward to super-national organizations and downward to sub-national authorities (Brenner, 1999). In addition, the rise of city-regions is elaborated as the outcome of the spatial selectivity of the state through creating a new locus to enable socio-economic regulation in coping with the negative externalities of urban entrepreneurialism, such as environmental pollution and fierce inter-city competition (Brenner, 2003; Jessop et al., 2008). Conventional studies on the regional governance of China sketch a half-baked framework, with a view to interpreting state spatiality as well. Li and Wu (2018) re-illustrate China's city-regionalism as a co-existence between top-down and bottom-up processes, based on up- and down-scaling. Luo and Shen (2009) enrich the non-linear scale-building perspective by adding the hybrid mode into the inter-scalar relation. Li et al. (2014) attempt to construct a full-scale framework of state spatiality by incorporating the horizontal dimension (destatization and statization), notably foregrounding the relation between the state and non-state domains. However, underpinning the city-regionalism is the relation among cities, which is inadequately illuminated by the case they selected. To some extent, the issue of a half-baked framework is partially offset by Xu and Yeh (2013), who scrutinize inter-jurisdictional cooperation through building a full-fledged and monolithic framework, involving inter-ministry politics, inter-scalar relations, inter-city politics, and state-market relations. However, the rationale of regional governance at the overall city-region scale is downplayed by the case-specific paradigm. This lacuna is the second theoretical gap that hampers our awareness of city-regionalism from the macro perspective. Although the case-specific paradigm details the intrinsic logic of city-regionalism considerably and offers profound theoretical elaboration, the lack of perspective throughout the regional practice undermines the understanding of the extent of institutional integration due to the scant theoretical dialogue across cases. Certainly, researchers are well aware that multi-dimensional relations emerge in MLG with reference to actor interaction (Liu et al., 2012; Yang, 2005; Zhang et al., 2019).

2.4.4 Research Gap

In seeking the assessment of the institutional integration of city-regions, the existing literature signifies that the word "institution" should be deployed as an organization-embraced approach, rather than institution-as-rule, and the specific forms of territorial organization have been meticulously identified as the territorial state and city. The institutional building process in the EU not only represents a successful example for institutional integration but also signals that the integration of territorial organization develops toward producing MLG at the city-regional scales. Correspondingly, under the MLG framework, three integrative pathways in the EU are identified: region-based institutions, inter-jurisdictional cooperation networks across local authorities, and public–private partnerships. Although this integration approaches sheds light on the assessment of institutional integration of a city-region, the conceptual framework that assesses institutional integration in China has two theoretical deficiencies. The first one is the half-baked framework that provides an incomplete picture to examine the relations between territorial organizations. The second is the case-specific paradigm, which fails to create a comprehensive interrogation of the institutional integration of a city-region at the macro level. Thus, further research is needed to question the institutional integration by examining MLG at the macro level to associate it with a host of city-regionalism praxis and by establishing a well-rounded conceptual framework wherein the relations across different levels of government, between cities, and between state and non-state domains are systemically integrated. In the context of examining MLG in China's city-region, the EU's national border issues and hurdles, however mitigated, do not apply to the nation-state of China.

2.5 Revisiting Governance Reshuffling Under the City-Regionalism of China: A Case-Specific Paradigm

According to the existing literature on regional studies, a case-specific paradigm has been widely employed to illustrate how governance structure is reshuffled in response to the functional integration of a city-region. To capture city-regionalism, the governance apparatus across different levels of government, between cities, and state and non-state domains are meticulously investigated based on the case-specific paradigm. These case studies probing the reshuffling of governance structure focus on two primary realms for the purpose of this book: (1) industrial integration propelled by cross-boundary economic cooperation zone, and (2) infrastructure integration driven by large-scale infrastructure projects. They are discussed in detail in this section.

2.5.1 Building the Integrative Governance Structure to Enable Cross-Boundary Economic Cooperation Zone

To facilitate the industrial integration of a city-region, a cross-boundary economic cooperation zone is often formulated to reconcile regional economic-geographic inequality and enhance urban competitiveness. Thus, many political economy studies place attention on the empirical case of the cross-boundary economic cooperation zone in decoding the specific process of city-regionalism in China. Generally, the case-specific paradigm employs the context-sensitive lens, indicating that the specific governance reshuffling process varies in accordance with different local socioeconomic environments. Within the making of a city-region, different levels of government attempt to create cross-boundary regions to effectively eliminate the administrative hurdles and restructure the functional space of a city-region. The existing literature identifies three types of spatial forms of cross-boundary economic cooperation zones as means to create an integrative governance structure (Table 2.3).

According to Liu et al. (2014), the three types of spatial forms (Table 2.3) have been anchored under four integration-oriented governance modes, namely, (1) MLG, (2) sub-regional differential governance, (3) local government-dominated governance mode, and (4) hosting governance.

MLG builds a two-level management entity with the explicit division of duties (Fig. 2.1). Specifically, the upper government empowers a specific agency to manage a cross-boundary economic cooperation zone, forming an upper committee that is affiliated to the upper government. Aside from the upper committee, the local government sets up subordinate committees to address daily issues of a cross-boundary economic cooperation zone. The subordinate committee is administratively managed by the local governments and upper committee concurrently. In terms of power configuration, the responsibility of the upper government is to coordinate the struggles between cities and supervise the daily operation of inter-city cooperation. The local government is the executive body in the governance mode. This mode is the conventional mode in the majority of cross-boundary economic cooperation zones. The "Tianfu New Area" and "Xi(*Xi-an*)-Xian(*Xianyang*) New Area" created the MLG to enable industrial integration between cities (Liu et al., 2014; Lu, 2015).

Table 2.3 Three spatial forms of cross-boundary economic cooperation zone

Type	Characteristics	Spatial form	Case
Type I	Integration by a common cross-boundary area Two cities develop in the opposite direction Cross-boundary area is located in the development axis of two cites and achieves spatial integration	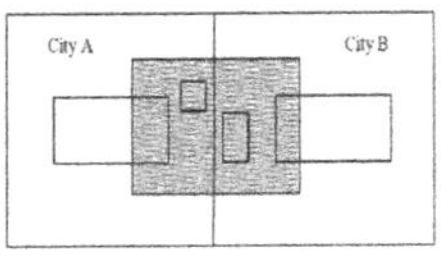	Gui-An New Area, Zheng-Bian New Area, Shenyang-Fushun New Town
Type II	Integration by constructing inter-related urban industrial zone, new town and new zone Every zone interconnects with others Achieving spatial integration	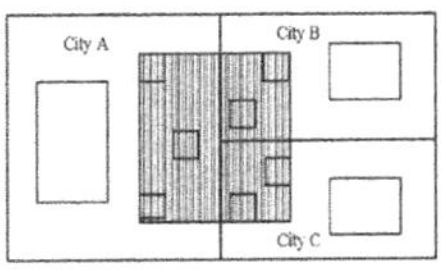	Tianfu New Area, Xi-Xian New Area
Type III	Integration by locating one cross-boundary industrial zone in one city Cross-boundary means cross-boundary development of economic element	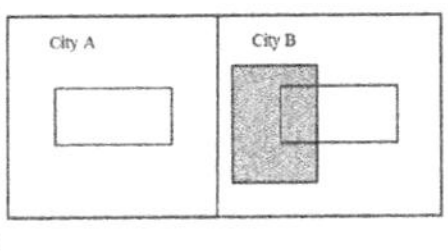	Jiangyin-Jingjiang, Industrial Park Su-Tong Science and Technology Park

Source Based on Liu et al., (2014)

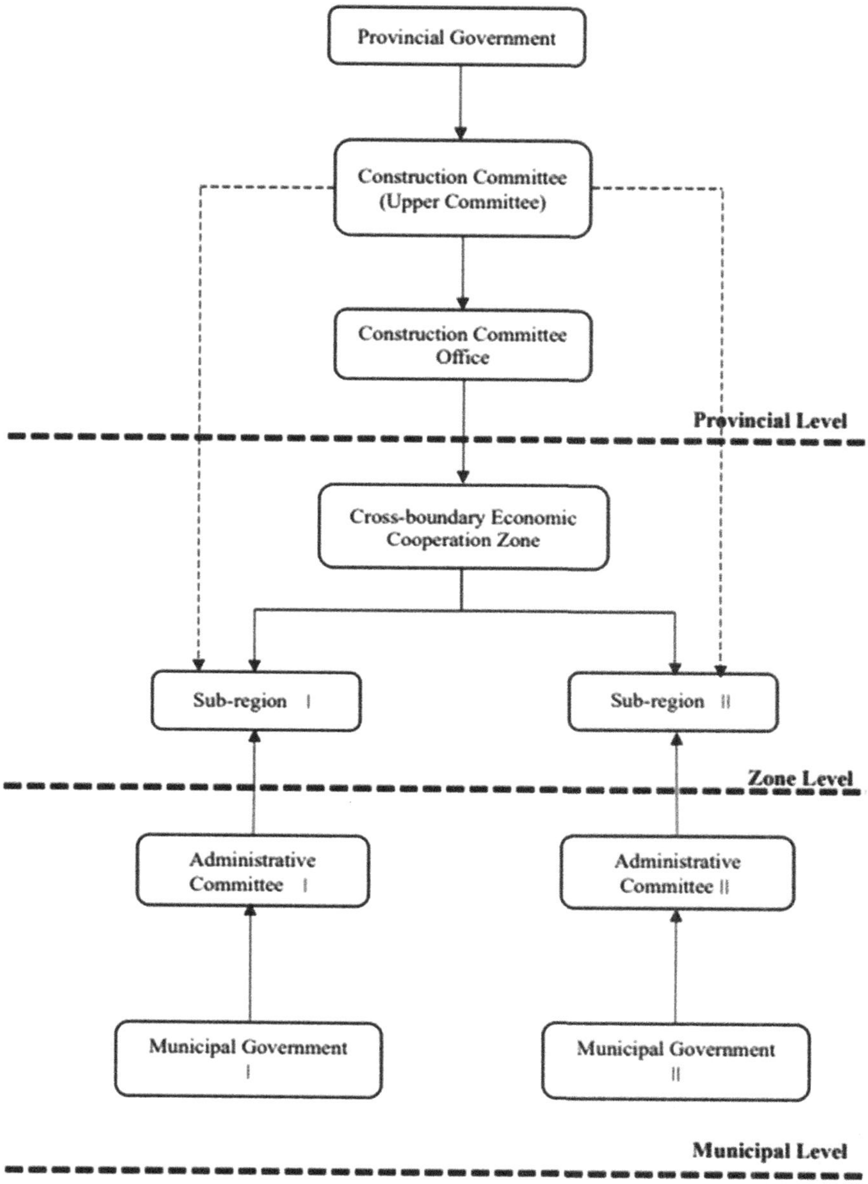

Fig. 2.1 Administrative system in MLG. *Source* Liu et al., (2014)

For sub-regional differential governance, cross-boundary economic cooperation zones are divided into direct-control areas administered by the upper government and indirect-control areas administered by the local government, reducing institutional organizations and enhancing administrative efficiency (Fig. 2.2). Regarding this mode, there is great power and institutional fragmentation in cross-boundary economic cooperation zones. The political intervention from the upper government inhibits the enthusiasm of the local government and undermines the economic-spatial and institutional integrity of the zone (Liu et al., 2014; Zhang et al., 2022a). The "Gui(*Guiyang*)-An(*Anshun*) New Area" adopts this governance mode (Liu et al., 2014; Wang & Li, 2015).

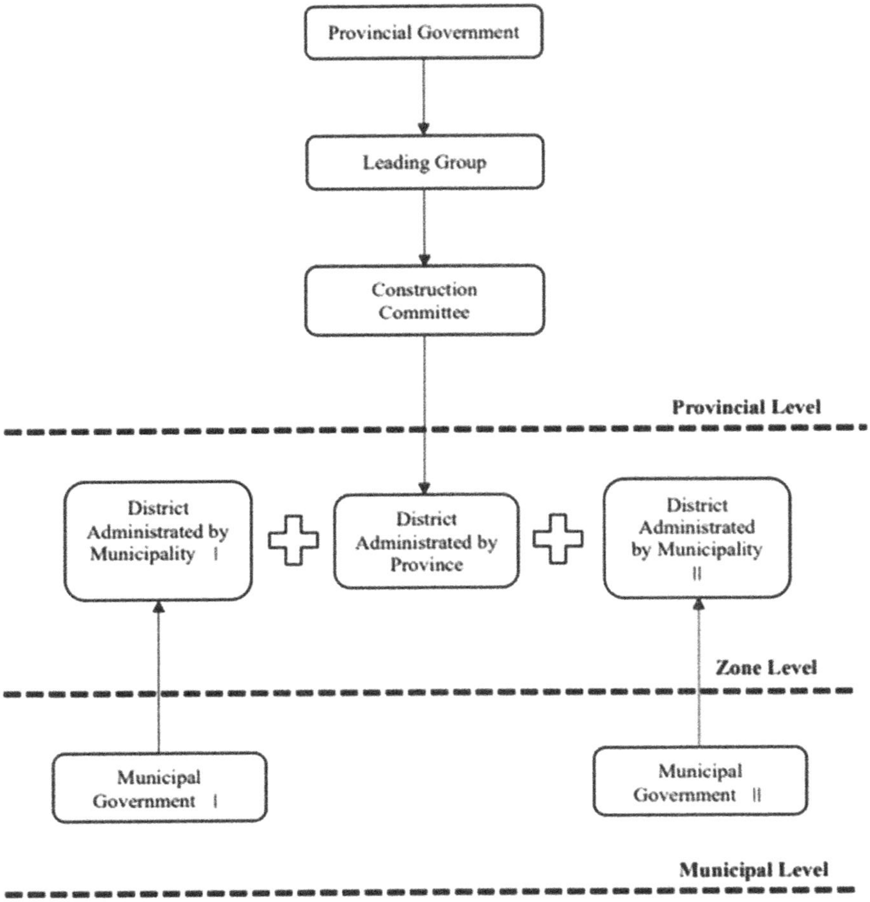

Fig. 2.2 Administrative system in the Sub-regional differential governance mode. *Source* Liu et al., (2014)

Regarding the local government-dominated governance mode, the local government is the absolute overarching actor, responsible for managing daily issues of cross-boundary economic cooperation zones, whereas upper governments are in charge of coordinating the conflicts of interest between cities (Fig. 2.3). In this mode, the upper government decentralizes the majority of administrative powers to local governments and merely retains the coordination power for inter-city struggles and tensions. Such a power configuration gives rise to the relatively flexible administrative configuration in the zone and great socioeconomic autonomy for the local government to conduct institutional innovation. The "Zheng(*Zhengdong*)-Bian(*Bianxi*) New Area" is a typical industrial district that adopts this governance mode (Liu et al., 2014; Zhao, 2009).

Hosting governance features a disjunction between sovereignty and the right to govern (Fig. 2.4). One municipal government transfers the administrative power to the target municipal government, and it retains merely the negotiation powers for some significant issues of the cross-boundary economic cooperation zone. Within this mode, on one hand, an exporting city expands its territory through injecting capital, labor, technologies, and other factors of production into the zone. On the other hand, an importing city transfers part of its administrative powers for its territory to an exporting city; however, it normally acquires a socioeconomic spillover effect from the exporting city (Luo & Shen, 2007; Zhang et al., 2022b). In terms of duty division, an importing city generally takes responsibility for social affairs such as land acquisition and housing demolition, and an exporting city takes charge of economic development and administrative management. The governance of the Jiangyin-Jingjiang industrial transfer park of Jiangsu Province follows this type (Liu et al., 2014; Luo & Shen, 2007; Zhang & Sun, 2019; Zhang et al., 2011).

To summarize, for the industrial integration of a city-region, the establishment of cross-boundary economic cooperation zones is not only an integration-oriented strategy for different levels of government, it is a focal point in studies to decipher city-regionalism. The governance structure in these cross-boundary economic cooperation zones is reshuffled toward an integrative mode, manifesting as the restructuring of two relations: (1) inter-scalar relations, and (2) inter-city relations. These territorial practices aim to create an integration-oriented governance structure among cities, and between the provincial government and local governments. To capture institutional building that enables industrial integration, the main task is to examine how integrative governance structure is triggered by complicated politics within the state apparatus contextually. However, these existing governance modes seem inadequate in terms of the complicated politics of city-regions and how the ad hoc and haphazard local socioeconomic environments reshuffle the governance structure to boost city-regionalism, which is in dire need of further investigation.

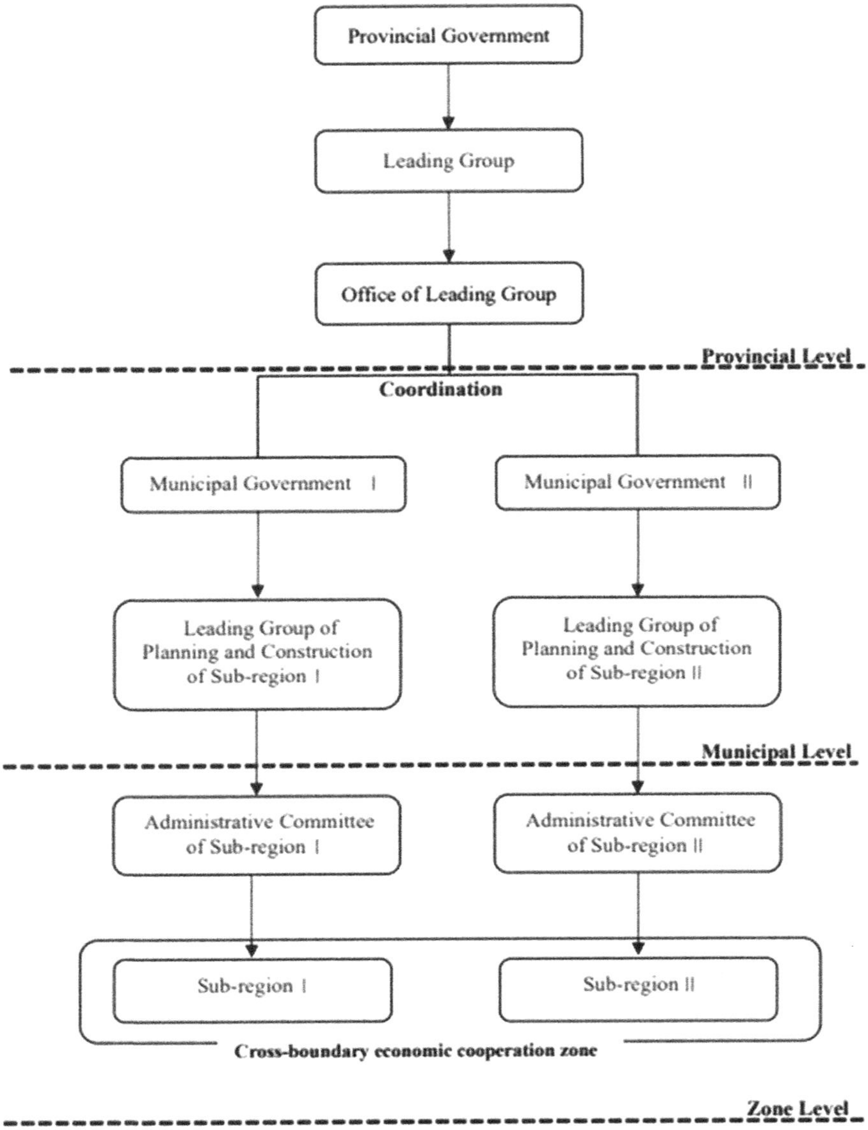

Fig. 2.3 Administrative system in local government-dominated governance mode. *Source* Liu et al., (2014)

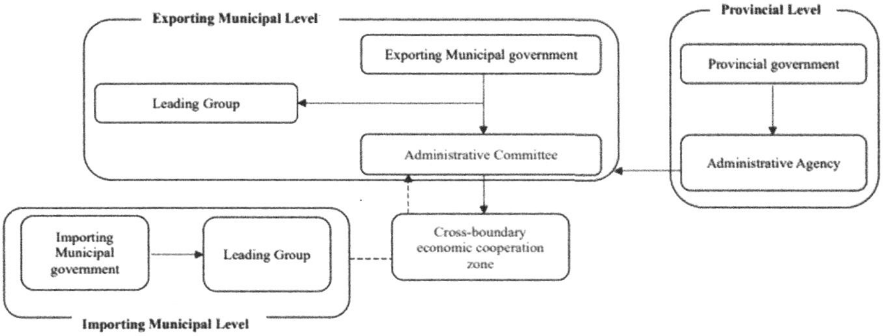

Fig. 2.4 Administrative system in the hosting governance mode. *Source* Liu et al., (2014)

2.5.2 Inter-scalar Politics in the Large-Scale Infrastructure Project

Infrastructure as the foundation for city-regional integration accelerates the mobility of capital, labor, and information flows across different territories. The integration of infrastructure is viewed as a precondition and an impetus for facilitating city-regional economic integration. Hence, the construction of large-scale infrastructure is essentially the dominant field in the making of city-regions, involving different stakeholders such as provincial governments, local governments, development enterprises, and state-owned corporations. The existing literature probing large-scale infrastructure projects focuses on the inter-scalar politics between different levels of authority (Li et al., 2014; Xu & Chen, 2014; Xu & Yeh, 2013; Zhang et al., 2023). Inter-scalar politics is an analytical lens to investigate the tensions and struggles across different levels of authority. The key dimensions in inter-scalar politics include not only the relations between different levels of government, but also the relations between the vertical hierarchy (*tiao*) and horizontal organizations (*kuai*) within each level of government in China (Lieberthal & Oksenberg, 1988; Mertha, 2005; Xu & Yeh, 2013).

Conceptually, *tiao* denotes functional administrative bureaucracies that consist of both high- and low-level government institutions. *Kuai* refers to the exclusive power of local governments, which holds the comprehensive functions and jurisdictions for its territory (Mertha, 2005). To decipher the relation of *tiao-kuai* in China, Mertha (2005) investigates the attempts of the central government to counteract local protectionism in China. He argues that *tiao* has been strategically employed as the political tool of the central government to recentralize powers in selected fields from the local to the provincial level (Mertha, 2005).

According to Lieberthal and Lampton (1992), the relation of *tiao-kuai* has manifested as follows:

> In China's unitary system, *tiao-tiao* lines of authority tie each unit vertically to superior organs of power at the Center, whereas *kuai-kuai* lines of authority tie them horizontally to local

> organs of power. In the Chinese scheme of things, both *tiao-tiao* and *kuai-kuai* authorities are to share power cooperatively according to a system of dual rule, or *shuangchong lingdao*. (p. 286)

The persistent power disequilibrium and continual jostling between vertical bureaucracies and horizontal coordinating bodies in China's state apparatus have been conceptualized as "fragmented authoritarianism" by Lieberthal and Oksenberg (1988). The central goal of *tiao-kuai* interaction is to determine which entity has the authority related to a specific issue in urban and regional development (Christoffersen, 1996). Essentially, within China's administrative system, *tiao* aims to ensure the authority of the central government by controlling localities in different functions and implementing the decrees of high-level government at the top-down uniformly (Mertha, 2005). *Kuai* has been created to enable the independence of local government in coping with external influence and making use of local conditions and advantages (Mertha, 2005). Thus, the intricacy of *tiao-kuai* relation speaks of the varied trajectory of power reshuffling between decentralization and (re)centralization in the regional governance in post-reform China.

In terms of inter-scalar politics across different levels of government, substantial cases focus on PRD, including the Guangzhou-Zhuhai railway (Xu & Yeh, 2013), PRD inter-city railway (Li et al., 2014; Xu & Chen, 2014) and Guangzhou-Foshan Metro (Li et al., 2014). Xu and Yeh (2013) utilize the Guangzhou-Zhuhai railway as an example to construe the institutional configuration in cross-boundary infrastructure projects. The construction of the Guangzhou-Zhuhai railway was conducted on a case-to-case basis through a considerable body of bargaining without the formal institutional arrangement of horizontal cooperation. Due to the lack of formal institutional arrangement, the successful implementation of an inter-city railway depends largely on the politics at multiple spatial scales. The legitimacy of local government can be supported by strengthening development and going beyond pure economic logic (Xu & Yeh, 2013; Zhang et al., 2021b). Li et al. (2014) examine the construction process of the inter-city railway of PRD and the Guangzhou-Foshan Metro by modifying the rescaling of the state in China. Within the horizontal and vertical cooperation processes, the physical and economic landscapes are shaped by state-led initiatives that varied from different levels of government, and destatization of power fosters a diverse platform for various stakeholders that compete yet cooperate with each other. In the research of Xu and Chen (2014), the result shows by examining the planning of inter-city railways in PRD that within the process of cooperation, the state somehow redefines itself and is no longer a passive entity. On the contrary, the state orchestrates integration in a more sophisticated manner. For inter-city relationships, in response to the lack of a formal institution for the adjacent cities to coordinate directly, the provincial government and its Development and Reform Commission evolve into the main agents to solve chaotic conflicts. However, due to the consideration of the provincial government, the core city tends to be protected and the benefit gambling among adjacent cities becomes severe and overriding political agendas. Under these circumstances, flexible informal institutions can successfully address any negotiation crisis. The problem is that the existing literature remains inadequate, and additional cases should be examined and explored, especially different institutional arrangements regarding city-regional politics across different levels of authority, through deciphering how they affect the construction of regional infrastructure projects.

2.5.3 *Research Gap*

The existing literature gives substantial attention to the case-specific paradigm in deciphering city-regionalism. The governance structure shuffles in response to the socioeconomic restructuring, specifically cross-boundary economic cooperation zones and large-scale infrastructure projects, are preliminarily explored by a series of empirical studies. Governance structure reshuffling is situated in different local socioeconomic environments, and the research paradigm is context-sensitive, crystallized by the case studies of cross-boundary economic cooperation zones and large-scale infrastructure projects. To capture the city-regionalism of China, additional case studies concerning cross-boundary economic cooperation zone and large-scale infrastructure projects are needed to decode the restructuring of governance structures.

2.6 Conclusions

Through a meticulous review of existing literature on city-regional integration, this chapter provides a solid theoretical foundation for portraying city-regional integration from an economic and institutional lens and the intrinsic logic of the formation of city-regions through the case-specific research paradigm. The four conclusions drawn are as follows:

First, in terms of the concept of the city-region and city-regional integration, a city-region is defined as the economic, social, and institutional interactions among an urban core, semi-urban areas, and rural hinterlands (Rodríguez-Pose, 2008). City-regional integration is seen as an integration in which the final target is the comprehensive goal, namely, developing a systemic and orderly, but differentiated economic system and establishing integration-oriented institutional alignment to strengthen the cross-boundary connection. The key dimensions of city-regional integration are categorized into functional and institutional integrations. Shen (2014) also classifies city-regional integration into economic, institutional, and social integrations. The existing literature sheds light on the nature of functional integration, particularly socioeconomic interaction (Sohn et al., 2009). The economic integration of a city-region is understood as a deliberately designed process that eliminates the barriers of goods and factors of production across territorial organizations. In other words, the ultimate landscape of an integrated economy at a city-regional scale occurs by forming a regional unified market across jurisdictions to facilitate the flow of goods and factors of production. Correspondingly, institutional integration is defined as "integration through regional integrative organization and comprehensive legal agreements" (Peng, 2000, p. 177).

Second, the assessment of the economic integration of a city-region is usually conducted by one of two approaches. The first approach is network analysis through employing the data on advanced production services or multinational corporations and their branches. The second approach is dividing economic integration into different parts, such as market, infrastructure, and industrial integration. This approach is based on the context-sensitive lens. According to the specific local issues in economic integration, different types of methodologies and indicators are selected to quantify the abovementioned aspects. Given the availability of data and the single dimension of network analysis, the second approach is prevalent and appropriate in measuring the economic integration of city-regions. However, scant literature undertakes a comprehensive investigation that fully considers the local socioeconomic milieus to decode the economic integration of a city-region.

Third, the institutional integration of a city-region is referred to as the integration of territorial organizations toward the production of MLG at the city-regional scale. Three specific approaches have been identified, namely, the region-based institutions, the inter-jurisdictional cooperation networks across local authorities, and the public–private partnerships. Existing studies that examine institutional integration in the Chinese context create a half-baked framework, obscuring the understanding of the de facto extent of institutional integration. In addition, the case-specific paradigm that they employ fails to provide an overall landscape of institutional integration and is devoid of the macro lens, which focuses on the whole practices of city-regionalism.

Fourth, the case-specific paradigm has been extensively deployed to construe the reshuffling of the governance structure in response to socioeconomic restructuring. The cases of cross-boundary economic cooperation zones and large-scale infrastructure projects have been normally devised to capture the city-regionalism of China. The conventional literature of case studies of the cross-boundary economic cooperation zone focuses on the building of integrative governance structures across jurisdictions to enable industrial integration. In examining cases of large-scale infrastructure projects, conventional studies place more attention on the politics of city-regions across different levels of jurisdiction. However, because the narrative of the political-economic tradition occurs case by case, context-sensitive empirical cases should be conducted to give a fuller discussion of city-regionalism in China. The above conclusions inform the design of this book, as detailed in Chap. 4.

References

Alderson, A. S., & Beckfield, J. (2004). Power and position in the world city system. *American Journal of Sociology, 109*(4), 811–851.

Al-Mohannadi, D. M., & Linke, P. (2016). On the systematic carbon integration of industrial parks for climate footprint reduction. *Journal of Cleaner Production, 112*(5), 4053–4064. https://doi.org/10.1016/j.jclepro.2015.05.094

Anderson, M. (1996). *Frontiers: Territory and state formation in the modern world* (Vol. 34, Issue 1, pp. 34–055). Polity Press. https://doi.org/10.5860/choice.34-0555

Andrew, C., & Doloreux, D. (2012). Economic development, social inclusion and urban governance: The case of the city-region of Ottawa in Canada. *International Journal of Urban and Regional Research, 36*(6), 1288–1305. https://doi.org/10.1111/j.1468-2427.2011.01025.x

Balassa, B. (1961). *The theory of economic integration*. Routledge.

Bennett, R. J. (1985). Central city-city region fiscal disparities in Austria: Estimates for 1979. *Urban Studies, 22*(1), 69–81. https://doi.org/10.1080/00420988520080071

Blatter, J. (2004). From 'spaces of place' to 'spaces of flows'? Territorial and functional governance in cross-border regions in Europe and North America. *International Journal of Urban and Regional Research, 28*(3), 530–548. https://doi.org/10.1111/j.0309-1317.2004.00534.x

Braun, E. (2008). *City marketing: Towards an integrated approach*. Erasmus Research Institute of Management (ERIM).

Breathnach, P. (2010). From spatial keynesianism to post-fordist neoliberalism: Emerging contradictions in the spatiality of the Irish State. *Antipode, 42*(5), 1180–1199. https://doi.org/10.1111/j.1467-8330.2010.00798.x

Brenner, N. (1998). Between fixity and motion: Accumulation, territorial organization and the historical geography of spatial scales. *Environment and Planning D: Society and Space, 16*(4), 459–481. https://doi.org/10.1068/d160459

Brenner, N. (1999). Globalisation as reterritorialisation: The re-scaling of urban governance in the European Union. *Urban Studies, 36*(3), 431–451. https://doi.org/10.1080/0042098993466

Brenner, N. (2003). Metropolitan institutional reform and the rescaling of state space in contemporary Western Europe. *European Urban and Regional Studies, 10*(4), 297–324. https://doi.org/10.1177/09697764030104002

Brenner, N. (2004). Urban governance and the production of new state spaces in Western Europe, 1960–2000. *Review of International Political Economy, 11*(3), 447–488.

Brenner, N. (2009). Open questions on state rescaling. *Cambridge Journal of Regions, Economy and Society, 2*(1), 123–139. https://doi.org/10.1093/cjres/rsp002

Breslin, S., & Higgott, R. (2000). Studying regions: Learning from the old, constructing the new. *New Political Economy, 5*(3), 333–352. https://doi.org/10.1080/713687784

Brie, M., & Stölting, E. (2012). Formal institutions and informal institutional arrangements. *International Handbook on Informal Governance*, 19–39. https://doi.org/10.4337/9781781001219.00008

Brunet-Jailly, E. (2005). Theorizing borders: An interdisciplinary perspective. *Geopolitics, 10*(4), 633–649. https://doi.org/10.1080/14650040500318449

Bryant, C. E., & Javalgi, R. G. (2016). Global economic integration in developing countries: The role of corruption and human capital investment. *Journal of Business Ethics, 136*(3), 437–450.

Castells, M. (1989). *The informational city: Information technology, economic restructuring, and the urban-regional process* (p. 24). Basil Blackwell Press.

Castells, M. (1996). *The rise of the network society* (Vol. 1). Basil Blackwell Press.

Chen, Y. (1998). The measurement of industrial structure convergence, desired convergence and undesired convergence. *China Industrial Economics, 12*(4), 37–43 (in Chinese). https://doi.org/10.19581/j.cnki.ciejournal.1998.04.007

Cheung, P. T. (2015). Toward collaborative governance between Hong Kong and Mainland China. *Urban Studies, 52*(10), 1915–1933. https://doi.org/10.1177/0042098014548139

Cheung, S. Y., & Phillimore, J. (2014). Refugees, social capital, and labour market integration in the UK. *Sociology, 48*(3), 518–536. https://doi.org/10.1177/0038038513491467

Chong, E., Huet, F., Saussier, S., & Steiner, F. (2006). Public-private partnerships and prices: Evidence from water distribution in France. *Review of Industrial Organization, 29*(1–2), 149.

Christoffersen, G. (1996). Governing China: From revolution through reform. *China Review International, 3*(2), 487–489. https://doi.org/10.1353/cri.1996.0037

Coombes, M. (2014). From city-region concept to boundaries for governance: The English case. *Urban Studies, 51*(11), 2426–2443. https://doi.org/10.1177/0042098013493482

Davoudi, S. (2003). European briefing: Poly-centricity in European spatial planning: From an analytical tool to a normative agenda. *European Planning Studies, 11*(8), 979–999. https://doi.org/10.1080/0965431032000146169

Delwaide, J. (2011). Introduction: Globalization, governance, crisis. *European Review, 19*(1), 1–4. https://doi.org/10.1017/S1062798710000359

Dematteis, G. (1997). Globalisation and regional integration: The case of the Italian urban system. *GeoJournal, 43*(4), 331–338. https://doi.org/10.1023/A:1006833200307

Derudder, B. (2006). On conceptual confusion in empirical analyses of a transnational urban network. *Urban Studies, 43*(11), 2027–2046. https://doi.org/10.1080/00420980600897842

Derudder, B., Witlox, F., Faulconbridge, J., & Beaverstock, J. (2008). Airline data for global city network research: Reviewing and refining existing approaches. *GeoJournal, 71*(1), 5–18.

Ehlers, N., & Buursink, J. (2000). Binational cities: People, institutions and structures. In M. van der Velde & H. van Houtum (Eds.), *Borders, regions, and people*. Pion.

Ehlers, N., Buursink, J., & Boekema, F. (2001). Introduction: Binational cities and their regions: From diverging cases to a common research agenda. *GeoJournal, 54*(1), 1–5. https://doi.org/10.1023/A:1021132212768

Fang, C., & Zhang, Y. (2014). Dynamic mechanism, spatial organization pattern and structure of integrated urbanization areas in China. *Urban Planning Forum, 6*, 5–12. (in Chinese).

Feiock, R. C. (2009). Metropolitan governance and institutional collective action. *Urban Affairs Review, 44*(3), 356–377. https://doi.org/10.1177/1078087408324000

Frank, L. K. (1925). The significance of industrial integration. *Journal of Political Economy, 33*(2), 179–195.

Friedmann, J. (1986). The world city hypothesis. *Development and Change, 17*(1), 69–83. https://doi.org/10.1111/j.1467-7660.1986.tb00231.x

Friedmann, J. (1995). Where we stand: A decade of world city research. In P. Knox & P. Taylor (Eds.), *World cities in a world-system* (pp. 21–47). Cambridge University Press.

Friedmann, J., & Miller, J. (1965). The urban field. *Journal of the American Institute of Planners, 31*(4), 312–320. https://doi.org/10.1080/01944366508978185

Friedmann, J., & Wolff, G. (1982). World city formation: An agenda for research and action. *International Journal of Urban and Regional Research, 6*(3), 309–344. https://doi.org/10.1111/j.1468-2427.1982.tb00384.x

García-Álvarez, J., & Trillo-Santamaría, J. M. (2013). Between regional spaces and spaces of regionalism: Cross-border region building in the Spanish 'state of the autonomies.' *Regional Studies, 47*(1), 104–115. https://doi.org/10.1080/00343404.2011.552495

Gass, R. J., Rickenbach, M., Schulte, L. A., & Zeuli, K. (2009). Cross-boundary coordination on forested landscapes: Investigating alternatives for implementation. *Environmental Management, 43*(1), 107–117. https://doi.org/10.1007/s00267-008-9195-2

Germain, A. (2011). *The fragmented or cosmopolitan metropolis?: A neighborhood story of immigration in montréal*. Centre Métropolis du Québec-Immigration et métropoles (CMQ-IM).

Goldberg, P. K., & Verboven, F. (2005). Market integration and convergence to the law of one price: Evidence from the European car market. *Journal of International Economics, 65*(1), 49–73. https://doi.org/10.1016/j.jinteco.2003.12.002

Gottmann, J. (1957). Megalopolis or the urbanization of the northeastern seaboard. *Economic Geography, 33*(3), 189–200. https://doi.org/10.2307/142307

Greif, A. (2000). The fundamental problem of exchange: A research agenda in historical institutional analysis. *European Review of Economic History, 4*(3), 251–284. https://doi.org/10.1017/S1361491600000071

Greif, A. (2006). *Institutions and the path to the modern economy: Lessons from medieval trade*. Cambridge University Press.

Grundy-Warr, C., Peachey, K., & Perry, M. (1999). Fragmented integration in the Singapore-Indonesian Border Zone: Southeast Asia's 'Growth Triangle' against the global economy. *International Journal of Urban and Regional Research, 23*(2), 304–328. https://doi.org/10.1111/1468-2427.00197

Gu, C., Zhang, M., Zhang, C., Zhang, X. M., Wang, C., & Chen, L. (2007). Prospects of urban agglomeration in the Changjiang (Yangtze) River Delta. *Scientia Geographica Sinica, 27*(1), 1. (in Chinese).

Hall, P. (1984). *The world cities*. Weidenfeld and Nicolson.

Hall, P. (2002). *Global city-regions in the twenty-first century*. Oxford University Press.

Hall, P. (2009). Looking backward, looking forward: The city region of the mid-21st century. *Regional Studies, 43*(6), 803–817. https://doi.org/10.1080/00343400903039673

Hall, P., & Pain, K. (Eds.). (2006). *The polycentric metropolitan region: Learning from mega-city regions In Europe*. Earthscan.

Harrison, J. (2010). Networks of connectivity, territorial fragmentation, uneven development: The new politics of city regionalism. *Political Geography, 29*(1), 17–27. https://doi.org/10.1016/j.polgeo.2009.12.002

Harrison, J. (2012). Life after regions? The evolution of city regionalism in England. *Regional Studies, 46*(9), 1243–1259. https://doi.org/10.1080/00343404.2010.521148

Harvey, D. (1989). From managerialism to entrepreneurialism: The transformation in urban governance in late capitalism. *Geografiska Annaler. Series B. Human Geography, 71*(1), 3–17. https://doi.org/10.1080/04353684.1989.11879583

Hooghe, L., & Marks, G. (1996). "Europe with the regions": Channels of regional representation in the European Union. *Publius: The Journal of Federalism, 26*(1), 73–92. https://doi.org/10.1093/oxfordjournals.pubjof.a029841

Hoyler, M., Kloosterman, R. C., & Sokol, M. (2008). Polycentric puzzles-emerging mega-city regions seen through the lens of advanced producer services. *Regional Studies, 42*(8), 1055–1064. https://doi.org/10.1080/00343400802389377

Hudalah, D., Firman, T., & Woltjer, J. (2014). Cultural cooperation, institution building and metropolitan governance in decentralizing Indonesia. *International Journal of Urban and Regional Research, 38*(6), 2217–2234. https://doi.org/10.1111/1468-2427.12096

Hveem, H. (2000). Explaining the regional phenomenon in an era of globalization. *Political Economy and the Changing Global Order, 2*, 70–81.

Jessop, B. (1996). Post-Fordism And the state. *In Comparative Welfare Systems* (pp. 165-183). Palgrave Macmillan.

Jessop, B. (2002). The political economy of scale. In *Globalization, regionalization and cross-border regions* (pp. 25-49). Palgrave Macmillan.

Jessop, B., Brenner, N., & Jones, M. (2008). Theorizing sociospatial relations. *Environment and Planning D: Society and Space, 26*(3), 389–401. https://doi.org/10.1068/d9107

Jonas, A. E., & Ward, K. (2007). Introduction to a debate on city-regions: New geographies of governance, democracy and social reproduction. *International Journal of Urban and Regional Research, 31*(1), 169–178. https://doi.org/10.1111/j.1468-2427.2007.00711.x

Keating, M. (1998). *The new regionalism in Western Europe: Territorial restructuring and political change*. E. Elgar. https://doi.org/10.1017/S000842390001711X

Keith, M. (2005). *After the cosmopolitan?: Multi-cultural cities and the future of racism*. Routledge.

Krätke, S. (2007). Metropolisation of the European economic territory as a consequence of increasing specialisation of urban agglomerations in the knowledge economy. *European Planning Studies, 15*(1), 1–27. https://doi.org/10.1080/09654310601016424

Kwak, Y. H., Chih, Y., & Ibbs, C. W. (2009). Towards a comprehensive understanding of public private partnerships for infrastructure development. *California Management Review, 51*(2), 51–78. https://doi.org/10.2307/41166480

Lang, R., & Knox, P. K. (2009). The new metropolis: Rethinking megalopolis. *Regional Studies, 43*(6), 789–802. https://doi.org/10.1080/00343400701654251

Lewis, M. W., & Wigen, K. (1997). *The myth of continents: A critique of metageography*. Oakland, California: University of California Press.

Li, X., & Yin, J. (2012). Research review on the effects of regional integration on industry. *City Planning Review, 36*(5), 91–96. (in Chinese).

Li, Y., & Wu, F. (2018). Understanding city regionalism in China: Regional cooperation in the Yangtze River Delta. *Regional Studies, 52*(3), 313–324. https://doi.org/10.1080/00343404.2017.1307953

Li, Z., Xu, J., & Yeh, A. G. (2014). State rescaling and the making of city regions in the Pearl River Delta, China. *Environment and Planning C: Government and Policy, 32*(1), 129–143. https://doi.org/10.1068/c11328

Lidström, A. (2013). Citizens in the city regions political orientations across municipal borders. *Urban Affairs Review, 49*(2), 282–306. https://doi.org/10.1177/1078087412457549

Lieberthal, K., & Lampton, D. M. (Eds.). (1992). *Bureaucracy, politics, and decision making in post-Mao China* (p. 19). University of California Press.

Lieberthal, K., & Oksenberg, M. (1988). *Policy making in China: Leaders, structures, and processes.* Princeton University Press.

Liesbet, H., & Gary, M. (2003). Unraveling the central state, but how? Types of multi-level governance. *American Political Science Review, 97*(2), 233–243. https://doi.org/10.1017/S0003055403000649

Liu, S., Luo, X., Tian, D., & Wang, Y. (2014). Preliminary study on China's cross-boundary new district in forming mechanisms, spatial organizations and governance model. *Economic Geography, 34*(12), 41–47 (in Chinese). https://doi.org/10.15957/j.cnki.jjdl.2014.12.007

Liu, Y. (2015). *The research on level measure and counter-measure of economic integration in Nanchang and Jiujiang* (Master's dissertation). Available from China Master's Theses Full-text Database (in Chinese).

Liu, L., Zhang, B., & Bi, J. (2012). Reforming China's multi-level environmental governance: Lessons from the 11th Five-Year Plan. *Environmental Science & Policy, 21*, 106–111. https://doi.org/10.1016/j.envsci.2012.05.001

Loo, B. P. (2010). Cross-boundary container truck congestion: The case of the Hong Kong-Pearl River Delta region. *Transportation, 37*(2), 257–274. https://doi.org/10.1007/s11116-009-9246-4

Lu, X. (2015). The comparison of manage system of western new national district. *City, 8*, 53–58. (in Chinese).

Luo, X., & Shen, J. (2007). Models of inter-city cooperation and its theoretical implications: An empirical study on the Yangtze River Delta. *Acta Geographica Sinica, 62*(2), 115–126. (in Chinese).

Luo, X., & Shen, J. (2009). A study on inter-city cooperation in the Yangtze river delta region, China. *Habitat International, 33*(1), 52–62. https://doi.org/10.1016/j.habitatint.2008.04.002

Lüthi, S., Thierstein, A., & Goebel, V. (2010). Intra-firm and extra-firm linkages in the knowledge economy: The case of the emerging mega-city region of Munich. *Global Networks, 10*(1), 114–137. https://doi.org/10.1111/j.1471-0374.2010.00277.x

Lv, D., & Zhang, Q. (2010). Analysis on Beijing-Tianjin integration under the background of economic globalization. China Population. *Resource and Environment, 20*(3), 162–167. (in Chinese).

Mansfield, E. D., & Milner, H. V. (1999). The new wave of regionalism. *International Organization, 53*(3), 589–627. https://doi.org/10.1162/002081899551002

Marks, G. (1993). Structural policy and multilevel governance in the EC. In A. Cafruny & G. Rosenthal (Eds.), *The state of the European community: The Maastricht debates and beyond* (Vol. II, pp. 391–410). Lynne Rienner.

Masek, J. G., Lindsay, F. E., & Goward, S. N. (2000). Dynamics of urban growth in the Washington DC metropolitan area, 1973–1996, from Landsat observations. *International Journal of Remote Sensing, 21*(18), 3473–3486. https://doi.org/10.1080/014311600750037507

McKee, M., Edwards, N., & Atun, R. (2006). Public-private partnerships for hospitals. *Bulletin of the World Health Organization, 84*, 890–896.

Mertha, A. C. (2005). China's "soft" centralization: Shifting tiao/kuai authority relations. *The China Quarterly, 184*, 791–810. https://doi.org/10.1017/S0305741005000500

Mongelli, F. P., Dorrucci, E., & Agur, I. (2005). What does European institutional integration tell us about trade integration? *ECB Occasional Paper, 40*, European Central Bank (ECB). Frankfurt a. M. http://dx.doi.org/10.2139/ssrn.807413

Moore, C. (2008). A Europe of the regions vs. the regions in Europe: reflections on regional engagement in Brussels. *Regional & Federal Studies, 18*(5), 517–535. https://doi.org/10.1080/13597560802351564

Mori, T. (1997). A modeling of megalopolis formation: The maturing of city systems. *Journal of Urban Economics, 42*(1), 133–157. https://doi.org/10.1006/juec.1996.2018

Moss, M. L., & Townsend, A. M. (2000). The Internet backbone and the American metropolis. *The Information Society, 16*(1), 35–47. https://doi.org/10.1080/019722400128310

Neal, Z. (2011). Differentiating centrality and power in the world city network. *Urban Studies, 48*(13), 2733–2748. https://doi.org/10.1177/0042098010388954

Nepal, R., & Jamasb, T. (2012). Interconnections and market integration in the Irish Single Electricity Market. *Energy Policy, 51*, 425–434. https://doi.org/10.1016/j.enpol.2012.08.047

Nicholls, W. J. (2005). Power and governance: Metropolitan governance in France. *Urban Studies, 42*(4), 783–800. https://doi.org/10.1080/00420980500060426

North, D. C. (1990). *Institutions, institutional change and economic performance*. Cambridge, MA: Cambridge University Press.

Oscar, J. (1994). *Border people: Life and society in the US-Mexico borderlands*. University of Arizona Press.

Pain, K. (2008). Examining 'core–periphery' relationships in a global city region: The case of London and South East England. *Regional Studies, 42*(8), 1161–1172. https://doi.org/10.1080/00343400701808857

Parr, J. B. (2003). Reinventing regions? The case of the polycentric urban region. *Regional Studies, 38*(3), 231–240.

Parr, J. B. (2005). Perspectives on the city-region. *Regional Studies, 39*(5), 555–566. https://doi.org/10.1080/00343400500151798

Partridge, M. D., & Rickman, D. S. (2008). Distance from urban agglomeration economics and rural poverty. *Journal of Regional Science, 48*(2), 285–310. https://doi.org/10.1111/j.1467-9787.2008.00552.x

Peng, D. (2000). The changing nature of East Asia as an economic region. *Pacific Affairs, 73*(2), 171–191. https://doi.org/10.2307/2672176

Perkmann, M. (2007a). Policy entrepreneurship and multi-level governance: A comparative study of European cross-border regions. *Environment and Planning c: Government and Policy, 25*(6), 861–879. https://doi.org/10.1068/c60m

Perkmann, M. (2007b). Construction of new territorial scales: A framework and case study of the EUREGIO cross-border region. *Regional Studies, 41*(2), 253–266. https://doi.org/10.1080/00343400600990517

Pinder, J. (1969). Problems of European integration. In G. Denton (Ed.), *Economic integration in Europe*. Weidenfeld and Nicolson.

Raunio, T. (2011). The changing world of EU governance. *International Studies Review, 13*(2), 314–317. https://doi.org/10.1111/j.1468-2486.2011.01030.x

Ravallion, M. (1986). Testing market integration. *American Journal of Agricultural Economics, 68*(1), 102–109. https://doi.org/10.2307/1241654

Robson, P. (2002). *The economics of international integration*. Routledge.

Rodríguez-Pose, A. (2008). The rise of the "city region" concept and its development policy implications. *European Planning Studies, 16*(8), 1025–1046. https://doi.org/10.1080/09654310802315567

Ruggie, J. G. (1993). Territoriality and beyond: Problematizing modernity in international relations. *International Organization, 47*(1), 139–174. https://doi.org/10.1017/S0020818300004732

Salet, W. G., & Thornley, A. (2007). Institutional influences on the integration of multilevel governance and spatial policy in European city regions. *Journal of Planning Education and Research, 27*(2), 188–198. https://doi.org/10.1177/0739456X07307207

Salet, W. G., Thornley, A., & Kreukels, A. (2003). *Metropolitan governance and spatial planning: Comparative case studies of European City regions*. Taylor & Francis.
Sassen, S. (1995).On concentration and centrality in the global city. In P. L. Knox & P. J. Taylor (Eds.), *World cities in a world system*. Cambridge University Press.
Sassen, S. (1991). *The global city: New York, London, Tokyo, Princeton*. Princeton University Press.
Sassen, S. (2006). *Cities in a world economy* (3rd ed.). Pine Forge Press.
Savini, F. (2012). Who makes the (new) metropolis? Cross-border coalition and urban development in Paris. *Environment and Planning A, 44*(8), 1875. https://doi.org/10.1068/a44632
Scanlon, C., & Adlam, J. (2011). Cosmopolitan minds and metropolitan societies: Social exclusion and social refusal revisited. *Psychodynamic Practice, 17*(3), 241–254. https://doi.org/10.1080/14753634.2011.587637
Scott, A. J. (2001). Globalization and the rise of city regions. *European Planning Studies, 9*(7), 813–826. https://doi.org/10.1080/09654310120079788
Scott, A. J. (2011). A world in emergence: Notes toward a resynthesis of urban-economic geography for the 21st century. *Urban Geography, 32*(6), 845–870. https://doi.org/10.2747/0272-3638.32.6.845
Scott, A. J., & Storper, M. (2007). Regions, globalization, development. *Regional Studies, 41*(S1), S191–S205. https://doi.org/10.1080/0034340032000108697
Shen, J. (2014). Not quite a twin city: Cross-boundary integration in Hong Kong and Shenzhen. *Habitat International, 42*, 138–146. https://doi.org/10.1016/j.habitatint.2013.12.003
Short, J. R., Kim, Y., Kuus, M., & Wells, H. (1996). The dirty little secret of world cities research: Data problems in comparative analysis. *International Journal of Urban and Regional Research, 20*(4), 697–717. https://doi.org/10.1111/j.1468-2427.1996.tb00343.x
Simmonds, R., & Hack, G. (2000). *Global city regions: Their emerging forms*. Taylor & Francis.
Smith, D. A., & Timberlake, M. (2002). Hierarchies of dominance among world cities: A network approach. In S. Sassen (Ed.), *Global networks, linked cities* (pp. 117–141). Routledge.
Sohn, C., Reitel, B., & Walther, O. (2009). Cross-border metropolitan integration in Europe: The case of Luxembourg, Basel, and Geneva. *Environment and Planning c: Government and Policy, 27*(5), 922–939. https://doi.org/10.1068/c0893r
Strange, S. (1996). *The retreat of the state: The diffusion of power in the world economy*. Cambridge University Press.
Streeten, P. (1964). *Economic integration: Aspects and problems (No. 5)*. Sythoff.
Sultana, S., & Weber, J. (2014). The nature of urban growth and the commuting transition: Endless sprawl or a growth wave? *Urban Studies, 51*(3), 544–576. https://doi.org/10.1177/0042098013498284
Swyngedouw, E. A. (1992). Territorial organization and the space/technology nexus. *Transactions of the Institute of British Geographers, 17*, 417–433.
Swyngedouw, E. A. (2000). Authoritarian governance, power, and the politics of rescaling. *Environment and Planning d: Society and Space, 18*(1), 63–76. https://doi.org/10.1068/d9s
Szczepański, M. (1999). Cultural borderland in sociological and political perspective (the case of Upper Silesia). In Local *power and modern community political life: Sociological essays*. Uniwersytet Slaski
Tabuchi, T. (1998). Urban agglomeration and dispersion: A synthesis of Alonso and Krugman. *Journal of Urban Economics, 44*(3), 333–351. https://doi.org/10.1006/juec.1997.2074
Taylor, P. J. (2001). Specification of the world city network. *Geographical Analysis, 33*(2), 181–194. https://doi.org/10.1111/j.1538-4632.2001.tb00443.x
Taylor, P. J., Evans, D. M., & Pain, K. (2008). Application of the interlocking network model to mega-city regions: Measuring poly-centricity within and beyond city regions. *Regional Studies, 42*(8), 1079–1093. https://doi.org/10.1080/00343400701874214
Tinbergen, J. (1954). *International economic integration*. Elsevier.
Tolone, W. J., Wilson, D., Raja, A., Xiang, W. N., Hao, H., Phelps, S., & Johnson, E. W. (2004). Critical infrastructure integration modeling and simulation. In *Intelligence and security informatics* (pp. 214-225). Springer.

Van Oort, F., Burger, M., & Raspe, O. (2010). On the economic foundation of the urban network paradigm: Spatial integration, functional integration and economic complementarities within the Dutch Randstad. *Urban Studies, 47*(4), 725–748. https://doi.org/10.1177/0042098009352362

Vermeijden, B. (2001). Dutch urban renewal, transformation of the policy discourse 1960–2000. *Journal of Housing and the Built Environment, 16*(2), 203–232. https://doi.org/10.1023/A:1017991519286

Wang, L., & Li, W. (2015). The interaction between the third sector and the government of 'The New Area between Guiyang and Anshun.' *Manager' Journal, 30*, 334. (in Chinese).

Ward, K., & Jonas, A. E. (2004). Competitive city regionalism as a politics of space: A critical reinterpretation of the new regionalism. *Environment and Planning A, 36*(12), 2119–2139. https://doi.org/10.1068/a36223

Wise, M. J. (1965). City region. In *Nature* (Vol. 207, No. 5001, p. 1063). Macmillan Magazines LTD.

Xu, J., & Chen, Y. (2014). Planning inter-city railways in China's mega-city regions: Insights from the Pearl River Delta. *China Review, 14*(1), 11–36.

Xu, J., & Yeh, A. G. (2013). Interjurisdictional cooperation through bargaining: The case of the Guangzhou-Zhuhai railway in the Pearl River Delta, China. *The China Quarterly, 213*, 130–151. https://doi.org/10.1017/S0305741013000283

Yang, C. (2005). Multilevel governance in the cross-boundary region of Hong Kong-Pearl River Delta, China. *Environment and Planning A, 37*(12), 2147–2168. https://doi.org/10.1068/a37230

Yang, C., & Li, S. M. (2013). Transformation of cross-boundary governance in the Greater Pearl River Delta, China: Contested geopolitics and emerging conflicts. *Habitat International, 40*, 25–34. https://doi.org/10.1016/j.habitatint.2013.02.001

Yeh, A. G., Yang, F. F., & Wang, J. (2015). Producer service linkages and city connectivity in the mega-city region of China: A case study of the Pearl River Delta. *Urban Studies, 52*(13), 2458–2482. https://doi.org/10.1177/0042098014544762

Yeung, H. W. C., & Olds, K. (2001). From the global city to globalising cities: Views from a developmental city-state in Pacific Asia. In *World forum on habitat-international conference on urbanizing world and UN human habitat II* (pp. 4–6). *Columbia University*.

Yuan, F., Sawaya, K. E., Loeffelholz, B. C., & Bauer, M. E. (2005). Land cover classification and change analysis of the twin cities (Minnesota) metropolitan area by multi-temporal landsat remote sensing. *Remote Sensing of Environment, 98*(2), 317–328. https://doi.org/10.1016/j.rse.2005.08.006

Zayati, A., Biennier, F., Moalla, M., & Badr, Y. (2012). Towards lean service bus architecture for industrial integration infrastructure and pull manufacturing strategies. *Journal of Intelligent Manufacturing, 23*(1), 125–139. https://doi.org/10.1007/s10845-010-0401-8

Zhang, J., Geng, L., Yin, J., & Luo, X. (2011). The regional cooperative governance from the perspective of regional spatial production—A case study of Jinjiang Park, Jiangyin Economic Development Zone. *Human Geography, 26*(1), 5–9 (in Chinese). https://doi.org/10.13959/j.issn.1003-2398.2011.01.016

Zhang. X., Cheung, D. M. W., Sun, Y., & Tan, J. (2019). Political decentralization and the path-dependent characteristics of the state authoritarianism: An integrated conceptual framework to understand China's territorial fragmentation. *Eurasian Geography and Economics, 60*(5), 548–581. https://doi.org/10.1080/15387216.2019.1667843

Zhang, X., Sun, Y., & Luan, X. (2023). Regulation flexibility and legitimacy building in governing intercity railways: The polymorphous role of the Chinese provincial government. *Journal of Urban Affairs*. https://doi.org/10.1080/07352166.2022.2141641

Zhang, X., Chen, S., Luan, X., & Yuan, M. (2021a). Understanding China's city-regionalization: Spatial structure and relationships between functional and institutional spaces in the Pearl River Delta. *Urban Geography, 42*(3), 312–339. https://doi.org/10.1080/02723638.2019.1710399

Zhang, X., Li, Y., Zhou, C., Luan, X., & Yuan, F. (2022a). Rescaling of the land regime in the making of city-regions: A case study of China's Pearl River Delta. *Urban Studies*. https://doi.org/10.1177/00420980221101781

Zhang, X., Shen, J., & Gao, X. (2021b). Towards a comprehensive understanding of intercity cooperation in China's city-regionalization: A comparative study of Shenzhen-Hong Kong and Guangzhou-Foshan city groups. *Land Use Policy, 103*, 105339. https://doi.org/10.1016/j.landusepol.2021.105339

Zhang, X., Shen, J., Sun, Y., Zhou, C., & Yang, Y. (2022b). Formation of city regions from bottom-up initiatives: Investigating coalitional developmentalism in the Pearl River Delta. *Annals of the American Association of Geographers*. https://doi.org/10.1080/24694452.2022.2130868

Zhang, X., & Sun, Y. (2019). Investigating institutional integration in the contexts of Chinese city-regionalization: Evidence from Shenzhen-Dongguan-Huizhou. *Land Use Policy, 88*, 104170.

Zhao, Y. (2009). Consideration on the development of pushing forward Zheng-Bian new region. *Journal of Anyang Normal University, 11*(1), 55–57 (in Chinese). https://doi.org/10.16140/j.cnki.1671-5330.2009.01.036

Chapter 3
Empirical Background: Policies, Practices and Evolutions to City-Regional Integration in China

3.1 Introduction

This chapter aims to identify acute challenges that impede the development of city-regions in China for the purpose of discerning issues to be investigated in the case study of SDH. In the first three sections, this chapter reviews the integration of city-region at the national, regional (PRD), and sub-regional (SDH) scales, including the evolution and formation of spatial layout, socioeconomic development tendency, and institutional arrangements for city-regionalism. It is important to note that for city-regional development in China, socioeconomic restructuring is entangled with inter-city and regional political–institutional reconstruction to re-configure the city-regional multilateral system.

Accordingly, the first section discusses the evolution of a city-region in the post-reform era, dividing it into five inter-connected phases. It scrutinizes the transformation of spatial patterns and the underlying driving forces in every phase, such as national development strategies, regional spatial plans, large-scale infrastructure projects, and urban economic growth. The second section centers on PRD integration and articulates long-term polycentric initiatives and strategy (e.g., the making of three sub-regions and the assistance strategy[1]). Moreover, it illustrates integration-oriented institutional designs in restructuring the spatial layout of PRD. It also systematically presents the status quo of the polycentric strategy in PRD by a range of empirical cases. The third section exposes the empirical issues of the major cities in SDH, challenging the city-regionalism of PRD, and how the initiatives for integration are translated into the concrete everyday practices of SDH integration, such as institutional arrangement and industrial and infrastructure collaboration in order to overcome the crisis of economic cycles.

[1] The assistance strategy refers to the regional strategy of Guangdong province under the initiatives of regional coordinated development. Specifically, to revitalize the backward cities in eastern, western and northern regions of Guangdong province, the Guangdong Provincial Government requires the core cities in the PRD region, such as Guangzhou and Shenzhen, to aid the outlying cities, such as Heyuan, Shanwei, and Yunfu, socioeconomically and institutionally.

X. Zhang, *Regional Integration and Governance Reshuffling in the Making of China's City-Regions*, Urban Sustainability, https://doi.org/10.1007/978-981-99-2792-0_3

The final section concludes the discussion of the evolution of city-regional integration under the city-regionalism in China, PRD, and SDH. Furthermore, the section presents the challenging issues of SDH integration to support the significant claims of this book.

3.2 Macro-context: City-Regional Integration in China Since 1978

Market forces have become the underlying drivers to propel regional economic integration in socialist China since the launch of the Opening-Door Policy in 1978. However, in the wake of repositioning to market-oriented governance, the powerful governmental intervention also played a dominant role in city-regional integration. The process of city-regional integration in the post-reform period has undergone five phases in light of regional policy, characteristics of city-regional integration, and other related issues. The criteria needed to divide the development process transformation are the regional spatial pattern and the driving forces that underlie these transformations including economic growth, policy incentives, government intervention, and global–local relations. Furthermore, in the creation of a competitive city-region, given the strong motivation of central and local governments, some weakly integrated or partly integrated areas are designated as the city-region factitiously. The five phases of city-regional development deserve detailed attention.

3.2.1 First Phase, 1978–1990: Eastern Area—First Strategy Under the Eastern–Central–Western Spatial Layout

China strategically embarked on the reform and the Opening-Door Policy in 1978, and the central government implemented "The Strategy of First Developing the Coastal Area" from 1978 to 1985. In December 1978, Deng Xiaoping proposed that "we should let some people become rich first through honest work and lawful operations in some areas".[2] After the announcement of the "6th Five-year Plan (1981–1985)", the central government undertook such differentiated development strategies to separate the coastal area from the inland area (Wei, 1992). However, from 1986 to 1990, the central government divided the territory into the eastern, central, and western economic belts, superseding the coastal-inland layout division in the early 1980s (Fig. 3.1). The "7th Five-year Plan (1986–1990)" further emphasized the acceleration of the developmental pace of the eastern coastal area (Cai & Zhang,

[2] CPC news. (2014, July 22). The outline of Deng Xiaoping's speech in closing meeting of the central work conference on December 13, 1978. Retrieved from http://dangshi.people.com.cn/n/2014/0722/c85037-25317997.html.

2015). The "National Program for General Planning of the National Land (1985–2000)" paved the way for regional development in the next stage from 1991 to 1998. Notably, the central government did not simply organize the country as western, central, and eastern economic belts but designated key development nodes and areas, such as PRD, YRD, the Liaozhongnan area, the Shandong Peninsula, the Wuhan area, and the Chongqing–Yichang area, where infrastructure projects were developed to stimulate the local economy (Hu, 2006).

From 1978 to 1990, the central government invested in a series of large-scale infrastructure projects in the eastern areas to facilitate physical integration with the purpose of intensifying inter-urban connections. Moreover, the government set up various Special Economic Zones to endow great economic autonomous power to enable the opening of the market in the eastern area, including Shenzhen, Zhuhai, Shantou, and Xiamen, established in 1979, and similar set-ups in Shanghai in 1982, Hainan in 1988, and Pudong New District in 1992. Special Economic Zones were set up to attract foreign capital and to develop an export-oriented mode of economic development. Substantial integration-first authorities or agencies also emerged to encourage inter-jurisdictional economic cooperation. For example, the "Shanghai Special Economic Zone Planning Office" was established by SC in December 1982 to organize and manage the joint conference of ten city mayors in YRD. The central government decentralized the power of overseeing investment projects to the local governments and allowed overseas financial institutions to locate their headquarters and branch offices in the eastern areas.

Eastern areas witnessed tremendous economic growth in this phase, and the disparity in internal income narrowed. However, the socioeconomic gap between the eastern area and the inland area enlarged (Wang & Shi, 2011) because of the influx of foreign capital and the rise of the urban economy in the eastern coastal area (Li, 2011). Furthermore, a high level of geographical agglomerations of manufacturing industries, capital-intensive industries, and provision of professional services appeared in the eastern area. Such industrial agglomeration signals that towns were closely connected to one another, and the liquidity of capital was strengthened geographically on the basis of the rise of the urban economy (Wu & Li, 2010). Despite the fast-tracked development in the economic sectors, the networks of railways and highways were not upgraded to the level that matched the economic development, and the role of infrastructure network was not recognized by governments in enhancing the regional transport efficiency from 1986 to 1994 (Wu et al., 2010). As such, the productivity of different regions was uneven, thereby triggering the weakness of intra-regional cooperation in the eastern area (She et al., 1995). In addition, the municipal government lacked the willingness and policy incentives for inter-city cooperation under the newly introduced market economy and the initial stage of the city-based accumulation regime in socialist China. Decentralization of administrative power further gave rise to inter-jurisdictional competitions devoid of concrete collaborative institutional designs (Yang, 2011). These situations resulted in a series of integration-based institutional reforms in the following phases.

The local autonomy did not elicit substantial attention from the central government, and the rapid growth of the urban economy and accumulation of urban assets

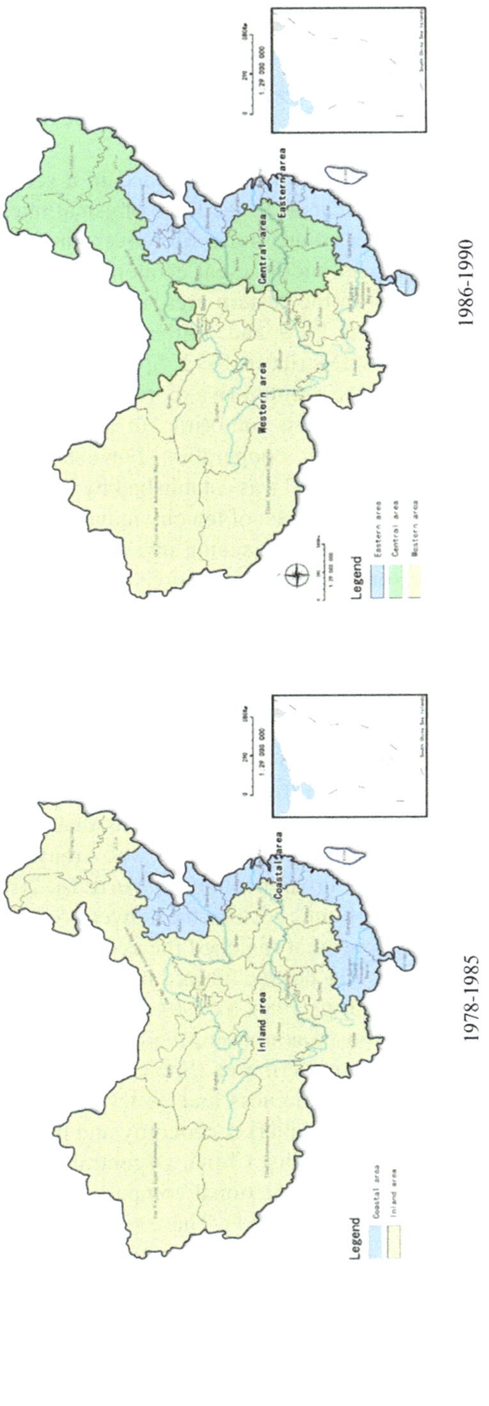

Fig. 3.1 Regional development strategy from 1978 to 1990. *Source* Author

was primarily driven by national preferential policy and top-down investment of regional infrastructure projects in this phase. Essentially, the actual consequences included not only the liberation and development of productivity in the eastern area and the formation of three plates (i.e., western, central, and eastern regions) but also the emergence of uncoordinated regional development. However, the underlying socioeconomic mechanism of the city-region was not yet mature and the policy focused primarily on the individual cities as well as regional plates.

3.2.2 Second Phase, 1991–1998: The Opening Layout Centered on Three Megacity-Regions

Based on the strategy of building three economic belts, the central government issued a host of specific regional strategies in this stage, especially the place promotion strategies for three megacity-regions, involving the Circum–Bohai–Sea Economic Zone, PRD, and YRD (Fig. 3.2). In addition, the central government proposed the strategy of opening border cities, riverside cities, and inland cities and approved the opening of a series of cities, such as Bole, Hekou, Pingxiang, Dongxing, Dandong, and Suifenhe. In 1992, the central government approved 14 national and cross-border economic cooperation zones, which served as a significant motor for economic integration. With regard to the economic coordination mechanism, some economic cooperation platforms were requested to accommodate additional municipalities. In the same year, the joint conference among the economic cooperation administration of fourteen cities was established in YRD. The central government formulated a development strategy to open five riverside cities (i.e., Chongqing, Yueyang, Wuhan, Wuhu, and Jiujiang) and five inland provincial capital cities in 1998 to further deepen the opening layout.

The tax-sharing system reform in 1994 was conducived to relieve the plight of central finance, reshuffling the vassal relation between central and local governments which treated the local government as the subsidiary body of the central government. The centralization of the fiscal power and the decentralization of developmental liabilities across different levels of government brought by the tax-sharing system reform introduced in 1994 enabled the local government to operate as an entrepreneurial government and deliver economic growth by land finance (Zou & Ren, 2015). After the tax reform, the increasing amounts of autonomous spending provided a great impetus for the local government to opt for urban land expansion by land finance. Moreover, intensive land development effectively enhanced the integration between urban and rural areas through the construction of industrial parks and infrastructure promotion. The "Forum for the Coordination of Urban Economy of the YRD" was convened in 1997 and encompassed 15 cities. The relationship between the urban and rural areas in PRD was reshuffled from a disengaged to an integrated connection and triggered the rise of the regional transportation network, where Guangzhou was viewed as the hub (Zheng et al., 2003). The density of the

Fig. 3.2 Regional development strategy from 1991 to 1998. *Source* Author

traffic network in the Circum–Bohai–Sea Economic Zone increased dramatically, and that of the regional highway even reached 0.3 km/km^2 (Luo & Zhou, 2006).

However, the land finance mode was entangled with the GDP-based cadre evaluation system, posing great administrative barriers among cities, and the administrative fragmentation of these megacity-regions triggered rock-ribbed conflicts among different territorial authorities (Yin & Feng, 2014). Given the fierce, yet chaotic intercity competition driven by the entrepreneurial government and the strong governmental intervention which partly superseded the role of market actors in resource allocation, the industry isomorphism became severe and hindered the formation of a regional industrial chain (Luo & Zhou, 2006). Additionally, regional cooperation as contemplated but not implemented because of the lack of region-based institutions and the blurred division of responsibility and power among ministries or departments in the central government (Ma & He, 2008). The regional policies promulgated by the central government incentivized city-regional development and encouraged bottom-up development.

3.2.3 Third Phase, 1999–2005: Four Plates to Enable Regional Coordinated Development Strategy

A significant adjustment of national strategy has occurred since 1999, shifting from a regional uneven development strategy to a regional coordinated development strategy. In September 1999, the "Decision concerning the Adjustment of State-owned Economy Strategy" was approved, and a develop-the-west strategy was launched in the Fourth Plenary Session. Specifically, in September 2003, the Political Bureau of the Central Committee approved the "Several Opinions about the Implementation of the Northeast China Revitalization Strategy." In October, 2005, the fifth plenary session of the 16th central committee put forward an overall strategy for China's regional coordination development (i.e., promoting the West Development strategy, revitalizing Northeast China, facilitating the rise of central China, and encouraging the eastern region to lead the development of the whole country) (Fig. 3.3). Premier Wen Jiabao chaired an "SC Executive Meeting" to approve the "Plan of Promoting the Rise of Central China" in 2006. To serve as an impetus to regional integration, in March 2001, the central government decided to build a large cross-provincial border economic belt, known as the Western Longhai–Lanxin Railway Economic Belt, the Upper Reaches of Yangtze River, Chengdu–Kunming, and Nanning–Guiyang. In 2003, the SC set up the "Leading Group for the Revitalization of Northeast Old Industrial Base" and founded the "Northern China Department" in the National Development and Reform Commission (NDRC). In addition, a consensus on inter-city economic cooperation, such as the "Langfang Consensus" in 2004, was reached with the expectation that it could effectively stimulate local economic vitality.

Thus, the differences in socioeconomic development among the four regions reduced over time compared with the last phase (Xu & Li, 2006). The gaps of per capita GDP among the eastern, central, and western areas after 2004 narrowed as well (Liu et al., 2009). Local protectionism was alleviated in these regions, and the flow of internal resources for the regional industry was accelerated with the increase in infrastructure projects (Fan, 2010). Nevertheless, the growth-oriented evaluation system for government officials further reinforced the administrative hurdles for the flow of human resources and essential productive factors (Yang, 2007). Such a system encouraged inter-city competition, because the local officials at the horizontal level were evaluated primarily by the performance of economic growth rather than the regional coordination (Edin, 2003). In regard to a regional plan, no mature cross-border coordinated plan was formulated in this phase (Yang, 2011). The regional common market was not established and fostered, thereby triggering the low efficiency of resource allocation among cities. Thus, although the central government has plaid increasing attention to the geographical scale of the city-region in formulating place-based policies, the defects of the existing administrative system, and the cadre evaluation system, and the absence of a regional coordination plan impeded the building of effective regional governance across territories.

Fig. 3.3 Regional development strategy from 1999 to 2005. *Source* Author

3.2.4 Fourth Phase, 2006–2013: Coordinated Regional Layout Through Development of City-Region

A series of urban agglomeration and metropolitan area plans were formulated in this phase to further forge the coordinated regional layout, and such plans were a primary characteristic of the city-regionalism of China. These integrated geographical sites were commonly composed of neighboring cities and provinces, thereby manifesting as polycentric or monocentric regions that were constituted by a range of different scales of jurisdictions. These regions included Changsha–Zhuzhou–Xiangtan urban agglomeration, Wuhan urban agglomeration, Central Henan urban agglomeration, and the Chengdu–Chongqing Economic Zone. This phase was the diffusion period for regional integration, and the layout of regional development shifted from uneven development to coordinated and stabilized development anchored by the mounting number of city-regions. The adjustment to the administrative division was widely deployed to overcome the negative externalities of the administrative division-based economic mode and integrate the fragmented urban governance for fostering the regional coordinated layout. For instance, in Tianjin, the Tanggu, Hangu, and Dagang Districts were revoked and replaced by the Binhai New Area in Tianjin in November 2009. In February 2010, the SC approved the adjustment for the administrative division of Shenyang and merged Dongling District and Hanggao base into Hunnan New District. In July 2010, SC approved the urban expansion of Shenzhen municipality from 327.5 to 1991 km^2.

A range of new regional plans from 2009 to 2012 was enacted to foster the polycentric layout at the national scale. Such as "Changsha-Zhuzhou-Xiangtan Urban Agglomeration Plan," "Regional Plan for Yangtze River Delta," "Regional Plan for Chengdu-Chongqing Economic Zone," "Regional Plan for Hu-Bao-E Urban Agglomeration," "Plan for Central China Economic Zone," and so forth. In the construction of a balanced regional layout, different levels of government unanimously emphasized the equal dispersion of social service facilities. In the "12th Five-year Plan (2011–2015)" that was promulgated by NDRC, the central government further focused on setting up a coordination mechanism in city-regions, especially in the field of industrial collaboration. In the views of the interior city-regions, in addition to relieving the development pressure for a regional core city, constructing reasonable upstream and downstream industrial divisions should also consolidate the industrial functions of medium and small cities. As the overarching geographical scaffolding for socioeconomic development, urban agglomeration should focus on the building of a reasonable layout, uniform planning, a well-rounded functional structure, and a development mode, which was referred to as "large city guiding small city" (NDRC, 2011). To foster the polycentric layout, in the working conference on the urbanization of the central government, it was confirmed that urban agglomeration would be regarded and constructed as the basic and primary form of regional development and urbanization in China.

With the global financial crisis in 2008, the export-oriented economic mode was destroyed and China reoriented towards the city-region as the overarching scale in enhancing the national economy. The regional policy continued to attach the most importance to polycentric development at the national scale through the large-scale making of city-regions. In some full-blown city-regions, the agglomeration effect of post-Fordist economic sectors enabled the tight intercity connection and facilitated the reshuffling of the social life of laborers to ensure economic-geographic restructuring. Regional policy and planning have increasingly documented and guided social life and social development, such as the prevailing "One Hour Commuting Circle" in urban and regional planning, and its relevant arrangements for infrastructure and social service facilities.

In the present situation, 32 urban agglomerations were proposed in China. These urban agglomerations were divided into three categories, namely, developed urban agglomeration, developing urban agglomeration, and potential urban agglomeration[3] (Fig. 3.4).

- *Developed urban agglomeration*—PRD, YRD, Beijing-Tianjin-Hebei city-region, urban agglomeration in the Middle Reaches of the Yangtze River, Shandong Peninsula city group, Liaoning peninsula urban agglomeration, Central Henan urban agglomeration, Guanzhong urban agglomeration, Chengyu urban agglomeration, urban agglomeration in the West Coast, and urban agglomeration in the East Coast.

[3] Sina News. (2013, December 19). Building 32 urban agglomerations in 2030. Retrieved from http://news.dichan.sina.com.cn/2013/12/19/989132.html.

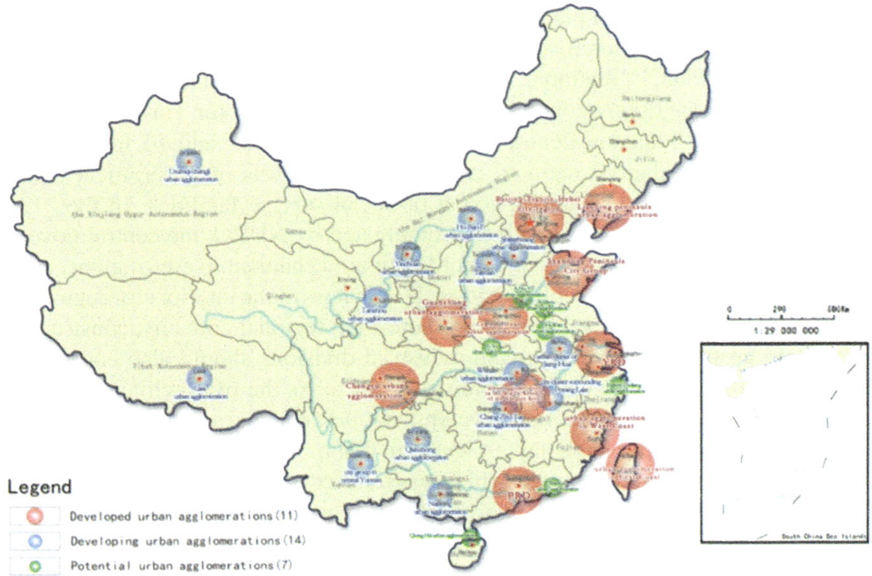

Fig. 3.4 Regional development strategy from 2006 to 2011. *Source* Author

- *Developing urban agglomeration*—Wuhan urban agglomeration, Changsha-Zhuzhou-Xiangtan urban agglomeration, the urban cluster of Jiang-Huai, Hu-Bao-E urban agglomeration, Lanzhou urban agglomeration, Urumqi-Changji urban agglomeration, Qianzhong urban agglomeration, Yinchuan urban agglomeration, Lasa urban agglomeration, Taiyuan urban agglomeration, Shijiazhuang urban agglomeration, city group in central Yunnan, city cluster surrounding Poyang Lake, and Nanning urban agglomeration.
- *Potential urban agglomeration*—Yu-Wan urban agglomeration, Ji-Lu-Yu urban agglomeration, E-Yu urban agglomeration, Xuzhou urban agglomeration, Eastern Zhejiang urban agglomeration, Shantou urban agglomeration, and Qiong-Hai urban agglomeration.

These urban agglomerations will be built strategically toward an overarching spatial form to enhance regional development. NDRC promulgated the "Main Functional Area Plan" in 2010 in light of the polycentric layout to provide differentiated guidelines for different regions in relation to intensive land use and protection of the environment and land resources. State land was further split into four categories, namely, an optimized development area, important development area, restricted development area, and prohibited development area. Furthermore, in light of the spatial distribution of the four land categories, the plan proposed three patterns: (a) an urbanization strategy pattern of "Three-horizontal & Two-longitudinal", (b) an agricultural strategy pattern of "Seven-area & Twenty-three-belt", and (c) an ecological security strategy pattern of "Two-screen & Three-belt." These ideal patterns are intended to be found in all the above-mentioned urban agglomerations, and the new development tasks for the city-regional integration were established virtually in terms of agricultural development, urbanization, and ecological and environmental protection (SC, 2010).

Overall, China strategically created a series of new metropolitan areas, urban agglomerations, and economic zones in this stage. Some of these city-regions had a solid socioeconomic foundation for regional integration. However, the socioeconomic links within these regions and the industrial relationship had to be further fostered by institutional arrangements (Xiao, 2008; Xu & Shi, 2010; Zhao et al., 2012). Inter-jurisdictional competition still lacked constraints and regulations both top-down and bottom-up in terms of institutional design. The production element market could not grow in step with the service market, and such a structural mismatch further led to a shortage of educational and infrastructure investments due to repeated investment, wasted resources and finance, and a lack of regional interest entities. Additionally, the construction of region-based agencies lagged behind *vis-à-vis* the rapid socioeconomic linkages in city-regions, and power reshuffling from the territorial government toward regional authority remained difficult due to the rigid administrative system in China. Some public affairs across jurisdictions, such as river control and inter-city infrastructure construction, remained difficult to address and led to damage to regional interests.

3.2.5 *Fifth Phase, 2014 to Present: Three New National Strategies for Enhancement of Competitiveness and Regional Coordinated Development*

On the basis of the four-plate strategy and the making of the city-region strategy in the previous stages, the central government put forward another three new regional and international strategies (i.e., "Coordination Development of Beijing-Tianjin-Hebei city-region," "Yangtze River Economic Belt," and "Belt and Road Initiatives"). In 2014, Chairman Xi Jinping said, "the capital economic cycle needs to depend on coordinated development of Beijing-Tianjin-Hebei city-region and continue to propel the industrial connection through forging the cross-border cooperation mechanism." From September to October 2013, Chairman Xi Jinping advocated the construction of the "New Silk Road Economic Belt" and "Maritime Silk Road" in Kazakhstan and Indonesia. The two strategies were the main components of the "Belt and Road Initiatives," which were elaborated and reinforced by several allied state visits, the China-Asian Expo, and the Asia–Pacific Economic Cooperation. In March 2015, the NDRC, Ministry of Foreign Affairs, and Commerce Department enacted "Vision and Actions on Jointly Building the Silk Road Economic Belt and Twenty-first Century Maritime Silk Road," which served as the outline of building "Belt and Road Initiatives" and included primary principles, key regions, and implementation mechanisms.

Meanwhile, such strategic transformation was viewed as an important approach to reconciling the recent economic crisis in China. The loss of advantages in resource endowment, price of labor, and institutional environment had given rise to the unavoidable economic transformation from an export-oriented economy to a domestic demand-oriented economy mode. This transformation urged the building of a city-regional integration mechanism through policy and institutional arrangements to create competitive city-regions at the global scale. The crisis of excess production capacity in the domestic market generated the initiative of the output of production capacity in formulating regional policies and strategies. Some coastal city-regions have received overriding attention from policymakers, aiming to create competitive city-regions under the national and international opening layout, such as the making of the Guangdong-Hong Kong-Macao Greater Bay Area and the recurrence of YRD integration in governmental agendas. However, in this round of regional development, the fever of city-regionalism also generated some weakly-integrated city-regions and artificial city-regions without underlying economic integrating mechanisms.

The national and international development strategies have restructured the spatial layout of the city-region and reinforced the coordinated development mode, which induced the fostering of the polycentric layout at the national scale (Fig. 3.5). Moreover, regarding the economic integration of city-region, SC promulgated "The New Urbanization Plan" in 2014 to proffer integration-oriented specific guidelines for city-regions due to the importance of forging inter-jurisdictional coordination mechanisms, promotion of coordination of industrial division, and construction of infrastructure among cities. The plan also concentrated on breaking administrative hurdles and monopolies, enabling the production factors to flow freely in favor of resource allocation. Based on this plan, in the integration process, different administrative levels of a city followed their own pathways for socioeconomic development. Regional core cities, such as direct-controlled municipalities, provincial capital cities, cities with independent planning and crucial hub cities, normally served as cornerstones for the city-regional integration socioeconomically and institutionally. Coastal cities must actively accelerate their industrial upgrading process and participate in the global industrial competition to upgrade their position in the global economy. Inland cities must establish a high-quality industrial system, including advanced manufacturing, strategic new industries, and modern services. However, speeding up the development of medium-sized and small-sized cities was the key to regional integration. These cities must consolidate their industrial basis by developing key industries and improving the quality of industrial operations and products. Small towns must absorb more from local agriculture and develop a key industry to gradually form a satellite town (SC, 2014).

China has been following the "13th Five-year Plan (2016–2020)". In terms of national strategy, based on the existing polycentric layout for regional development, China must further reinforce the role of city-regions by developing benign integration. The reinforcement of city-regions requires the building of large-scale infrastructure and effective institutional alignments for regional coordination. Additionally, the central government should intensify the leading role of geographical axis and belt throughout China from the eastern area to the western area, including "the Road

Fig. 3.5 Regional development strategy from 2014 to present. *Source* Author

and Belt Initiatives", the "Yangtze River Economic Belt", and other crucial railway lines.[4]

The preceding review investigates two overt tendencies with regard to the evolution of national strategies and policies. The first tendency is the "making of city-regions" in socialist China, which transforms the spatial layout of national development from a dichotomy between the inland area and coastal area to a city-region-based polycentric layout. From 1978 to the present, the national strategies that resulted from the severe conflicts under the regional repositioning from an uneven development strategy to a coordinated development strategy have broken the monopolized regional advantages of eastern coastal areas by creating economic benefit-sharing mechanisms across regions at the national scale. Such transformation manifested in the trend of developing the "Coastal Area plus Inland Area," which evolved into a spatial layout of four plates and formed the "four plus three" strategy. The polycentric regional layout consisted of 32 urban agglomerations, which were predicted to accommodate a population of 0.8 billion in 2030.[5] The second tendency is the geographical down-scaling of regional development in China from the traditional four-plate strategy to the specific urban agglomeration that requires a focus on

[4] Xiao, J. & Wang, Y. (2015, September 23). The regional strategic thinking in the era of 13th five-year plan. Retrieved from http://epaper.gmw.cn/gmrb/html/2015-09/23/nw.D110000gmrb_20150923_1-15.htm?div=-1.

[5] Xinhuanet. (2013, December 17). The appearance of 32 urban agglomeration development path, established in 2030 as prediction. Retrieved from http://cq.winshang.com/news-206197.html.

the coordination mechanism among different levels of a city and town. Therefore, a series of institutional arrangements were developed to target the establishment of an effective regional coordination mechanism to reinforce city-regional integration.

The central government began to devolve power to localities, and this power included administrative power, such as administrative approval authority and planning approval; power for formulating policies and schemes to propel socioeconomic development, such as central-local governmental tax-sharing reform; and regulatory power, such as formulating and enacting regional planning, environmental and fiscal regulations. Such decentralization made the city government less dependent on the central government and more flexible to foster the city-region by integration-first institutional alignments, such as a regional spatial plan, a joint conference on development, and inter-jurisdictional economic agreements. Despite the strengthening of the role of the territorial government, the central government retained institutional and administrative control in city-regional integration for building a regional coordination networking mechanism. The mechanism was to integrate top-down and bottom-up institutional approaches and to enable the roles of different levels of government, social organizations, and enterprises. With encouragement from upper-level governments, territorial practices in institutional arrangement reshaped the local governance mode to foster city-regional integration. Thus, a new concept was determined and developed. However, whether existing institutional arrangements can successfully acquire city-regional integration from economic and institutional perspectives remains an ambiguous and perplexing question. In addition, comprehensive integration should be conducted to grasp the inherent institutional logics of city-region, especially for certain areas officially documented as city-regions; however, these areas are weakly connected in terms of socioeconomic and institutional linkages.

3.2.6 Summary

The analysis above indicates the transformation of the spatial layout of regional development and the evolution of city-regional integration at the national scale after the launch of economic reform and the Opening-Door Policy in 1978. Overall, the spatial characteristics of this transformation are summarized as spatial repositioning, shifting from the division into inland areas and coastal areas to a polycentric spatial layout at the overall national scale. However, the driving forces underlying this long-term evolution are multi-dimensional and not just reshaped by the intricate central-local relations but remolded by globalization, decentralization, and marketization (Wei, 2001). Table 3.1 systemically presents the features of regional evolution in five phases.

According to Table 3.1, during the evolution process during which a city-region gradually superseded the regional plate to serve as the overarching spatial unit for regional development, there are three inter-locked strands of re-configuring the trajectory of regional development in post-reform China. The first strand is the changing national spatial strategy in keeping with the transformation of governmental

Table 3.1 Features of regional evolution from 1978 to now

Phase	One	Two	Three	Four	Five
Time period	1978–1990	1991–1998	1999–2005	2006–2013	2014 to present
Governmental initiatives	Regional uneven development	Regional uneven development	Regional coordinated development	Regional coordinated development	Regional coordinated development
Regional spatial layout	Eastern-central-western region	Eastern-central-western regions centered on three megacity-regions	Eastern-central-western-Northeastern region	32 urban agglomerations	National and international opening layout
Spatial unit	Regional plate	Regional plate	Regional plate	City-region	City-region
Economic landscape	• Revitalization of economy of eastern regions • Rise of export-oriented economic mode	• Flourishing urban economy • Rise of inter-city economic linkages	• Reducing the economic gaps among four plates • Preliminary making of city-region • Reinforcement of Administrative division-based economy	• Creating polycentric economic-geographic layout • Export-oriented economy destroyed by global financial crisis	• Transforming from export-oriented economy to domestic demand-oriented economy mode • Output of production capacity
Institutional landscape	• Power decentralized to localities • Policy experimentation through designating the opening cities • Marketization	• Rise of urban entrepreneurialism • Rock-ribbed conflicts between jurisdictions • Preliminary integration-oriented institutional design	• Place-based development strategy • Lack of cross-border coordinated mechanism	• Building of diversified regional coordinated mechanism • Rise of a range of regional plans	• Substantial integration-oriented institutional alignment for city-region • Designating city-region, even for weakly or partly integrated region

Source Author

initiatives from uneven regional development to coordinated regional development. Specifically, the regional spatial layout was crafted according to the national development strategies initiated by the central government, which focused on regional efficiency fostered by an uneven regional development strategy. These efforts are exemplified by the "Development Strategy of Coastal Area"; the setting up of a "Special Economic Zone"; and the opening of border cities, riverside cities, and inland cities. Afterwards, the focus shifted to regional equity by implementing regional coordinated development strategies, such as the encouragement of 32 urban agglomerations, spearheading the "Belt and Road Initiative" from China's side, and the Yangtze River Economic Belt.

In the wake of the transformation of the developmental initiatives, the central government decentralized power to the territorial government to reinforce the local socioeconomic autonomy and provided regional preferential policies and integration-first institutional arrangements for facilitating city-regionalism. Specifically, top-down empowerment boosted the high vitality of the urban economy and strengthened the functional linkages between cities through allowing the localities to select their development paths independently. For example, the rise of the Sunan and Kunshan models was chiefly driven by the economic innovations of local governments. In addition, due to the directories from the central government and the aspirations of the local government to forge city-regionalism, some city-regions were only nominally listed in the governmental agendas. It is worth noting that the above-mentioned endeavors in the making of the city-region have encountered difficulties as the new regional authorities were not empowered by the incumbent administrative system to drive the formal regional plans (see Sect. 3.2.4). The contradictions between the new regional authority and the original administrative institutions were difficult to reconcile; their cooperation heavily depended on institutional innovations, such as the creation of joint conferences, the establishment of case-specific regional authorities, and doing regional spatial planning.

The second strand is the reshuffling of economic driving forces in promoting regional development in China. In the early period of the Opening-Door policy, the economic driving forces were developing the export-oriented economic mode in some special open cities through decentralization of power and experimentation with policy to attract global capital. In the next phase, from 1991 to 1998, the rise of the urban economy was triggered by diversified pathways for enhancing it, such as the Wenzhou model, Sunan model, and Tianjin model. Three typologies of the urban economy were recognized by the central government and differentiated by the governance structure related to the share of state-owned capital. One category attracted foreign capital and investment to achieve industrialization, accepting international principles, such as the Kunshan and Guangdong models (Chien, 2007). Another category utilized local resources or locally-developed businesses. Notable examples were the Wenzhou and Sunan models, which fully utilized local natural resources to develop town and township enterprises or local family-owned businesses in achieving first industrialization and then marketization. There are subtle differences between these two models (Wei, 2010). Specifically, the Sunnan model was a localized urban economy driven by state-directed township and village enterprises

(Wei & Gu, 2010). The Wenzhou model followed family-owned small businesses characterized by "market-oriented flexible production and endogenous development" (Wei et al., 2007, p. 427). The last category was based on state-owned businesses to deliver economic growth, such as the Tianjin models (Zeng et al., 2015). The early stages of regional integration for YRD and PRD were driven by classical models of urban economies, such as in the Guangdong, Wenzhou, Kunshan, and Sunan models, whereas some of them were merely thought of as "Privilege Economies." These export-oriented economies and local economies, built on cheap labor forces and resource advantages, cannot be sustained under the global economic transformation moving from Fordism to post-Fordism (Zeng et al., 2015). This situation was exemplified by the economic depression of some export-oriented cities, such as Dongguan, Shenzhen and Suzhou. Traditional urban economy models and regional industrial divisions were devastated by new economic trends (i.e., risk-induced economic transformation), especially after the global economic crisis in 2008. Pillar manufacturing in China has experienced a tremendous decline at the local scale, such as in Dongguan, and the upgrading of industrial structures would be conceived of as a significant task for urban and regional development. Furthermore, the export-oriented economic model has been gradually repositioned as a domestic demand-based economic model. Transforming the national economy and improving the industrial capacity was urgent for the central government in a rapid changing and globalizing world economy. The introduction of the "Belt and Road Initiative" is a case in point. One of its functions is to address the excessive outdated capacity so as to optimize the economic and industrial structures, especially after 2014 (see Table 3.1). This tremendously changing economic landscape and environment compelled the repositioning of the urban economy and industry in a more globalized context and a global urban system and also left new challenges for the city-regions in implementing the governmental initiatives of regional coordinated development.

The last important strand of the evolution of governance structure, started with the introduction of tax-sharing system reforms formally in 1994. Under the new fiscal policy, local governments gradually became entrepreneurial, characterized by inter-jurisdictional competitions to attract mobile capital and forge business-style urban agendas for acquiring additional developmental resources from the central government. The exacerbated fiscal burden on the local government induced by the fiscal reform resulted in a strong willingness to collaborate with business elites and enterprises and to operate as urban coalitions for stimulating economic growth. Under this circumstance, city-regionalism acts as a new spatial economic strategy for local government to achieve not only economic growth but also political promotion for local officials. Specifically, the municipal governments are actively involved in the inter-jurisdictional competition to re-territorialize the capital flow and stimulate economic growth, by maligning the land price competition and others. In contrast, regional resources, such as the labor force, mineral resources, capital, information and technology, and national preferential policies and fiscal aids, were integrated at a lower level. Moreover, to overcome the fierce inter-city competition and other negative externalities of urban entrepreneurialism, city-regionalism has been strategically deployed by the central government to create a new regulatory arena for both regional economic growth and regional coordinated development. A series of integration-oriented institutional alignments has been formulated to coordinate regional competition. Local governments have been enthusiastic about inter-jurisdictional cooperation and have taken advantage of city-regionalism as a strategy to seek advantage to hasten urban development.

In addition to vertical interaction across different levels of government, the statization and destatization processes in socialist China reflect that the relationships across the government, market, and civil society are necessarily restructured with reference to achieving city-regional integration. Economic factors were increasingly playing a prominent role in the governance structure of city-regions. With regard to China's economic reform in 1978 and the transformation from a planned economy to a market-oriented economy, dozens of initial regional development policies stimulated the economic growth of coastal areas, and the majority of the Special Economic Zones in China entrenched an export-oriented economy in the early stage. After several decades of institutional loosening and preferential policies from the central government, the emergence of some urban economies accelerated the internal linkage among different cities through building uniform regional industries and forming region-based divisions of labor. However, in recent years, connected to the rapid restructuring of the global economy and the decline of the export-oriented economic model, city-regionalism has turned into an imperative strategy for local officials to break the local economic deadlock, even for transnational corporations. The economic demand turned to the main impetus for localities to collaborate with one another through acquiring industrial upgrading as a means of repositioning themselves in the world city system.

Overall, Table 3.1 reveals that the underlying economic force is entangled with the institution-building process in different phases, and the intersection of the economic evolution and the reshuffling of governance structures reshapes the economic-spatial layout of regional development in China. Recently, city-regional integration has gained increasing attention from the central government and the local government, and city-regionalism has been strategically conducted at the overall national scale. Furthermore, some weakly integrated regions or regions that were deeply affected by the restructuring of the global economy are officially designated as city-regions and proffered a host of integration-oriented institutional alignments and large-scale infrastructure projects from different levels of the government. To provide integration-oriented institutional alignments in securing city-regionalism, this phenomenon has already arisen not only in the central and western regions but also in the eastern coastal region.

However, under the restructuring of the macro-economic context, does existing integration-oriented institutional praxis effectively apply in the localities seeking to create a path for economic repositioning or industrial upgrading? In other words, within the creation of city-regions, how the regional governance structure that the governments endeavor to build enables or encumbers the socioeconomic development of a region is less than transparent. Thus, how the institutional alignments or governance structure reshuffles itself in response to the economic integration of city-region, especially in some weakly integrated regions or regions that confront the tremendous effects from changing global economy, is under-researched. The present study aims to fill this gap by investigating the city-regional development of SDH. The general evolution of PRD integration is reviewed in the next section to identify major issues to be examined by this book.

3.3 Meso-context: Evolution of the PRD Integration and Making of Three Sub-regions

The embeddedness of SDH in the integration process of PRD and the positions of the three sub-regions can be found in the framework of the PRD's regional plans. After the promulgation of the *Outline Plan* in 2008, a series of formal and informal institutions of SDH was fostered by different territorial organizations to create an institutionalized governance apparatus and tackle cross-boundary public issues.

The Guangdong Provincial Government initiated the first version of spatial planning, which is titled as the "PRD Urban System Plan (1991–2010)". To repair the territorial fragmentation and enable clustering development, the Construction Committee of Guangdong Province (CCGP) proposed the strategy of integrative development of neighboring towns, thereby creating ten integrated regions. In particular, Shenzhen, Dongguan, and Huizhou formed three integrated regions with their neighboring towns, respectively. The "PRD Urban System Plan (1991–2010)" uses the term "urban system" to refer to urban agglomeration (CCGP, 1991). This plan primarily aimed to generate an orderly and rational division of the urban system.[6] The regional spatial structure of PRD has been restructured by the flow of capital since 1994 owing to the Hong Kong enterprises and opening platforms of Shenzhen and Zhuhai. Consequently, the "PRD Economic Region Urban Cluster Plan" was promulgated in response to this socioeconomic spatial reconstruction, thereby defining the polycentric structure of PRD as "one core (i.e., Guangzhou) and two sub-cores (i.e. Shenzhen and Zhuhai)" and designating three sub-regions, namely, East Pearl River, West Pearl River, and Middle Pearl River (Fig. 3.6).

Based on these core cities, PRD initiated three sub-regions (i.e., SDH, Guangzhou-Foshan-Zhaoqing sub-region, and Zhuhai-Zhongshan-Jiangmen sub-region) (CCGP, 1994). This plan first put forward officially SDH and the initiatives for the polycentric development of PRD (Fig. 3.7).

The Construction Department of Guangdong Province (CDGP) issued the "PRD Urban Cluster Coordinated Development Plan (2004–2020)". The plan followed the division of three sub-regions in PRD to obtain displaced development and fulfill the leading role of regional core cities (CDGP, 2004). Given the polycentric structure of PRD, this regional plan focused on the establishment of a regional coordination mechanism and common policies to achieve a balanced power equilibrium. Shenzhen was fostered as the core of a sub-region composed of Shenzhen, Dongguan, and

---

[6] Liang, W. (2006, January 6). The big change from urban system to urban agglomeration of PRD during 7 years. Retrieved from http://finance.sina.com.cn/review/20060106/16572255437.shtml.

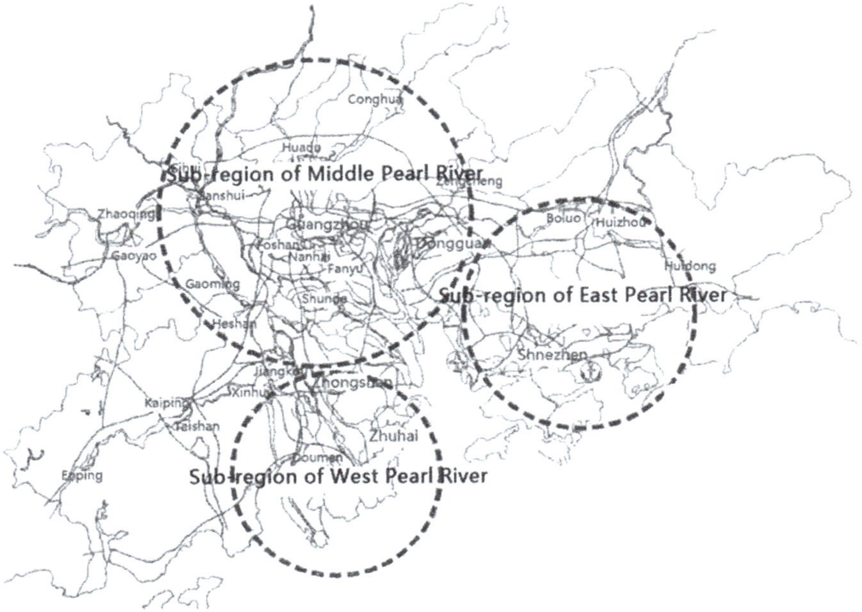

Fig. 3.6 Spatial structure of PRD. *Source* CCGP (1994)

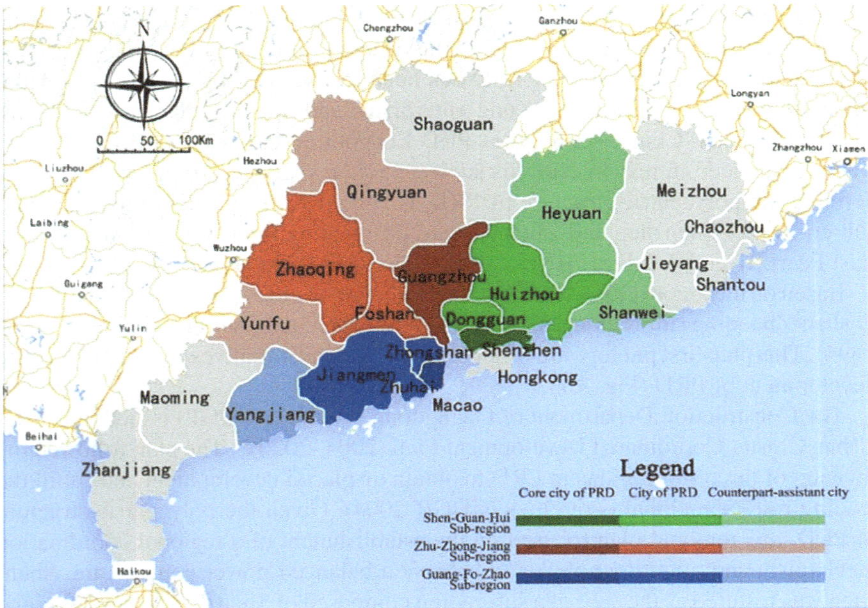

Fig. 3.7 Three sub-regions in the PRD region. *Source* Author

Huizhou, and shared the same socioeconomic status as Guangzhou. Eventually, PRD formed "two cores (Guangzhou and Shenzhen) and one sub-core (Zhuhai)." Therefore, this version of a regional plan was conceived as a milestone for SDH, not only for the enforcement of SDH integration, but also for member cities in SDH to achieve industrial integration. The industrial guidelines for the integration were three: Shenzhen must develop a creative industry and top service as the open port for Southeast Asia and Hong Kong; Dongguan should focus on industrial upgrading and developing high-tech industry and modern manufacturing; and Huizhou must reinforce its traditional industries, such as electronic information, electrical machinery, textile, and clothing. In addition, the regional plan stipulates the policies and principles to economic and infrastructure integrations of PRD and brings other elements, such as ecological, resource, population, and spatial regulation, into the action framework of regional integration.

In 2008, NDRC promulgated the *Outline Plan*, which provided the guidelines of the integration of PRD in detail; PRD integration was thereby split into six sections, namely, industrial distribution, urban and rural planning, infrastructure, public service, environment protection, and tourism development (NDRC, 2008). Furthermore, the chapter on the promotion of regional coordination development specifies the role of industrial division in SDH's integration and mandates the industrial position of SDH in the PRD region, namely, as the core of electronic information and top manufacturing products and a global industrial base of electronic information, reinforcing the modern service industry and establishing the regional service and creation core by industrial upgrading.

Importantly, this *Outline Plan* was not simply a matter of putting forth a plan for industrial development. Compared with the previous versions of the regional plan, the *Outline Plan* set up a comprehensive framework to enhance regional integration and provides explicit guidance for the economic integration of SDH. Afterwards, a series of city-regional plans were proposed in accordance with this *Outline Plan*: the PRD Urban and Rural Integration Plan (2009–2020), the Guangzhou-Foshan city-region Plan (2009–2020); and the SDH Coordinated Development Plan (2012–2020). Thus, the *Outline Plan* set a milestone in fostering the integration of PRD. More plans crystallizing other contents of the *Outline Plan* were enacted subsequently. The "PRD Megalopolis Integration Plan (2009–2020)" was formulated by the Guangdong Provincial Government and focused on the integration of urban and rural planning systems, industrial layout, public service provision, infrastructure construction, and environmental protection (Fig. 3.8).

Three sub-regions launched their own regional integration plans in 2009, 2010, and 2012. The fifth version of the PRD plan, known as the "PRD Region Plan," aiming to cover all the territory of PRD, was initiated in 2014, incorporating all the towns in the nine cities. Importantly, this regional plan was conceived as a crucial program to integrate PRD spatially, socially, and economically, driven by the initiatives of regional coordinated development.

In 2011, NDRC officially promulgated the "Main Functional Area Planning" to provide specific development guidelines for the whole country's land, dividing

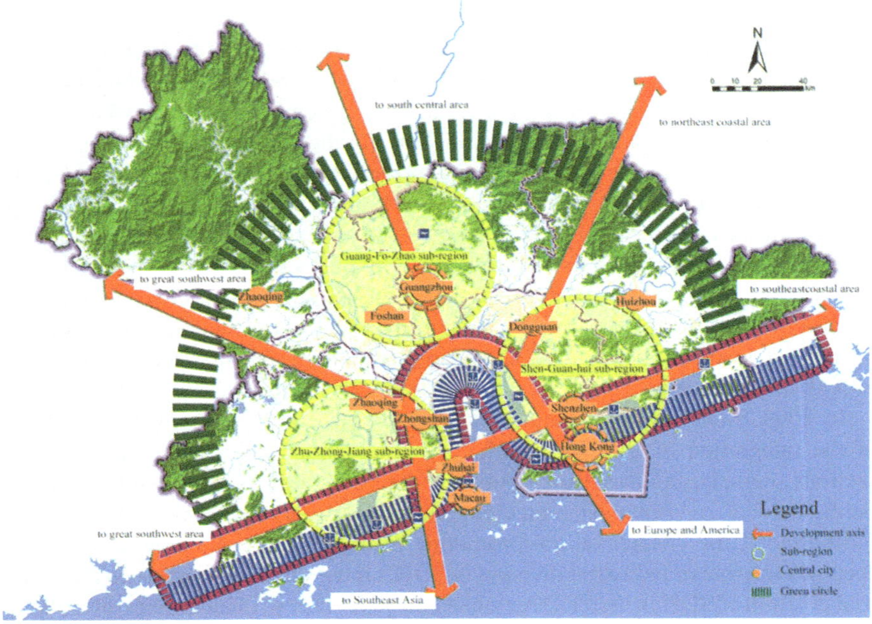

Fig. 3.8 Spatial structure in the PRD Megalopolis Integration Plan (2009–2020). *Source* Guangdong Provincial Government (2009)

land into four types based on the land use intensity, namely, an optimized development area, important development area, restricted development area, and prohibited development area. According to this plan, PRD was classified as a development area. Accordingly, the spatial structure, town distribution, population distribution, industrial structure, development mode, infrastructure layout, and the ecological system must be optimized. In the urbanization strategy of this plan, PRD was positioned to establish modern manufacturing and service bases. Further, as the open gateway to Southern China, the main city-region of PRD would be involved in global economic competition and national technological creation and serve as a research base and a crucial engine for national economic growth, through internal economic integration with Hong Kong and Macao. Regarding SDH, Shenzhen is mandated as a national-level innovation city and an international hub of technological R&D and high-level business services. Dongguan and Huizhou were expected to reinforce their geographical advantage to drive industrial division in the PRD region and foster regional competitiveness. The SDH region, which had been positioned as the center of technological development, high-tech industry, and modern service and manufacturing, was expected to continue to strengthen infrastructure integration (SC, 2010).

Apart from the strategy of creating three sub-regions, the promotion of a "Double Transfer" in 2008 (i.e., labor relocation and industry relocation) began to strengthen the socioeconomic connections between PRD and other regions in Guangdong

Province. On the basis of such an integration-oriented industrial promotion strategy, approximately 28 provincial industrial transfer parks were constructed to facilitate industrial and social integration within the entire territory of Guangdong Province. The Guangdong Provincial Government issued an assistance strategy to help the western, eastern, and northern areas of Guangdong Province in 2014. As such, three sub-regions enlarged their scales spatially as follows: Qingyuan and Yunfu were incorporated into the Guangzhou-Foshan-Zhaoqing sub-region, Shanwei and Heyuan were incorporated into SDH, and Yangjiang was integrated into the Zhuhai-Zhongshan-Jiangmen sub-region. This endeavor was an institutional attempt to achieve polycentric development and repair fragmented regional governance, through city-regionalism.

The actual spatial structure of PRD has been evaluated under these plans from the perspective of urban networks and inter-regional connections. There is a competing viewpoint concerning the polycentric degree of PRD in terms of socioeconomic space. Li (2008) graphically studies the evolution of the spatial structure of PRD, particularly the transformation from a growth point to a growth axis via a growth pole. Yu et al. (2009) employ the data from the 5th National Census in 2000 to identify the spatial structure of PRD and clarify a "core–periphery–fringe" spatial pattern in PRD. Based on such three divisions, PRD has become a functional polycentric city-region. After 2000, the spatial structure of a "core–periphery–fringe" was reinforced; however, they argue that the polycentric spatial structure was unremarkable in PRD (Yu et al., 2009). Conventional studies maintain that after the 2010s, the spatial structure of PRD has increasingly exhibited an apparent polycentric layout for socioeconomic space. This conclusion is gaining ground through a host of empirical studies. Chen et al. (2013) employ the data of inter-city passenger traffic flow to elaborate the spatial pattern of PRD. The results reveal an imbalanced spatial variation between center and peripheral areas, eastern and western coastal areas, and among the three sub-regions. In addition, the gradient hierarchy and polycentric layout coexist in PRD (Chen et al., 2013). Feng et al. (2014) use the data of inter-city railway traffic flow to ascertain the spatial structure of PRD and conclude that PRD has spatially manifested a high level of functional polycentricism. Furthermore, high-intensive transportation connection between eastern and western regions is lacking, thereby negatively influencing the functional polycentricism of PRD.

Empirically speaking, the polycentric structure has been strengthened with the increasingly tight inter-city socioeconomic connection. The three sub-regions of PRD as a divided spatial strategy under city-regionalism have attracted substantial attention from scholars and policymakers. Lai et al. (2015) call the strategy a decentralization strategy and use the inter-city connection to measure the internal linkage within the three sub-regions. The result indicates that the reality is not totally consistent with the ideal layout of the decentralization strategy. Chen et al. (2013) conducted a comprehensive comparison among three sub-regions and concluded that the internal integrated extents of the Guangzhou-Foshan-Zhaoqing sub-region, SDH, and the Zhuhai-Zhongshan-Jiangmen sub-region decreased; the relative equilibrium of urban function, urban size, and urban hierarchy gave rise to weak complementarity. Mei et al. (2012) affirm the considerable development of PRD in the past two decades and measured the potential interactions of prefecture-level cities and counties in the PRD region. The major findings show that PRD witnessed a tendency toward unbalanced development in the recent two decades, especially between central and peripheral areas. Moreover, they identified Guangzhou–Foshan twin city and SDH as having high urban potential value, and external radiation to neighboring areas is clearly stronger than any other regions in PRD.

In summary, city-regionalism as a spatial economic strategy was vigorously conducted by the Guangdong Provincial Government for regional coordinated development and improvement of the competitiveness of PRD. The creation of the city-region included at least two steps. The first stage designated the Guangzhou-Foshan-Zhaoqing sub-region, SDH, and the Zhuhai-Zhongshan-Jiangmen sub-region as city-regions within the cities of PRD, which were grouped to achieve agglomeration centered on the core cities. The second stage applied the assistance strategy (see Sect. 3.1) to achieve integration of PRD with the remaining regions in the Guangdong Province, that is the western, eastern, and northern regions. To reinforce PRD integration with three sub-regions, the provincial government promulgated several rounds of PRD regional spatial plans and offered a series of preferential policies, such as regional infrastructure projects, a cross-border economic cooperation zone, and a transfer payment system to different types of jurisdictions. Thus, PRD as a megacity-region in Guangdong was largely secured by integration-oriented institution alignments, and the city-regionalism of PRD is directed by provincial and municipal governments. However, the central government was not directly involved in the regionalization of PRD.

3.4 Micro-context: Industrial and Infrastructure Integrations Under the Integration-Oriented Institutional Arrangements of SDH

SDH in the eastern PRD administratively includes five cities, namely, Shenzhen, Dongguan, Huizhou, Shanwei, and Heyuan (Fig. 3.9). Although the socioeconomic status of SDH was documented and stipulated in the "PRD Urban System Plan (1991–2010)", its integration action started after the promulgation of the *Outline Plan*. From the territorial praxis of the SDH integration, the focal points of these endeavors focused on three domains, involving institutional arrangement, industrial collaboration, and infrastructure development.

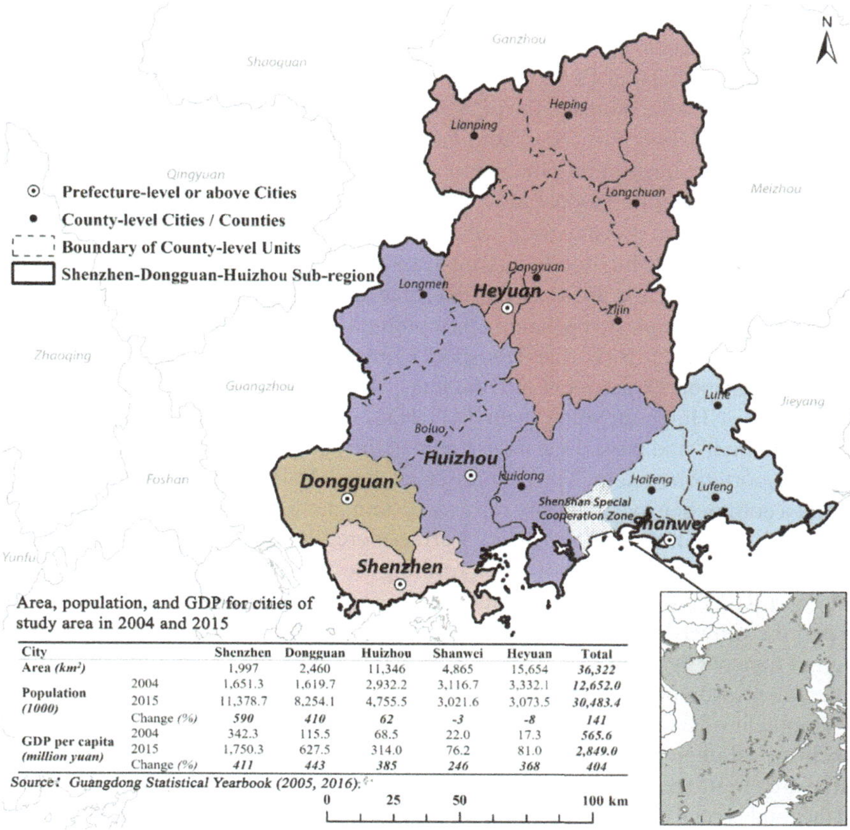

Area, population, and GDP for cities of study area in 2004 and 2015

City		Shenzhen	Dongguan	Huizhou	Shanwei	Heyuan	Total
Area *(km²)*		1,997	2,460	11,346	4,865	15,654	***36,322***
Population *(1000)*	2004	1,651.3	1,619.7	2,932.2	3,116.7	3,332.1	***12,652.0***
	2015	11,378.7	8,254.1	4,755.5	3,021.6	3,073.5	***30,483.4***
	Change *(%)*	***590***	***410***	***62***	***-3***	***-8***	***141***
GDP per capita *(million yuan)*	2004	342.3	115.5	68.5	22.0	17.3	***565.6***
	2015	1,750.3	627.5	314.0	76.2	81.0	***2,849.0***
	Change *(%)*	***411***	***443***	***385***	***246***	***368***	***404***

Source: Guangdong Statistical Yearbook (2005, 2016).

Fig. 3.9 Location of SDH. *Source* Author

3.4.1 Institutional Arrangement in SDH: Toward Integrative Governance Structure

To facilitate the SDH integration, different levels of government provided various formal and informal institutional arrangements to enable socioeconomic integration by creating an SDH-wide integrative governance structure. These institutional arrangements included regional planning, joint conferences, and a cooperation agreement. This integrative institutional building process aimed to create a region-based governance structure to guide a series of themes for regional development; such themes include industrial collaboration, construction of large-scale infrastructure projects, environmental and resource protection, and provision of public service.

The promulgation of the "SDH Coordinated Development Plan (2012–2020)" positioned SDH as a significant global modern industrial base in accordance with the

Outline Plan. The plan presents a host of specific measures to obtain infrastructure, economic, and social integrations of SDH. For infrastructure, this plan put forward the concept of the "One Hour Economic and Life Cycle" by reconstructing some transportation lines. For the industry, the plan provided the imagery of the spatial structure as "One Core and Two Sub-cores" and intended to nurture five cross-boundary new districts as the demonstration plot in economic integration, encompassing Pingshan and Huiyang Economic Cooperation Zone, Shenzhen–Huizhou around Daya Bay Coastal Tourism and Leisure Belt, Shenzhen–Dongguan Jiaoyi Bay Cooperation Zone, Dongguan and Huizhou Industrial Transfer Park, and Dongguan and Huidong Industrial Transfer Park. In addition, this plan first put forward the integration of housing estates through strengthening the communication and collaboration between cities (NDRC, 2012). Next, the Transport Commission of Shenzhen Municipality (TCSM), Transport Department of Dongguan Municipality (TDDM), and Transport Department of Huizhou Municipality (TDHM) enacted the "SDH Transportation Integration Plan" and passed the assessment at the end of 2014 to establish an SDH-wide transportation system. The primary objective of the three cities in this plan was to build a comprehensive inter-city railway system, named "Four Vertical Lines, Two Horizontal Lines, and One Sub-line." Furthermore, the plan proposed the comprehensive transportation system of SDH, forming the "One Hour Commuting Cycle of East Pearl River" to lay a solid foundation for regional integration. The plan incorporated the railway system, the highway system, the inter-city bus system, and the external transport system (TCSM et al., 2014). In 2014, the Guangdong Provincial Government promulgated the agenda "Significant Tasks of Further Promoting the Development of Eastern, Northern, and Western Areas of Guangdong Province" to launch the strategy of aid activity.[7] To respond to this strategy, SDH enlarged its spatial area to further incorporate two outlying cities, namely, Shanwei and Heyuan, and the total area of SDH has doubled *vis-a-vis* the previous area,[8] forming the spatial structure of "Three plus Two."

In addition to the regional plan, other formal and informal institutional arrangements play a significant role in the integration process as well. On February 27, May 16, and September 24 in 2009, the municipal governments of Shenzhen, Dongguan, and Huizhou signed the "Framework Agreement of Promoting Closely Cooperation of Eastern Pearl River Region" and "Cooperation Agreement of Planning Integration among Shenzhen, Dongguan, and Huizhou," to formulate the action framework for some specific affairs concerning SDH integration. In 2010, the "East Pearl River Forum" was held in Shenzhen, and the three cities' urban planning departments energetically participated in this forum, aiming to implement the initiatives of the

---

[7] Guangdong Provincial Government. (2015, April 16). Significant tasks of further promoting the development of eastern, northern, and western areas of Guangdong Province. Retrieved from http://zwgk.gd.gov.cn/006939748/201504/t20150430_578853.html.

[8] Sina News. (2014, October 17). Extension of SDH, adding Shanwei and Heyuan as the form of three plus two. Retrieved from http://gd.sina.com.cn/dg/social/2014-10-17/082610360.html.

Outline Plan and identify the economic position of SDH in the global economy.[9] As a typical informal institutional arrangement, joint conferences among the three cities were successfully held eight times from 2009 to 2014. Under the framework of joint conferences, working groups were instructed by the deputy mayor, city leader coordination conferences, and ad hoc groups to cope with the issues of urban planning, transport infrastructure, industrial collaboration, and environmental protection (Kou, 2014). These joint conferences signed a multitude of agreements, involving a host of cross-boundary matters, such as the construction of border roads, governing of the Border River, integration of planning system, implementation of major projects, automotive industry cooperation, and so forth. Moreover, intergovernmental collaborations had fruitful achievements in some special issues involving tourism, planning integration, informatization cooperation, police issues, and industrial collaboration.

Based on the abovementioned review, different levels of government devoted attention to the integration of SDH. The "SDH Coordinated Development Plan (2012–2020)" paved the way for economic, infrastructure, and social integration, which laid a solid foundation for the implementation of the SDH integration strategy. Apart from regional planning, some informal institutional arrangements also coped with city-regional issues under the guidelines of regional planning. These informal institutional arrangements not only involved the adjustment of relations from the provincial to the local levels but also incorporated the reconstruction of horizontal inter-government relations. Importantly, in the "SDH Coordinated Development Plan (2012–2020)," industrial integration and transportation integration were treated as the most significant tasks in SDH integration. New collaborative actions concerning the industrial and transportation integrations were formulated in this plan; such actions included fostering cross-border new districts and joint conferences between cities,[10] constructing a comprehensive transportation system, building industrial alliances, and constructing large-scale infrastructure projects.

Overall, the efforts of institutional alignments attempted to create a region-based governance structure, marked by the building of an integration-oriented institutional framework through formal and informal institutions to enable the socioeconomic integration of SDH. Thus, institutional integration has been vigorously propelled by the different levels of government in SDH on the basis of formal and informal institutional praxis. However, whether these institutional alignments can successfully reconstruct an integration-oriented institutional framework or a region-based governance structure remains unknown. Furthermore, the mechanism through which the existing governance structure is shaped by the integration-oriented institutional arrangements from governments can enable the socioeconomic integration of SDH. Although the SDH underwent a series of integration-oriented institutional alignments

---

[9] Huizhou Housing. (2010, November 2). The Pearl River east coast BBS opened in Shenzhen in 2010, Elite gathering to discuss the position. Retrieved from http://huizhou.house.sina.com.cn/news/2010-11-02/09324643.shtml.

[10] It is important to note that the building of joint conference between cities has received substantial attention from policymakers in Guangdong. The joint conference served as one of important layers within the three-layer implementing system of Outline Plan and was increasingly treated as the primary institutional form to enhance the PRD integration (Lai et al., 2015).

across broader themes for regional development, the less-than-transparent extent of institutional participation calls attention to the abovementioned two critical issues, which remain under-researched.

3.4.2 Industrial Integration in SDH: City-Based Crises in Industrial Development and Severe Industrial Isomorphism

After the promulgation of the *Outline Plan*, the integration of SDH achieved some crucial goals with reference to infrastructure integration, industrial development, public service, and environmental protection. Nevertheless, some potential risks are related to industrial integration. Traditionally, the local governments of SDH have attracted cheap labor to develop labor-intensive, low-end industries. Overall, SDH had a solid manufacturing-based industrial foundation with sufficient industry types. Nevertheless, industrial integration of SDH presented a significant limitation, which refers to the serious industry isomorphism between cities, resulting in the potential crises of socioeconomic development in member cities and challenging city-regionalism (Hu, 2011; Liu, 2011; Zhou et al., 2012). These difficult problems signify that the SDH region may be partially integrated in terms of socioeconomic linkages but is officially treated as a city-region driven by the strategy of regional coordinated development.

Regarding the individual cities in SDH, Shenzhen established the top industrial system in the beginning of the 2000s, manifested in the comprehensive modern industrial types. The system includes the electronic information industry, the bio-industry, the new energy industry, equipment manufacturing, and the cultural creative industry.[11] Although Dongguan served as the centerpiece of traditional manufacturing in PRD through developing five provincial-level industrial clusters (i.e., electronics, clothing, furniture, wool, and hardware), traditional manufacturing had higher labor costs than manufacturing in southeastern Asia. Dongguan was keen on industrial upgrading, shifting from low-end manufacturing to top services and knowledge-intensive industries.[12] In addition, Huizhou strongly depended on its low-end manufacturing, such as the electronics and petrochemical industries. Owing to the weak competitiveness of its traditional industries, such as the clothing and textile industries, the shoemaking industry, and industrial transformation, industrial upgrading was an urgent target for Huizhou in SDH economic integration (Ouyang, 2012).

---

[11] Yang, L. (2014, May 9). Cultivating five billions of the leading industry. Retrieved from http://www.sz-qb.com/html/2014-05/09/content_8575.htm.

[12] Zhongguo Jingying Wang. (2016, June 6). Dilemma for Dongguan's industrial transformation. Retrieved from http://www.cb.com.cn/index.php?m=content&c=index&a=show&catid=26&id=1136753&all.

Given the nonequivalent urban administrative and socioeconomic status and different crises of capital accumulation of these cities, a serious mismatch existed in the local demands for the industrial development of the SDH integration. First, the Shenzhen Municipal Government paid more attention to the industrial integration with Hong Kong to reposition its industry to modern high-end industrial species, such as financial services, R&D, and artificial intelligence. However, the rise of Shenzhen's economy has elicited an overflow of socioeconomic activities, such as the supply chain of Huawei relocating from Shenzhen to Dongguan, because of the shortage of land resources for construction in Shenzhen. According to governmental statistics, the planned supply of construction land of Shenzhen merely reached 1600 and 1350 ha in 2015 and 2016, respectively.[13] However, the planned supply of construction land of Beijing reached 4600 ha in 2015[14] and 4100 ha in 2016,[15] which nearly tripled that of Shenzhen. Such a serious crisis of land resources has propelled Shenzhen to prefer SDH over others as the overarching geographical scaffolding to reconcile the conflicts between the provision of land resources and fast-tracked industrial development. Shenzhen was reluctant to commit to industrial integration with Dongguan and Huizhou except for the demand to relocate the low-end industries to achieve valuable industrial lands. However, owing to the aspiration for industrial transformation and upgrading, Dongguan anchored its hope on the SDH integration to achieve a benign integration with Shenzhen with reference to transportation and industrial upgrading. Within the integration process, Huizhou sought to undertake more industrial transfers in cooperation with Shenzhen and Dongguan (Hu, 2011). Therefore, Dongguan and Huizhou had an insurmountable hurdle in promoting industrial upgrading toward the high-end sectors and boycotting the low-end industry from Shenzhen. However, the new member cities, Shanwei and Heiyuan, were the backward regions of Guangdong Province; in particular, the former was labeled "collapsing Shanwei (*Taxian de Shanwei*)." According to governmental statistics, the GDP of Shanwei reached merely about 76 billion Yuan in 2015, ranking the second worst of all cities in Guangdong Province, and approximately 83 billion Yuan in 2016, ranking the second worst as well. These two outlying cities had a strong desire to accept the industries that moved out from Shenzhen to revitalize their backward urban economy. This phenomenon provided a new pathway for Shenzhen, Shanwei, and Heyuan to break the local deadlock in development.

In addition to this disparate city-based crisis in industrial development, the severe industrial isomorphism among cities challenged the industrial integration of SDH. According to some empirical studies, the average of the industry isomorphism index reached 0.88, whereas the Zhuhai-Zhongshan-Jiangmen sub-region and Guangzhou-Foshan-Zhaoqing sub-region only had 0.72 and 0.49, respectively, in 2007 (Liu, 2011). For the internal industrial isomorphism, the index of Shenzhen–Dongguan,

[13] Shenzhen, China, Retrieved from http://www.sz.gov.cn/cn/xxgk/zfxxgj/ghjh/csgh/jq/.

[14] Xinhuanet. (2015, April 7). Beijing: the planned supply of construction land reaches 4600 hectare in 2015. Retrieved from http://www.xinhuanet.com/fortune/2015-04/07/c_1114889802.htm.

[15] Askci. (2017, February 21). Policy analysis: the plan of the supply of construction land in Beijing in 2017. Retrieved from http://www.askci.com/news/finance/20170221/16395991225.shtml.

Shenzhen–Huizhou, and Dongguan–Huizhou reached 0.88, 0.86, and 0.82, respectively (Liu, 2011). The industries with the highest isomorphic indices were plastic products, communication devices, computers, and other electronic equipment.[16] This high-level industrial isomorphism resulted in cut-throat competition among cities for the foregoing industries. Under such circumstances, scholars conducted a series of preliminary investigations to determine the economic integration of SDH. Xing et al. (2013) focus on the measurements of logistic integration in SDH and demonstrate that the degree of logistic integration lags behind the degree of development of the volume of freight traffic in SDH. Lai et al. (2015) employ the data on headquarters and branch offices to find that the economic integration of SDH is partly successful, and the integration still needs to be strengthened, especially for the connections between Shenzhen and Huizhou and between Dongguan and Huizhou.

The different levels of government in SDH have undergone preliminary approaches to reconcile the disparate city-based crisis in industrial development and cope with the issues of severe industrial isomorphism. The "SDH Coordinated Development Plan (2012–2020)" proposed the idea of creating cross-boundary economic cooperation zones, including the Pingshan and Huiyang Economic Cooperation Zone, the Shenzhen–Huizhou around Daya Bay Coastal Tourism and Leisure Belt, the Shenzhen–Dongguan Jiaoyi Bay Cooperation Zone, the Dongguan and Huizhou Industrial Transfer Park, and the Dongguan and Huidong Industrial Transfer Park (NDRC, 2012). These six cross-boundary economic cooperation zones were regarded as the "Bridgehead (*Qiaotoubao*)" in the regional agenda for the integration of SDH. After 2011, these zones started to pursue land development, construction of infrastructures, and attraction of investment. Moreover, the Guangdong Provincial Government approved the "Framework of SSSCZ" in February of 2011, and the "Working Committee of SSSCZ" and "Management Committee of SSSCZ" were established in May of 2011.[17]

As the typical industrial transfer park in SDH, the "Dongguan and Huizhou Industrial Transfer Park" was approved in 2009, and the Guangdong Urban and Rural Planning and Design Institute formulated the master planning based on the "Dongguan–Qiaotou Industrial Transfer Park," which was the predecessor of the "Dongguan and Huizhou Industrial Transfer Park," extending the area from 4 to 10.8 km^2. The "Dongguan and Huizhou Industrial Transfer Park" is located in the junction of Longcheng, Longhua, and Pingling towns in Huizhou and has convenient traffic conditions with Guanghe, Heqing, Renshen, and Huizhou Shugang highways meeting here (Guangdong Urban and Rural Planning and Design Institute, 2007). The primary objective of the "Dongguan and Huizhou Industrial Transfer Park" was to develop electronic appliances, new building materials, and automotive components by attracting investments from Dongguan. As one of fifteen key zones in Guangdong Province, this park provides settled enterprises in the 14 preferential policies of the free service charge offered by the Guangdong Provincial Government. Based on these policies,

[16] Sheng, W. (2010, March 16). Shenzhen–Dongguan–Huizhou develops into the leading area of PRD. Retrieved from http://www.bioon.com/organization/shenzhen/436599_3.shtml.

[17] Official website of SSSCZ. Retrieved from http://www.shenshan.gov.cn/home/about#survey.

"Dongguan and Huizhou Industrial Transfer Park" introduced 13 new programs, and the total investment reached 11.5 billion Yuan in 2014.[18]

Compared with the "Dongguan and Huizhou Industrial Transfer Park," the "Dongguan and Huidong Industrial Transfer Park" had a longer history and was approved in 2006 by the Guangdong Provincial Government. From 2006 to 2012, the area of the "Dongguan and Huidong Industrial Transfer Park," which is in the Daling Town Huidong County of Huizhou, expanded from 426.6 to 746.6 ha.[19] According to the "SDH Coordinated Development Plan (2012–2020)," the leading industries of the "Dongguan and Huidong Industrial Transfer Park" were advanced machinery and equipment manufacturing, new electronic information, and LED to develop the important origin of strategic emerging industries and a top industry hub. Until May 2014, the investment of infrastructure in the "Dongguan and Huidong Industrial Transfer Park" reached 1.81 billion Yuan and formed the infrastructure network. In addition, the "Dongguan and Huidong Industrial Transfer Park" constructed social amenities gradually, encompassing Yongji ecological park and golf course. To enhance its attractiveness, the "Dongguan and Huidong Industrial Transfer Park" proposed ten service platforms, involving financial services, quality testing services, administrative services, public logistics information services, law services, the collaboration between industry and university, mobile network services, project intermediary services, and security services.[20] At the end of 2013, the scalable industrial output value reached 12.6 billion Yuan and accounted for 33% of the scalable industrial output value of Huidong County in Huizhou. The taxes on the "Dongguan and Huidong Industrial Transfer Park" reached 0.15 billion Yuan and increased by 40.15% compared with that in 2012.[21]

Compared with the foregoing two parks, SSSCZ was endowed with prefecture-level city administrative power for personal appointments and removals. SSSCZ is located in southern China, in the east of PRD, bordering the South China Sea, and connected to Huizhou and east to Haifeng County, Shanwei. SSSCZ has an area of 468.3 km^2 and covers Ebu, Xiaomo, Chishi, and Houyu towns in Haifeng County of Shanwei. In terms of administrative authority, the "Working Committee of SSSCZ" and "Management Committee of SSSCZ" were the agencies of the Guangdong Provincial Government, and SSSCZ was co-managed by Shenzhen and Shanwei municipalities. Shenzhen was responsible for the economic construction and management, and Shanwei was in charge of land requisition, demolition, and social affairs. Shenshan Highway, 324 National Line, Xiashen Railway, and Chaoguan

[18] Official website of Huizhou Industrial Transfer Park. Introduction of Huizhou Industrial Transfer Park. Retrieved from http://www.jscyy.gov.cn/digital.asp?genusid=1119.

[19] Official website of Dongguan–Huidong Industrial Transfer Park. Introduction of Dongguan–Huidong Industrial Transfer Park. Retrieved from http://www.hdpark.com.cn/about/.

[20] Official website of The Economic and Information Commission of Guangdong Province. Introduction of Dongguan-Huidong Industrial Transfer Park. Retrieved from http://www.gdei.gov.cn/flxx/cyzy/gwjk/201407/t20140704_112874.htm.

[21] Official website of The Economic and Information Commission of Guangdong Province. Introduction of Dongguan-Huidong Industrial Transfer Park. Retrieved from http://www.gdei.gov.cn/flxx/cyzy/gwjk/201407/t20140704_112874.htm.

Highway passed through SSSCZ to provide convenient traffic conditions.[22] To foster the test area of SDH cross-border cooperation, SSSCZ included five types of leading industries, namely, information technology, coastal tourism, international logistics, new energy materials, and bio-pharmaceuticals.[23] A total of 37 projects were under construction, the total area reached 7.71 million m^2, the total investment was 25 billion Yuan, and 36 programs were developed in 2014.[24]

In sum, the underlying crises of the industrial integration of SDH emerged from the confluence of disparate city-based crises in industrial development and serious industrial isomorphism. Under the existing integration-oriented institutional alignments, the cross-boundary economic cooperation zone has been deployed as the crisis-management strategy to overcome such underlying crises and facilitate SDH integration. However, some urgent questions remain to be scrutinized in this book. (1) Under the several years of endeavor to integrate SDH, to what extent is SDH integrated as a regional industrial entity with particular reference to the actual degree of industrial isomorphism? (2) How has the above-mentioned governance structure reshuffled itself to cope with the underlying crisis of industrial integration in the case of the cross-boundary economic cooperation zone? Scrupulous investigation of these two correlative questions will contribute to a comprehensive and profound grasp of the SDH integration under China's city-regionalism.

3.4.3 *Infrastructure Integration in SDH: Building Regional Infrastructure Network*

According to the guidelines from the "SDH Transportation Integration Plan" the regional highway network, railway network, and inter-city public transport system were constructed to enable the regional infrastructure network. A total of 11 inter-city bus lines operated in the adjacent towns and townships in Shenzhen, Dongguan, and Huizhou until the end of 2013.[25] The first bus lines (*Shenhui sanxian*) linking the three cities were built in 2014.[26] The sharing rate of the taxis' dynamic information exceeded 90%, and the toll stations of ordinary highways in border areas were demolished to accelerate regionalization. Several boundary connecting lines was accomplished. Among the eight expressways, Huishen Coastal Highway, Guangshen Riverine Expressway, Boshen Highway, and Meiguan Highway (Qinghu–Liguang

[22] Official website of SSSCZ. Retrieved from http://www.shenshan.gov.cn/home/about#survey.

[23] NetEase News. (2015, February 9). Shenshan Special Economic Cooperation Zone: Shenzhen builds a new to east. Retrieved from http://help.3g.163.com/15/0209/09/AI0IPNKQ00964J4O.html.

[24] Shenzhen Evening News. (2015, April 9). Shenzhen-Shanwei Economic Cooperation Zone have 21 new projects. Retrieved from http://wb.sznews.com/html/2015-04/09/content_3190494.htm.

[25] TCSM. (2015, July 27). Planning of SDH integration and its reading. Retrieved from http://jt.sz.bendibao.com/news/2015727/715020.htm.

[26] Shenzhen Evening News. (2014, June 9). SDH integration has the symbol beginning. Retrieved from http://news.gmw.cn/newspaper/2014-06/09/content_3370277.htm.

Section) were extended, and the second phase of Qingping Highway was opened to traffic in recent years. The out-ring expressway and the eastern border highway were also constructed. Additionally, going through the Lanzhou Eastern Road and Huizhou Xinan Road, Shenzhen Jingxiu Eastern Road and Huizhou Shihua Road, Shenzhen Ruyi Road and Dongguan Longfeng Road, the western section of Qinglin Road in Longgang District and the eastern section of Danzi Eastern Road are ongoing projects.[27] Consistent with the prediction, Huidashugang Highway was accomplished in 2015, combining seven other highways to build the "Two Hours Life Cycle among Shenzhen, Huizhou, and Hong Kong".[28] In July 2015, the Shenzhen, Dongguan, and Huizhou municipal governments signed an agreement to construct 36 butt-joint projects on road networks in border regions.[29] From the perspective of the transportation infrastructure network, the construction of highway network, railway, and bus lines is ongoing and will lay the foundation for SDH to achieve social, economic, and institutional integration. However, although the initiative of building the regional infrastructure network was raised, the actual degree of infrastructure integration remains less than transparent, and the dynamics of governance structure to enable the large-scale infrastructure project needs to be further investigated.

3.4.4 *Summary*

According to the preceding analysis, the degrees of the socioeconomic and institutional integrations of SDH remain ambiguous with reference to the recent endeavor of the government to facilitate city-regionalism. In terms of institutional arrangement, the different levels of government in SDH devoted themselves to create an integrative governance structure to enable the initiatives of integration. In seeking a region-based governance structure, the formal and informal institutional forms have been entrenched to create an action framework between governments of member cities, and specific guides and blueprints for industrial and infrastructure integration were promulgated. In the realm of industrial integration, different city-based crises of industrial development and serious industrial isomorphism led to disparate local demands for industrial paths and orientations that were difficult to coordinate across city borders. Thus, industrial integration across the five municipalities in SDH remains thorny. Consequently, the cross-boundary economic cooperation zone as a significant part of the existing institutional arrangements has been popularized in the

[27] TCSM. (2015, July 27). The planning of SDH integration and its reading. Retrieved from http://jt.sz.bendibao.com/news/2015727/715020.htm.

[28] Shenzhen Evening News. (2015, October 29). Hui–Da Highway will concatenate seven other highway in Guangdong Province, forming "Two Hours Life Cycle among Shenzhen, Huizhou, and Hong Kong". Retrieved from http://union.china.com.cn/qyjj/txt/2015-10/26/content_8325121.htm.

[29] Guangzhou Daily. (2016, January 10). Promoting integration of transportation, SDH will smooth break five cross-border End Road within two years. Retrieved from http://gz.ifeng.com/a/20160110/4181118_0.shtml.

integration praxis between cities to overcome the underlying hurdles. Correspondingly, in the field of infrastructure integration, the regional infrastructure network was strategically formulated by the "SDH Transportation Integration Plan." The transportation infrastructure was central to the strategy of the regional infrastructure network, manifesting as the enhancement of the connectivity between cities.

Given the complicated situation of the SDH integration, some urgent issues need to be further investigated. First, do the government-imposed institutional arrangements successfully forge a region-based governance structure to accomplish city-regionalism in SDH? This question is prompted by the rise of varied integration-oriented institutional practices and aims to induce a comprehensive investigation of the degree of institutional integration of SDH. Second, to what extent is SDH integrated as a regional industrial entity, given the degree of industrial isomorphism? This question is raised by the vagueness of the status quo of industrial integration under the underlying crises in SDH. Third, will the governance structure created for cross-boundary economic cooperation zones overcome the underlying hurdles to enable industrial integration; if so, how? This question arises from the phenomenon of the widespread projects of cross-boundary economic cooperation zones in SDH integration. Fourth, to what extent is SDH integrated as a regional infrastructure network through strengthening the connectivity between and among the three cities? Last, how does the governance structure created for constructing large-scale regional infrastructure projects enable infrastructure integration in SDH? Addressing these five inter-connected empirical questions will contribute to the micro-level understanding of city-regionalism in the southeastern coastal area of China.

3.5 Conclusions

This chapter investigates the empirical background of city-regional development in China and identifies the major issues that emerge in the SDH integration process. The review identifies key empirical issues of city-regional development that need to be addressed. First, two intertwined tendencies are emerging in the socioeconomic–spatial evolution of city-regions in the post-reform era of China. The first tendency is the geographical down-scaling of regional development, shifting with the ideological changes from regional unbalanced development to regional coordinated development. Such down-scaling of regional development manifests as the geographical reshuffling of a national strategy, restructuring from a coastal–inland division to an eastern–central–western division, then to thirty-two urban agglomerations, and finally, to four plates plus three belts. Therefore, creating a polycentric layout throughout the national territory has emerged as the primary strategy to implement the initiatives of regional coordinated development. In addition, the city-regionalism in China has been forged under the confluence of state decentralization and recentralization. The political decentralization from the central government to local governments gives rise to greater socioeconomic and political autonomy in localities, triggering the diversified bottom–up institutional approaches for inter-jurisdictional coordination. The recentralizing process from the local government to provincial and central governments is a politico–strategic response to the rise of urban entrepreneurialism by deploying the city-region as a crisis-management strategy to cope with the fierce and chaotic inter-city competition.

Second, under the prevalent city-regionalism, some weakly and partly integrated regions have also been geared toward productive city-regions not only in central and western but also in eastern coastal regions. PRD city-regionalism involving the setting up of three sub-regions was orchestrated by the Guangdong Provincial Government in order to facilitate regional development in the PRD region and the entire Guangdong Province. The creation of three sub-regions has been enabled through three progressive stages of development. The first stage is to define the regional core cities in regional planning. Shenzhen and Guangzhou were documented as the core cities, and Zhuhai acts as the sub-core city. The second stage is designating three sub-regions within the PRD region. These are SDH which incorporated Shenzhen, Dongguan, and Huizhou; the Guangzhou-Foshan-Zhaoqing sub-region which consisted of Guangzhou, Foshan, and Zhaoqing; and the Zhuhai-Zhongshan-Jiangmen sub-region, which was made up of Zhuhai, Zhongshan, and Jiangmen. The third stage is driven by the strategy of revitalizing the eastern, western, and northern regions of Guangdong Province. Some outlying cities in the backward regions were incorporated into these regions. The three sub-regions are essentially a polycentric strategy of the Guangdong Provincial Government to enable regional coordinated development across strong and weak regions.

Third, regarding SDH integration, different levels of government impose a series of institutional alignments to enable the building of institutional and socioeconomic linkages within the designated regions. Although existing institutional arrangements attempt to forge a region-based governance structure to enable socioeconomic integration, industrial integration has confronted the underlying hurdles that are co-produced by the disparate city-based demand for industrial development and severe industrial isomorphism between and among cities. To overcome such emerging hurdles in industrial integration, different levels of government have strategically set up cross-boundary economic cooperation zones to provide a platform for facilitating integration. In addition, the regional infrastructure network has been formulated by the governments in SDH to accelerate infrastructure integration.

SDH in fact represents a polycentric strategy to realize the initiatives of creating coordinated regional development in the Guangdong Province. It still remains unclear whether such an endeavor will succeed and what lessons other city-regional development endeavors can learn from SDH. Thus, addressing the following questions is of great significance: (1) to what extent does SDH operate as an economically integrated entity? (2) to what extent does SDH operate as an institutionally integrated entity through redefining the governance structure? (3) to what extent does the governance structure enable or encumber the functioning of SDH as an integrated socio-economic entity through the cross-boundary economic cooperation zone and large-scale infrastructure project? These questions will be investigated in this book, applying concepts and analytical tools reviewed in the last chapter and integrated in the research design chapter.

References

Cai, Z., & Zhang, K. (2015). Retrospect of sixty years history for China's regional development strategy (1953–2013). *Gansu Social Sciences, 40*(2), 153–157. (in Chinese). https://doi.org/10.15891/j.cnki.cn62-1093/c.2015.02.033

CCGP (Construction Committee of Guangdong Province). (1991). *PRD urban system plan (1991–2010)*. CCGP. (in Chinese).

CCGP (Construction Committee of Guangdong Province). (1994). *PRD economic region urban cluster plan*. CCGP. (in Chinese).

CDGP (Construction Department of Guangdong Province). (2004). *PRD urban cluster coordinated development plan (2004–2020)*. CDGP. (in Chinese).

Chen, W., Ma, X., Cai, L., Luan, X., & Li, G. (2013). Characteristics of regional city connection's spatial pattern based on intercity passenger traffic flow in Pearl River Delta. *Economic Geography, 33*(4), 48–55. (in Chinese). https://doi.org/10.15957/j.cnki.jjdl.2013.04.017

Chien, S. S. (2007). Institutional innovations, asymmetric decentralization, and local economic development: A case study of Kunshan, in post-Mao China. *Environment and Planning C: Government and Policy, 25*(2), 269–290. https://doi.org/10.1068/c0558

Edin, M. (2003). State capacity and local agent control in China: CCP cadre management from a township perspective. *The China Quarterly, 173*, 35–52. https://doi.org/10.1017/S0009443903000044

Fan, Z. (2010). Territorial economic strategy and regional economic integration. *Comparative Economic & Social System, 26*(6), 64–73. (in Chinese).

Feng, C., Xie, X., Ma, X., & Cai, L. (2014). Functional polycentricity of the urban region in the Zhunjiang River Delta based on intercity rail traffic flow. *Scientia Geographica Sinica, 34*(6), 648–655. (in Chinese). https://doi.org/10.13249/j.cnki.sgs.2014.06.013

Guangdong Urban and Rural Planning and Design Institute. (2007). *The regulatory planning for industry garden of Longmen Jinshan*. Guangdong Urban and Rural Planning and Design Institute. (in Chinese).

Guangdong Provincial Government. (2009). *PRD megalopolis integration plan* (2009–2020). Guangdong Provincial Government. (in Chinese).

Hu, C. (2011). The connotation and characteristics analysis of "Shen-Guan-Hui Sub-region" integration. *Special Zone Economy, 29*(4), 15–16. (in Chinese).

Hu, X. (2006). Evolution and prospect of China's regional planning. *Acta Geographica Sinica, 61*(6), 585–592. (in Chinese).

Kou, D. (2014). The present situation, experience and direction of improvement for regional coordination of Guangdong province. *Urban Insight, 6*(5), 77–84. (in Chinese).

Lai, S., Yan, Y., Liu, G., Pang, X., & Li, Y. (2015). A review, evaluation and reflection of Pearl River Delta regional Plan. *Urban Planning Forum*, (4), 12–19. (in Chinese). https://doi.org/10.16361/j.upf.201504002

Li, H. (2011). Regional disparities and growth performance of economic openness: 1985–2008. *Journal of International Trade*, (9), 25–33. (in Chinese). https://doi.org/10.13510/j.cnki.jit.2011.09.007

Li, X. (2008). A study on spatial structure evolution of the city cluster of the Greater Pearl River Delta. *Urban Planning Forum, 2*, 49–52. (in Chinese).

Liu, P. (2011). *Research on iso-structuralizing of "Shen-Guan-Hui" industrial development* (Master's thesis). Available from China Master's Theses Full-text Database. (in Chinese).

Liu, S., Wang, Y., & Hu, A. (2009). The effect of western development program and regional economic convergence in China. *Economic Research Journal, 44*(9), 94–105. (in Chinese).

Luo, R., & Zhou, L. (2006). Study on the development of economic integration in the Littoral area of Bohai. *Journal of Tianjin Administration Institute, 8*(3), 33–38. (in Chinese). https://doi.org/10.16326/j.cnki.1008-7168.2006.03.007

Ma, J., & He, P. (2008). Studying on Bohai-Rim regional market integration advancement. *Journal of Hebei University of Science and Technology (Social Sciences), 8*(2), 10–14. (in Chinese).

Mei, Z., Xu, S., Ouyang, J., & Shi, C. (2012). The spatio-temporal evolvement of spatial interaction among cities of Zhujiang River Delta in recent 20 years. *Scientia Geographica Sinica, 32*(6), 694–701. (in Chinese). https://doi.org/10.13249/j.cnki.sgs.2012.06.007

NDRC (National Development and Reform Commission). (2008). *The outline of the plan for the reform and development of the PRD (2008–2020)*. NDRC. (in Chinese).

NDRC (National Development and Reform Commission). (2011). *The 12th five-year plan*. NDRC. (in Chinese).

NDRC (National Development and Reform Commission). (2012). *SDH coordinated development plan (2012–2020)*. NDRC. (in Chinese).

Ouyang, D. (2012). *Strategic studies on integration of "Shen-Guan-Hui" economic circle* (Master's thesis). Available from China Master's Theses Full-text Database. (in Chinese).

SC (State Council). (2010). *The main functional area planning*. SC. (in Chinese).

SC (State Council). (2014). *The new urbanization plan (2014–2020)*. SC. (in Chinese).

She, Z., Shen, D., Tang, Z., & Gong, C. (1995). Production distribution, infrastructure and urban development in Yangtze River Delta. *Resources and Environment in the Yangtze Basin, 4*(2), 97–105. (in Chinese).

TCSM (Transport Commission of Shenzhen Municipality), TDDM (Transport Department of Dongguan Municipality), & TDHM (Transport Department of Huizhou Municipality). (2014). *SDH transportation integration plan*. TCSM, TDDM, & TDHM. (in Chinese).

Wang, Z., & Shi, J. (2011). The imbalanced development of China's regional economy: Cause analysis. *Journal of Zhejiang University (humanities and Social Sciences), 41*(6), 91–103. (in Chinese).

Wei, H. (1992). The changing of income difference among regions in China. *Economic Research Journal, 27*(4), 61–65. (in Chinese).

Wei, Y. D. (2001). Decentralization, marketization, and globalization: The triple processes underlying regional development in China. *Asian Geographer, 20*(1–2), 7–23. https://doi.org/10.1080/10225706.2001.9684073

Wei, Y. D. (2010). Beyond new regionalism, beyond global production networks: Remaking the Sunan model, China. *Environment and Planning C: Government and Policy, 28*(1), 72–96. https://doi.org/10.1068/c0934r

Wei, Y. D., & Gu, C. (2010). Industrial development and spatial structure in Changzhou City, China: The restructuring of the Sunan model. *Urban Geography, 31*(3), 321–347. https://doi.org/10.2747/0272-3638.31.3.321

Wei, Y. D., Li, W., & Wang, C. (2007). Restructuring industrial districts, scaling up regional development: A study of the Wenzhou model, China. *Economic Geography, 83*(4), 421–444.

Wu, S., & Li, S. (2010). Market integration, industrial agglomeration and regional specialization: adopting two-digit manufacturing abate of China's 29 provinces. *Industrial Economic Research, 49*(6), 7–16. (in Chinese). https://doi.org/10.13269/j.cnki.ier.2010.06.006

Wu, W., Cao, Y., & Liang, S. (2010). Temporal and spatial evolution of integrated transport accessibility in the Yangtze River Delta: 1986–2005. *Progress in Geography, 29*(5), 619–626. (in Chinese).

Xiao, Q. (2008). Industrial integration of western strait economic zone. *Journal of the Party School of the Central Committee of the C.P.C., 12*(6), 64–66. (in Chinese).

Xing, H., Liu, K., & Deng, Y. (2013). Evaluation research on the regional logistics cooperation and integration degree of Shenzhen, Dongguan and Huizhou. *Economic Geography, 33*(7), 109–114. (in Chinese). https://doi.org/10.15957/j.cnki.jjdl.2013.07.021

Xu, J., & Shi, L. (2010). Spatial motive mechanism and integrative development mode of Hohhot-Baotou-Ordos region. *Journal of Arid Land Resources and Environment, 24*(7), 52–57. (in Chinese). https://doi.org/10.13448/j.cnki.jalre.2010.07.029

Xu, S., & Li, S. (2006). Analysis on the trend of regional income disparity in China. *Economic Research Journal, 41*(7), 106–116. (in Chinese).

Yang, A. (2007). Intergovernmental cooperation under the context of regional integration: Driver, model and expectation. *Cass Journal of Political Science, 3*, 77–86. (in Chinese).

Yang, F. (2011). Study on deepening the regional economic integration in Yangtze River Delta. *Journal of Industrial Technological Economics, 5*, 8–14. (in Chinese).

Yin, L., & Feng, B. (2014). From urban competition to regional cooperation: The restructuring of governance system in urbanized areas. *Reform of Economic System, 32*(5), 38–42. (in Chinese).

Yu, T., Li, N., & Wu, Z. (2009). Structure and changes of the Pearl River Delta region since 2000. *Urban Planning Forum, 1*, 23–32. (in Chinese).

Zeng, G., Shang, Y., & Si, Y. (2015). The convergent evolution of China's regional economic development models. *Geographical Research, 34*(11), 2005–2020. (in Chinese).

Zhao, L., Han, Z., & Ma, H. (2012). Analysis of the inherent functional relations of central plains economic region. *Economic Geography, 32*(3), 57–62. (in Chinese). https://doi.org/10.15957/j.cnki.jjdl.2012.03.024

Zheng, J., Zhao, T., & Zhu, Y. (2003). Analysis and appraise on "gradient development strategy" of China's urban megalopolises-taking Pearl River Delta megalopolis as example. *China Soft Science, 18*(10), 120–123. (in Chinese).

Zhou, Y., Lu, Y., Kong, C., & Zhang, H. (2012). Regional integration and the cooperation between Nansha district and Hong Kong and Macao. *Urban Insight, 4*(6), 88–94. (in Chinese).

Zou, Y., & Ren, Q. (2015). The particularity of China's tax system from the perspective of tax sharing system reform. *Macroeconomics*, (7), 104–107. (in Chinese). https://doi.org/10.16304/j.cnki.11-3952/f.2015.07.015

Chapter 4
Analytical Framework and Research Method

4.1 Introduction

As noted in Chap. 1, this book aims to formulate recommendations for city-regional integration in the context of China's reorientation toward developing city-regions as a locus of economic development. Specifically, this book critically investigates the extent of the economic and institutional integrations of city-regions and reveals how the governance structure reshuffles itself to enable economic integration. Given its ad hoc and intricate local context in institutional alignment, socioeconomic transformation, and local economic growth initiatives, the SDH is selected as the empirical case in this work to shed light on the city-regionalism of China. Consequently, four research questions are posed. (1) To what extent does SDH operate as an economically integrated entity as reflected in price similarity, industrial structure similarity, and infrastructure development? (2) To what extent does SDH operate as an institutionally integrated entity through redefining the governance structure across different levels of government, between cities, and between state and non-state domains? (3) How does the governance structure reshuffle itself to enable or encumber the function of SDH as an integrated economic entity in China's reorientation toward city-regions as a locus of economic development? (4) How does the territorial praxis of SDH contribute to the theoretical debates on China's city-regionalism?

In this chapter, the analytical framework and research methodology for answering the research questions are presented on the basis of the theoretical foundation with reference to economic and institutional integrations and governance structure reshuffling under city-regionalism in Chap. 2 and the empirical context of city-regionalism in China in Chap. 3. The current chapter comprises two sections. The first section shows the three levels of the analytical framework used to address the research questions. These levels are (a) assessment of the extent of the macro-economic integration of SDH, (b) assessment of the extent of the institutional integration of SDH, and (c) examination of how governance structure reshuffles to enable the economic integration of SDH. The last level is analyzed by critically examining two case studies: one on a cross-boundary economic cooperation zone and the other on an inter-city railway

X. Zhang, *Regional Integration and Governance Reshuffling in the Making of China's City-Regions*, Urban Sustainability, https://doi.org/10.1007/978-981-99-2792-0_4

project. The second section explains the case selection, data collection methods, and data analysis methods.

4.2 Analytical Framework

The research framework presented in Fig. 4.1 is comprised of three stages of investigation in order to address the four research questions. First, this book quantitatively assesses the extent of the economic integration of SDH in terms of the aspects of market, industrial, and infrastructure integrations. In view of the definition of economic integration, the final state of economic integration, manifesting the rise of the regional economic and market unit, redefines the market from the urban sphere to a regionally integrative market among localities. The literature generally highlights that industrial development and infrastructure construction are crucial pillars in facilitating economic integration (Scott, 2001). The quantitative methods pertaining to the assessment of economic integration include the following: the relative price method for measuring market integration; the degree of industrial structure similarity and a modified Krugman's index for measuring industrial integration; and coefficients of variation, degrees of infrastructure development similarity, and an infrastructure connectivity index for measuring infrastructure integration. It is important to note that social integration is a significant component of city-regional integration and is closely associated with the economic and institutional integration of a city-region. However, economic and institutional orders serve as the driving forces for the social order of city-regionalism (Scott, 2001), and citizenship is normally viewed as the outcome of city-regionalism or a new type of political order and driven by economic and institutional factors (Jonas & Ward, 2007; Rodríguez-Pose, 2008), this book pays attention to the economic and institutional integration of a city-region and the social dimension needs to be investigated in future studies.

Second, this book assesses the extent of the institutional integration of SDH across different levels of government, between cities, and between state and non-state domains. This book formulates an MLG-based analytical framework to ascertain the institutional integration of SDH by referring to the territorial praxis of the EU and China. Institutional integration is assessed from the aspects of the inter-scalar relation, inter-city relation, and state–market–civil society relation. The assessment includes not only static power configuration in governance structure but also dynamic changes from the three perspectives. The qualitative methods pertaining to the assessment of the institutional integration of city-regions are the analyses of government documents and inter-city cooperation news. It is worth mentioning that institutional integration in China is a government-dominated integrating process whereby market and social actors stay at the edge position. Thus, in the third step of investigating two regional projects, this book probes primarily the inter-scalar relations across government hierarchies and inter-city relations.

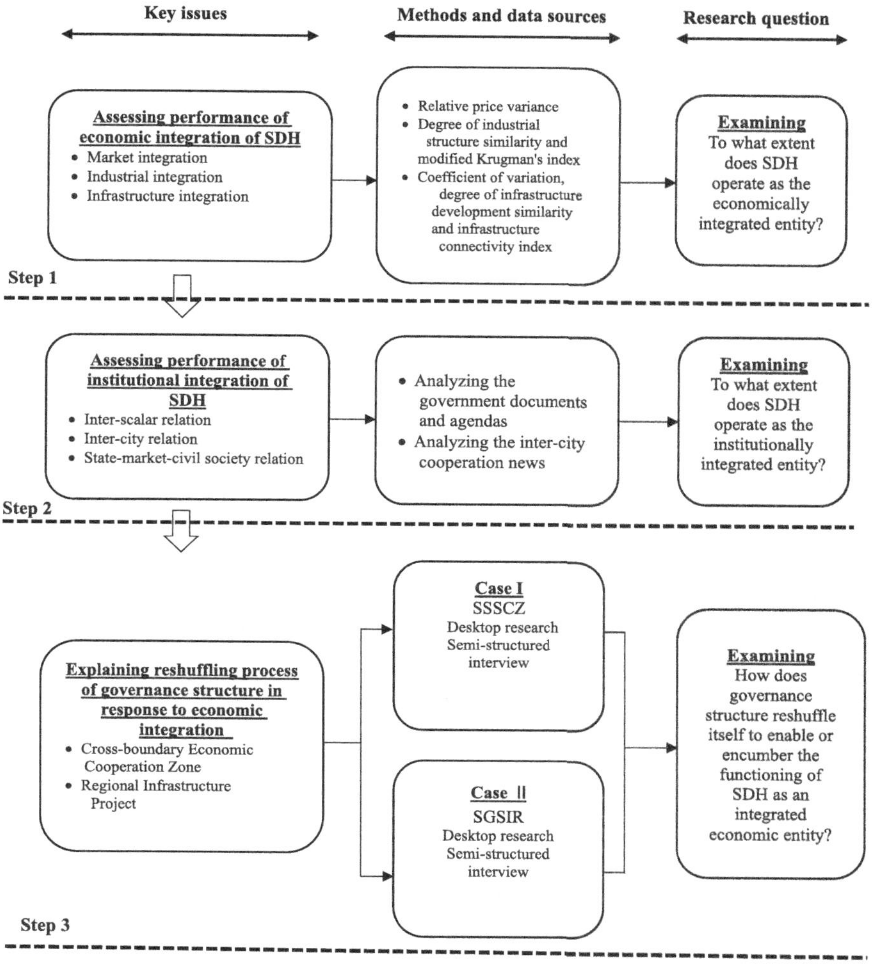

Fig. 4.1 Research framework. *Source* Author

Third, this book delves into the underlying mechanism of how the governance structure reshuffles itself to enable the economic integration of city-regions. Specifically, two empirical case studies are carried out to ascertain the city-regional politics under the prevalent city-regionalism. As industrial and infrastructure integrations are regarded as two crucial components of the economic integration of city-regions, this book selects SSSCZ and SGSIR as the cases to illustrate the city-regional politics of SDH under the integration-oriented development of China. The key issues and actors involved in the two regional development projects are identified initially. Then, the attributes of the players in every phase, particularly their aims in negotiation, participate level, valuation orientation, and relative advantages, are clarified through semi-structured interviews. Finally, governance structure reshuffling across different levels of government and between cities is analyzed by investigating the interaction processes between and among different stakeholders.

Finally, the contributions to the theories of China's city-regionalism, such as regional governance, inter-scalar, and inter-city politics, and the policy implications in formulating integration-oriented institutional design have been investigated and summarized in accordance with the empirical findings. The recommendations for facilitating the SDH integration under China's city-regionalism are proposed on the basis of the assessment of the economic and institutional integrations of SDH and how governance structure reshuffles itself in response to the functional integration of SDH in the cross-boundary economic cooperation zone and inter-city railway project. The results should shed light on the underlying forces that could propel city-regional development in China.

4.2.1 Assessing the Economic Integration of SDH

On the basis of the review of the mainstream literature on the long distance of economic integration (see Sects. 4.2.3.2 and 4.2.3.3), scholars (e.g., Friedmann, Wolff, and Sassen) have employed the data of multinational corporations and producer service firms, as well as infrastructure data, to measure the characteristics of global city networks. However, data availability is limited, and the statistical performance in China is inconsistent. In addition, the approach conceptualizes merely the one-fold dimension of economic integration and oversimplifies the complexity of the regional economy. In principle, the method should be highly connected to the local context. In other words, the method should use a context-sensitive lens to conceptualize the economic integration in different concrete city-regions.

In relation to the above-mentioned research lacuna, the local context of SDH is systematically introduced in Chap. 3 with particular reference to economic integration. The first one is the underlying crisis in the economic integration of SDH that manifests as industrial isomorphism. The emerging industrial isomorphism induces fierce industrial competition among cities, thereby attracting mobile capital. The second one is building the regional infrastructure network to enable the city-regionalism of SDH. Chapter 3 comprehensively presents these two issues in the context of economic integration. According to the summary of the economic integration of city-regions, the primary objective of economic integration is the free flow of goods and factors of production, and its core component is the creation of a region-wide unified market (see Chap. 2). Thus, on the basis of the discussion in the existing literature and the local context of SDH, this book deciphers the extent of the economic integration of SDH with an emphasis on the following three major aspects: (1) market integration, (2) industrial integration, and (3) infrastructure integration.

Figure 4.2 shows further details of the analytical framework for assessing the economic integration of SDH. This book employs the relative price variance to assess the extent of market integration, the coefficient of variation and degree of

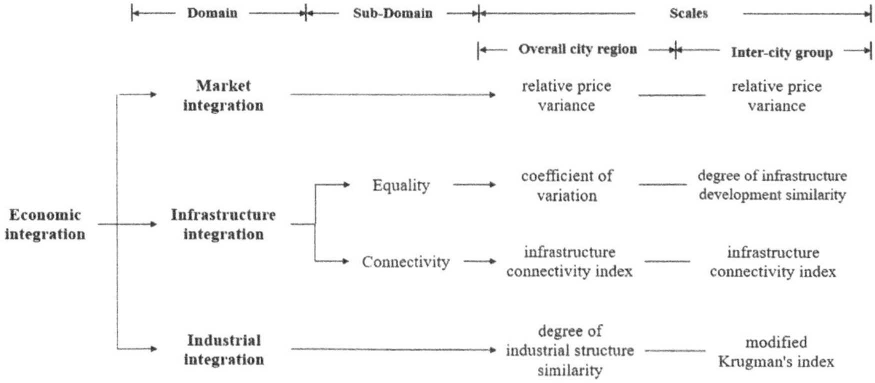

Fig. 4.2 Analytical framework for assessment of economic integration of city-region. *Source* Author

infrastructure development similarity to assess the extent of infrastructure equality, the infrastructure connectivity index to assess the extent of infrastructure connectivity, and the degree of industrial similarity and modified Krugman's index to assess the extent of industrial integration.

First, the aspect of market integration attempts to answer the question, "to what extent does SDH create a regional unified market to accelerate the flow of goods?" In the case of SDH, market integration is regarded as the final state of regional economic integration. Considering that the factors of production, such as land, have difficulty flowing between cities, this book primarily examines the extent of integration of the market for consumer goods. Second, with regard to the emerging issue of serious industrial isomorphism in SDH (see Sect. 3.4.2), this book assesses the extent of industrial integration by using the measurement of the similarity of industrial structure as the central task. Third, for the territorial practice of building the regional infrastructure network to enable the flow of goods and efficient utilization of factors of production in SDH (see Sect. 3.4.3), the infrastructure integration/network is assessed in terms of equality and connectivity.

The setting up of SDH was conducted formally in 2004, when the Guangdong Provincial Government enacted the "PRD Urban Cluster Coordinated Development Plan (2004–2020)." Hence, this book sets the time period of the assessment as 2004–2015. Furthermore, this book assesses economic integration at two scales: (1) the overall SDH scale, and (2) the inter-city group scale.

The first scale reveals the macro-economic landscape of SDH from the perspectives of market, industrial, and infrastructure integrations. The second scale exhibits the economic integration between the two cities. In the assessment, the inter-city groups are divided into ten groups that are widely utilized in the following sections: Shenzhen–Dongguan, Shenzhen–Huizhou, Shenzhen–Heyuan, Shenzhen–Shanwei, Dongguan–Huizhou, Dongguan–Shanwei, Dongguan–Heyuan, Huizhou–Shanwei, Huizhou–Heyuan, and Heyuan–Shanwei. According to the concept of the city-region and positioning in several versions of PRD planning, Shenzhen is regarded as a regional core city, Dongguan and Huizhou are ordinary cities, and Shanwei and Heyuan are outlying cities in SDH. The inter-city group scale thereby is categorized into the following two types.

1. Shenzhen with its surrounding cities: Shenzhen–Dongguan, Shenzhen–Huizhou, Shenzhen–Heyuan, and Shenzhen–Shanwei.
2. Shenzhen–Dongguan–Huizhou: Shenzhen–Dongguan, Shenzhen–Huizhou, and Dongguan–Huizhou.

Consequently, this book primarily focuses on the following analytical tiers:

- First-tier analysis: the overall SDH area
- Second-tier analysis: the inter-city group
  - Shenzhen with its surrounding cities
  - Shenzhen–Dongguan–Huizhou

4.2.1.1 Assessing the Market Integration of SDH

The preceding considerations and reflections on the long-standing views about the nature of economic integration indicate that the ultimate state of economic integration is the formation of a regional integrative market system that enables the free flow of goods and efficient utilization of factors of production (Robson, 2002; Tinbergen, 1954). The notion of market integration is that the assets of the market with identical risks have the same expected returns (Bekaert & Harvey, 1995). As identified in the conventional literature, the approaches to assess market integration include (1) trade flow (Naughton, 2003), (2) production (Xu, 2002; Zheng & Li, 2003), and (3) price methods (Chen & Li, 2009; Li et al., 2002; Yu & Huang, 1998). Price is known to serve as a "barometer" of the supply–demand relationship in the market of goods. Ravallion (1986) systematically elaborates on the relation between the price of goods and market integration:

> Such an equilibrium will have the property that, if trade takes place at all between any two regions, then price in the importing region equals price in the exporting region plus the unit transport cost incurred by moving between the two. If this holds then the markets can be said to be spatially integrated. (p.103)

In this sense, the price of tradable goods have been extensively deployed to decode the degree of market integration (Asche et al., 1999; Chen & Li, 2009; Goldberg & Verboven, 2005; Ravallion, 1986). Some scholars have applied static price correlations, focusing on spatial variance price as the most common measure of market integration (Hurd, 1975) that triggers a number of inferential dangers (Ravallion, 1986). Static price correlations "test for the long-run convergence to zero or close to it. [......] if prices at different markets are generated by identical but independent stationary autoregressive processes then they will asymptotically converge to zero variance. Thus, nothing can be inferred about the interlinkage of markets from the results of such tests" (Ravallion, 1986, p.102).

On the basis of this recognition, Ravallion (1986) adopts spatial price differentials for tradable goods and uses the monthly price of rice to measure market integration. On the basis of this method, which primarily focuses on market adjustment speed to spatial price differentials, relative price variance has been widely employed as an analytical tool for measuring market integration and segmentation (Parsley & Wei, 1996, 2001a, 2001b). This price-based approach indicates that a certain region has a high degree of market integration if the region shows a rapid convergence of price differences among localities, and vice versa (Parsley & Wei, 2001a, 2001b). The relative price method (i.e., relative price variance) not only accurately depicts the final state of integration, which is strongly linked to the price movement of tradable goods, but also conveniently compares inter-localities based on other approaches. Hence, this book applies the relative price method to assess the market integration of SDH.

On the basis of the existing literature, this book notes that a negative correlation exists between relative price variance and the degree of market integration. In other words, the regional integrative market is likely to be formed, and the degree of market integration is high if the relative price variance is small, and vice versa.

Model selection

The theoretical foundation of the price method is Samuelson's iceberg transportation cost, specifically referring to the fixed proportion of goods wastage during transport. This cost can be conceptualized as

$$P_j = P_i e^{(f|y-y_i|)} \tag{4.1}$$

P_j Price of goods in the place of consumption (cost, insurance, and freight)
P_i Price of goods in the place of production
y Location of consumer
y_i Location of production
f Positive constant

As shown in the previous equation, if $P_j(1-f) > P_i$ or $P_i(1-f) > P_j$, then arbitrage, that is, the trade between two places, is generated. If the previous conditions are not established, then the two places form an integrated market. In other words, if P_j/P_i has a limited scope, then the loss between the places of production and consumption, including all the transportation and transaction costs, is minimal. The market between the two places can be seen as an integrated entity.

This book borrows the work of Gui et al. in 2006 and that of Chen and Li in 2009 to use the relative price variance between two places $\left(\mathrm{var}\left(\frac{P_j}{P_i}\right)\right)$ in assessing market integration. If $\mathrm{var}\left(\frac{P_j}{P_i}\right)$ is increasingly convergent with the time order, then the "iceberg cost" tends to be narrowed, and the trade barrier tends to be reduced.

For the t year and k goods, this book uses the first difference of relative price to deal with the relative price and conceptualizes it as follows:

$$\Delta Q_{ijt}^{k} = \ln\left(\frac{P_{it}^{k}}{P_{jt}^{k}}\right) - \ln\left(\frac{P_{i(t-1)}^{k}}{P_{j(t-1)}^{k}}\right) \tag{4.2}$$

In this model, if f is small, then this book can determine that Q_{ijt}^{k} and ΔQ_{ijt}^{k} are convergent. Furthermore, the market can be seen as an integrated market. Then, the equation can be continually changed as follows.

$$\Delta Q_{ijt}^{k} = \ln\left(\frac{P_{it}^{k}}{P_{i(t-1)}^{k}}\right) - \ln\left(\frac{P_{jt}^{k}}{P_{j(t-1)}^{k}}\right) \tag{4.3}$$

According to the equation, the number of ΔQ_{ijt}^{k} becomes a negative number due to the change in locations. Therefore, this book needs to use the absolute value ($\left|\Delta Q_{ijt}^{k}\right|$) to measure ΔQ_{ijt}^{k}. Moreover, this book uses the de-mean approach to reduce the system errors caused by the inner link with specific goods. The method calculates the average of $\left|\Delta Q_{ijt}^{k}\right|$ among five cities ($\left|\overline{\Delta Q_{t}^{k}}\right|$) and uses $\left|\Delta Q_{ijt}^{k}\right|$ to deduct $\overline{\left|\Delta Q_{t}^{k}\right|\left|\Delta Q_{t}^{k}\right|}$

$$\left|\Delta Q_{ijk}^{k}\right| - \overline{\left|\Delta Q_{t}^{k}\right|} = \left(a^{k} - \overline{a^{k}}\right) + \left(\varepsilon_{ijk}^{k} - \overline{\varepsilon_{ijt}^{k}}\right) \tag{4.4}$$

From the previous equation, ε_{ijk}^{k} refers to the price-changing part largely related to the special market environment in i and j. Then,

$$q_{ijt}^{k} = \varepsilon_{ijt}^{k} - \overline{\varepsilon_{ijt}^{k}} = \left|\Delta Q_{ijt}^{k}\right| - \left|\overline{\Delta Q_{t}^{k}}\right| \tag{4.5}$$

q_{ijt}^{k} indicates the changing part of relative price variance.

Indicator selection

The existing literature reveals that selected goods should focus on the daily life of urban citizens and ensure high mobility in cases of transfers to and from different places, that is, tradable goods (Chen & Li, 2009). Considering data availability, the daily life of urban citizens, and high mobility, this book selects seven categories of specific consumer goods, listed in Table 4.1.

4.2.1.2 Assessing the Industrial Integration of SDH

As found in the literature review, industrial development is a crucial topic in the economic integration of SDH. Industrial integration is illustrated as a vertically integrated industrial system and is described as "the functional coordination of one or more units in each of several successive stages of production" (Frank, 1925, p.179). According to Behrman (1972), industrial integration is "the extent to which the new

Table. 4.1 Indicator of relative price method to assess market integration

Domain	Indicator	Unit
Market integration	• Food	Consumer price indices
	• Clothing	
	• Tobacco, liquor, and articles	
	• Health care and personal articles	
	• Household facilities, articles, and services	
	• Transportation and communication	
	• Recreation, education, cultural articles, and services	

Source Author

industry is tied with other industrial segments—forward and backward" (p. 48). Therefore, the industrial integration of a city-region involves the industrial linkage among different industrial segments and the functional linkage among several successive stages of production across jurisdictions. As a result of the complexity of the assessment of industrial linkage, the assessment of industrial integration is embedded into the specific context of cases and follows the problem-solving paradigm. The empirical review of SDH indicates that industrial isomorphism is the main challenge existing in SDH industrial integration (see Sect. 3.4.2). Thus, in assessing the industrial integration of SDH, the key issue is to gauge the degree of similarity in industrial structure. Competing views are adopted with regard to the relation between the degree of industrial structure similarity and industrial integration (Chen, 1998; Fan, 2011). Generally speaking, undesired and desired convergences exist in the relation between the degrees of industrial structure similarity and industrial integration. The undesired convergence refers to the industrial convergence that is contrary to the regional industrial division, and it is the result of local protectionism (Chen, 1998). Given the status quo of SDH, local protectionism largely restrains industrial integration. Thus, the primary consideration for the SDH's industrial integration is to avoid undesired convergence. This book maintains that if the degree of industrial structure similarity is high, then the phenomenon of industrial isomorphism is serious, and the industrial integration degree is low, and vice versa.

Model selection

As previously outlined, the correlation of industries between cities is revealed quantitatively in this section. Industrial isomorphism, manifesting as the similarity and differentiation of industrial structures between cities, reflects the industrial division in the city-region. Although Liu (2011) measures the degree of industrial structure similarity of SDH, the data used was from 2009 and thus fails to uncover the recent industrial development of SDH. Moreover, Shanwei and Heyuan were not incorporated into SDH. Therefore, the research into industrial integration has great empirical significance with regard to deciphering industrial isomorphism.

Previous studies on the measurement of industrial structural similarity adopted the structural superposition (Finger & Kreinin, 1979) and Krugman's indices (Krugman, 1991). Considering the massive information required by these two approaches, a number of scholars modified Krugman's indices (Fan, 2013; Wang & Chen, 2006). In the pioneering work of Fan (2013) on the assessment of the industrial integration of city-regions, the degree of industrial structure similarity was adopted, and Krugman's index was modified, ultimately arriving at a minimal formula. Thus, the present study uses the model of Fan in 2013 to assess the industrial integration of SDH.

To measure industrial integration at the overall SDH scale, this book formulates the following formula. The city-region is *A*, which is composed of *m* cities. This book selected *n* industry categories as the sample. The degree of industrial structure similarity can be conceptualized as

$$I_c = 1 - \frac{1}{2(m-1)} \sum_{l=1}^{m} \sum_{k=1}^{n} \left| R_{lk} - \overline{R_k} \right| \tag{4.6}$$

I_c Degree of industrial structure similarity of city-region *A*
l City in city-region *A*, with the value varying from 1 to *m*
k Selected industry, with the value varying from 1 to *n*
R_{lk} Proportion of output value of *k* industry in city *l* accounting for the gross output value of all selected industries in city *l*
$\overline{R_k}$ Proportion of output value of *k* industry in all cities in city-region *A* accounting for the gross output value of selected industries in all cities in city-region *A*

According to the hypothesis, if $I_c = 0$, then the industries of city-region *A* are fully integrated. If $I_c = 1$, then the industries of city-region *A* are fully disjointed. Within the interval $0 < I_c < 1$, if I_c is close to 0, then the degree of industry integration of the whole city-region *A* is high, whereas the industry integration is low.

To measure the industrial integration at the inter-city group scale, i.e., between two cities, this book formulates the following formula. The city-region is *A*, which comprises *m* cities. This book selected *n* industry categories as the sample. The modified Krugman's index can be conceptualized as

$$B_{ijt} = 1 - \frac{1}{2} \sum_{k=1}^{n} \left| R_{ik} - R_{jk} \right| \tag{4.7}$$

B_{ijt} Modified Krugman's index between cities *i* and *j*
R_{ik} Proportion of output value of *k* industry in city *i* accounting for the gross output value of all selected industries in city *i*
R_{jk} Proportion of output value of *k* industry in city *j* accounting for the gross output value of all selected industries in city *j*

According to the hypothesis, if $B_{ijt} = 0$, then the industries between cities *i* and *j* are fully integrated. If $B_{ijt} = 1$, then the industries between cities *i* and *j* are fully disjointed. Within the interval $0 < B_{ijt} < 1$, if B_{ijt} is close to 0, then the degree of industry integration between cities i and j is high, whereas the industry integration is low.

Indicator selection

The existing literature reveals that the selected industry should have a close interaction with the other cities and serve as the basic sector of the urban economy (Bai, 2011; Fan, 2013; Liu, 2015). Considering data availability and industry characteristics, this book selects specific industries that include four branch industries in the primary industry and 27 branch industries in secondary industries, listed in Table 4.2.

4.2.1.3 Assessing the Infrastructure Integration of SDH

Regional infrastructure is a crucial pillar of the economic integration of city-regions as it accommodates economic activities and commuting flow. The theoretical and empirical research on infrastructure integration is scant. Schrijnen (2000) analyzes the influence of infrastructure networks on city-regional development and finds that five key factors, namely, polarity, decentrality, equality, continuity, and formality, explain why the influence is positive in the cases that he examined. According to his work, equality as one of the crucial elements of an infrastructure network reflects the infrastructure allocation policies over cities and urban and rural areas. Equality is of great importance for city-regional integration as it affects the creation of regional economic units and the development of an integral functional entity. In addition, given the role of infrastructure networks in regional development, the primary objective of infrastructure integration can be understood as the establishment of physical connectivity and mobility. Such connectivity between cities also plays a significant role in the economic integration of SDH. Thus, this book assesses the infrastructure integration of SDH from the perspectives of equality and connectivity. Specifically, infrastructure equality mainly refers to the distribution equality of infrastructure across jurisdictions, and infrastructure connectivity means the accessibility from one city to another through transportation infrastructure.

(1) Infrastructure equality

Model selection

This book adopts the coefficient of variation to assess infrastructure equality at the overall SDH scale. Considering the paucity of methods for assessing infrastructure equality between localities, this research draws on the logic of assessing industrial integration between localities in applying the degree of infrastructure development similarity to measure infrastructure equality at the inter-city group scale.

On the basis of the existing literature, this book notes that a positive correlation exists between infrastructure equality and the degree of infrastructure integration. In other words, if infrastructure equality is high, then the degree of infrastructure

Table. 4.2 Selected industries to assess industrial integration

Domain	Selected industries		Unit
Industrial integration	Primary industry	• Farming, forestry, animal husbandry, fishery	Gross output value (100 million Yuan)
	Secondary industry	• Processing of farm and sideline food • Manufacture of food • Leather, fur, feather, and related products • Manufacture of textile garments, footwear, and headgear • Manufacture of furniture • Textile industry • Manufacture of wine beverages and refined tea • Papermaking and paper products • Printing and recording medium reproduction • Timber processing, bamboo, cane, palm fiber, and straw products • Manufacture of cultural, educational, sports, and entertainment articles • Petroleum refining, coking, and nuclear fuel processing • Manufacture of raw chemical materials and chemical products • Manufacture of medicines • Manufacture of chemical fibers • Rubber and plastic products • Nonmetal mineral products • Smelting and pressing of ferrous metals • Smelting and pressing of nonferrous metals • Metal products • Manufacture of electrical machinery and equipment • Manufacture of general-purpose machinery • Manufacture of instruments and meters • Manufacture of railway, ship, aeronautics, and other transport equipment • Manufacture of automobile and special-purpose machinery • Manufacture of communication equipment and computers and other electronic equipment	

Source Author

development similarity, and degree of infrastructure integration are high, and vice versa.

To assess the infrastructure distribution equality at the overall SDH scale, this book adopts the coefficient of variation, which indicates the ratio of the sample standard deviation to the sample means. If the coefficient of variation is large, then the infrastructure equality of SDH is weak. This book conceptualizes the coefficient of variation as follows.

$$I = \sum_{k=1}^{n} V_k / n \tag{4.8}$$

$$V_k = SD_k / M_k \tag{4.9}$$

n Total number of categories of selected infrastructure in the city-region
k One category of selected infrastructure in the city-region, with the value varying from 1 to n
SD_k Standard deviation of k category of infrastructure in the city-region
M_k Mean of k category of infrastructure in the city-region
V_k Coefficient of variation of k category of infrastructure in the city-region
I Coefficient of variation of all selected infrastructure in the city-region

On the basis of the previous formula, I represents the infrastructure equality at the overall SDH scale; if I is large, then the infrastructure equality of SDH and the infrastructure integration of SDH are low, and vice versa.

To assess infrastructure equality at the inter-city group scale, this book adopts and conceptualizes the degree of infrastructure development similarity as follows:

$$B_{ijt} = 1 - \frac{1}{n} \sum_{k=1}^{n} \left| S_{ik} - S_{jk} \right| \tag{4.10}$$

n Total number of categories of selected infrastructure in the city-region
k One category of selected infrastructure in the city-region, with the value varying from 1 to n
S_{ik} Score of k category of infrastructure in the city i
S_{jk} Score of k category of infrastructure in city j
B_{ijt} Degree of infrastructure development similarity between cities i and j

On the basis of the previous formula, B_{ijt} represents the infrastructure equality between cities i and j; if B_{ijt} is large, then the infrastructure equality and infrastructure integration between cities i and j are high, and vice versa.

Indicator selection

This book establishes an indicator system to evaluate the infrastructure system of SDH. According to available data and the content of the "SDH Coordinated Development Plan (2012–2020)," the infrastructure projects include "road and transport, telecommunication service, and ecological environment." This book constructs the indicator system and uses the expert scoring method and analytic hierarchy process to confirm the weights for the indicators as shown in Table 4.3.

Table. 4.3 Indicator system assessing infrastructure integration in terms of infrastructure distribution equality

Domain	Indicator	Type of indicator	Unit
Road and transportation	Density of expressways and Class I to IV highway networks	Derive	1/km
	Road lambs per kilometer	Derive	1 km
	Ratio of road area	Derive	%
Telecommunication service	Internet user per 100 inhabitants	Basic	Per 100 inhabitants
	Subscribers of mobile telephones per 100 inhabitants	Basic	Per 100 inhabitants
	Number of post offices per 10 000 inhabitants	Basic	Per 10 000 inhabitants
Ecological environment	Ratio of sewage treatment	Basic	%
	Ratio of waste water and consumption	Basic	%
	Ratio of industrial solid waste utilized	Basic	%
Water supply	Ratio of urban water supply	Basic	%

Source Author

The first three indicators in the indicator system involve the derive indicator; they are calculated as follows:

- Density of expressways and Class I to IV highway networks = length of expressways and Class I to IV highways/total land area of the administrative region
- Road lambs per kilometer = length of road/number of road lambs
- Ratio of road area = surface area of roads/total land area of the administrative region

Among the previous formulas, the indicator "length of expressways and Class I to IV highways" is from the *Guangdong Statistical Yearbook (2005–2016)*; the indicator "total land area of administrative region" is from the *China City Statistical Yearbook (2005–2016)*; and the indicators "length of the road," "number of road lambs," and "surface area of roads" are from the *China Urban Construction Statistical Yearbook (2005–2016)*.

The last seven indicators are basic indicators. Specifically, the data for the indicator "ratio of sewage treatment" is from the *Guangdong Statistical Yearbook (2005–2016)*; data for the indicators "Internet user per 100 inhabitants," "subscribers of mobile telephones per 100 inhabitants," "number of post offices per 100 inhabitants," "ratio of waste water and consumption," and "ratio of industrial solid wastes utilized" are from the *China City Statistical Yearbook (2005–2016)*; and data for the indicators "length of road" and "ratio of urban water supply" are from the statistical yearbooks of five municipalities.

(2) Infrastructure connectivity

Model selection

This book adopts the infrastructure connectivity index to assess the infrastructure connectivity at the overall SDH and inter-city group scales. This book notes that a positive correlation exists between infrastructure connectivity and the degree of infrastructure integration. In other words, if the infrastructure connectivity is high, then the infrastructure connectivity index and degree of infrastructure integration are high, and vice versa.

Considering the importance of transportation in the regional infrastructure connection, this book selects the numbers of inter-city railways, expressways, and national highways as indicators to measure the infrastructure connectivity index and further assess infrastructure connectivity.

To assess infrastructure connectivity at the overall SDH scale, this book conceptualizes the infrastructure connectivity index as follows.

$$I_l = \alpha R_t + \beta E_t + \lambda H_t \tag{4.11}$$

I_l Infrastructure connectivity index of the overall city-region
R_t Number of inter-city railways of the overall city-region in year t
E_t Number of expressways of the overall city-region in year t
H_t Number of national highways of the overall city-region in year t
α Weight of inter-city railway
β Weight of expressway
λ Weight of national highway

To assess the infrastructure connectivity at the inter-city scale, this book conceptualizes the infrastructure connectivity index as follows.

$$B_{ijt} = \alpha R_{zt} + \beta E_{zt} + \lambda H_{zt} \tag{4.12}$$

B_{ijt} Infrastructure connectivity index of the overall city-region
R_{zt} Number of inter-city railways between cities i and j in year t
E_{zt} Number of expressways between cities i and j in year t
H_{zt} Number of national highways between cities i and j in year t
α Weight of inter-city railway
β Weight of expressway
λ Weight of national highway

Indicator selection

Considering the importance of transportation in the measurement of infrastructure connectivity, this book selects the following indicators:

- Number of inter-city railways
- Number of expressways
- Number of national highways

This book identifies and counts the existing roads through *Google Maps* and then determines the completion time of these roads due to the lack of historical data on inter-city railways, expressways, and national highways between cities. Finally, the numbers of inter-city railways, expressways and national highways completed in the span of 12 years can be obtained.

4.2.2 Assessing the Institutional Integration of SDH

The assessment of the institutional integration of SDH evaluates the extent to which the governance structure of SDH operates as the MLG paradigm from the perspective of inter-scalar, inter-city, and state–market–society relations. By deploying the organization-embraced approach instead of the institution-as-a-rule framework and by drawing on the territorial practices of MLG in the context of the EU integration (see Chap. 2), this book formulates an MLG-based analytical framework in which institutional integration can be appropriately assessed. Given the difference between China and the EU in terms of state intervention, the EU's MLG has been adapted to suit China's special characteristics—i.e., the dominance of the state. Three intrinsic dimensions are encapsulated: inter-scalar, inter-city, and state–market–society relations (Fig. 4.3 and Table 4.4).

4.2.2.1 Inter-scalar Relations

In capturing inter-scalar relations, the current administrative setting and the readjustment of the administrative division are used to demonstrate how territorial states are rescaled in response to the changing socioeconomic environment (Ma, 2005; Wu, 2016; Yang, 2005). Within these two integrative strands, this book demonstrates the four aspects in the formal administrative setting, namely, the current number of administrative levels of jurisdiction, the readjustment of administrative division, the local institutional innovation and policy experiment, and the existence of formal regional authority.

Yang (2005) meticulously articulated the MLG in the cross-border region between Hong Kong and PRD by disclosing the settings of multiple jurisdictions. Ma (2005)

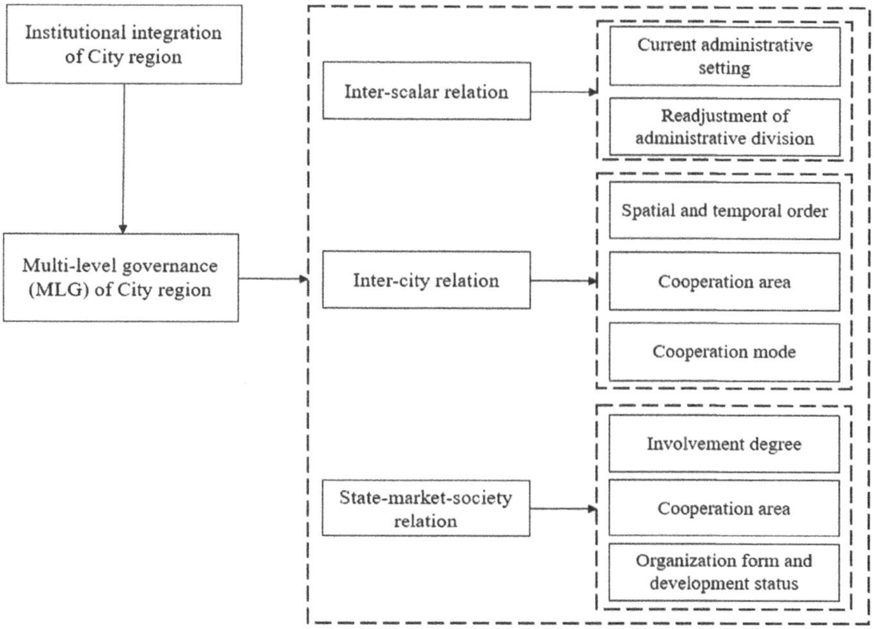

Fig. 4.3 Analytical framework for assessment of institutional integration of city-region. *Source* Author

analyzed the administrative setting in post-reform China, which is arguably characterized by several waves of administrative annexation, leading to the changing multilevel governmental relations. The central government monopolizes the power of restructuring the administrative ranks of different jurisdictions, which directly determines the administrative power and fiscal resources that localities can obtain (Ma, 2005). Thus, the current administrative setting and readjustment of administrative division are important lenses in deciphering the central constraint on localities. Specifically, in the current administrative setting, this book primarily investigates how many tiers these jurisdictions have and whether these jurisdictions administratively operate at the multi-level. According to the EU praxis, the setting of jurisdictions should have multi-tiers, especially at the sub-regional level. In the readjustment of administrative divisions, this book primarily investigates the changing of the administrative setting of SDH after 2000 and what the extent of central government enables local administrative annexation. Through investigating the changing of the administrative setting, this book can judge the controlling power of the central government and its enforcement on local development.

In addition to the central government-induced administrative setup, some local institutional innovations and policy experiments on administrative settings, such as the trans-jurisdictional industrial park, have received increasing attention from scholars (Liu et al., 2014; Jiang & Luo, 2016). Institutional innovation is treated as

Table. 4.4 Analytical dimension, research questions, and data sources on institutional integration

Analytical dimension	Research questions	Data source
Inter-scalar relations	What is governance structure from the perspective of a hierarchical government? • How many jurisdictions are in SDH and what are their administrative levels? • How does the number of jurisdictions change historically? • Does SDH have some informal institutional innovation in an administrative setting? • What are the readjustments of administrative division or jurisdictional practice in SDH? • What is the current situation of formal regional authority in SDH?	Official documents Official website of provincial, municipal, county, and town governments
Inter-city relations	What is governance structure from the city-to-city perspective? • What is the time–space distribution feature of inter-city cooperation in SDH? • What are the primary cooperation areas of different inter-city groups? • What are the primary institutional modes of inter-city cooperation? • What are the features of different institutional modes in terms of promoting collaboration actions?	Information from search engines and inter-city cooperation news
State–market–civil society relations	What is the governance structure from the perspective of relations between state and non-state actors? • What are the extents of state, market, and civil society involved in the institutional integration of SDH? • What types of cooperation areas do state, market, and civil society engage in? • What are the overarching organization forms of market and civil society in the SDH integration process?	Information from search engines and inter-city cooperation news

Source Author

an important manifestation of the rise of localities under the decentralization process of China. Thus, local practices of institutional innovation or policy experiments are selected in the current work to examine the central–local relations.

Regional authority is central to MLG in terms of inter-scalar relations. According to the territorial practices of MLG in the EU milieu (see Chap. 2), independent or semi-independent regional authorities have been widely established, aiming to provide public service at the city-regional scale. Thus, the investigation of the current administrative setting and the readjustment of administrative divisions in SDH in this book focuses on the establishment of regional authorities with the MLG structure to the build sub-regional fabric.

4.2.2.2 Inter-city Relations

Inter-city cooperation is generally posited as the primary standard by which to analyze inter-city relations (Douglass, 2002; Luo & Shen, 2009; Xu, 2008). According to Rodríguez–Pose (2008), a city-region refers to inter-jurisdictional connections that are underpinned by the dynamics among urban centers, semi-urban areas, and outskirts. Therefore, the spatial layout of inter-city cooperation is the key dimension to assess MLG in terms of inter-city relations. According to the EU praxis, the spatial and temporal order has manifested as a voluntary process of city grouping. Thus, within the creation of city-region in SDH, inter-city cooperation should reflect a core-periphery spatial layout centered on the core city in this regard. Specifically, to demonstrate the spatial and temporal order of inter-city relations, this book employs news on inter-city cooperation and divides the time period of news on inter-city cooperation into six groups: 2004–2005, 2006–2007, 2008–2009, 2010–2011, 2012–2013, and 2014–2015. Counting the number of inter-city cooperation news items can contribute to demonstrating the spatiotemporal layout of SDH in terms of inter-city relations.

The EU practices signify that building an inter-city cooperation network across local authorities grounded by a formal agreement is a new avenue with which to establish regional governance for enabling multi-objective tasks beyond the cooperation merely in the field of institution (Perkmann, 2007b). Thus, a cooperation area, which primarily refers to the types of issues of inter-city cooperation, has been investigated to assess whether the inter-city cooperation in SDH is multi-objective. Based on the news about inter-city cooperation, this book divided the primary cooperation areas into five groups, namely, comprehensive issues, economic development, institutional design, secondary function, and aid activity. The comprehensive issue group refers to comprehensive cooperation that covers all the fields of regional public issues. The economic development group refers to inter-city cooperation for the theme of economic growth and comprises two branches, namely, industrial development and infrastructure construction. The institutional design group denotes the inter-city cooperation for the building of an integrative institution and involves the three branches of institutional construction, urban management, and urban planning. The secondary function group means the inter-city cooperation for facilitating social development, including science and education, entertainment, disaster emergency, and resource and environment. The aid activity group refers to a special inter-city assistance activity driven by the strategy for the revitalization and development of eastern, western, and northern Guangdong. The Guangdong Provincial Government experiments with the pair-city strategy by selecting advanced cities to assist less-developed cities. This division can elaborate the overarching cooperation areas on which the inter-city cooperation focuses.

As for specific inter-city cooperation, the existing literature has paid close attention to cooperation modes to examine the performance of inter-city cooperative behavior, such as regional spatial planning, economic development zones, and coordination forums (Luo & Shen, 2008, 2009; Zhang, 2006). The features of these different types of cooperation modes, such as political mobilization, aims, and institutional tools, are keys in decoding inter-city cooperation efficiency and effectiveness (Luo & Shen, 2009). Formalized institutional arrangements, such as joint conferences on development, regional spatial planning, and cooperation agreements, tend to be more efficient than informal institutional arrangements (Xu & Yeh, 2013; Yang, 2005). Thus, by analyzing the spatial and temporal order and cooperation areas of these inter-city cooperation news items, this book identifies the primary forms of cooperation in these news items to demonstrate the cooperation modes in terms of inter-city relations.

Hence, this book now primarily focuses on three features of an inter-city cooperation network: (1) spatial and temporal orders treat a sub-region as a voluntary process of city grouping and manifests as a core-peripheral layout; (2) the cooperation area has been used to judge whether the inter-city cooperation network operates with multi-objectives; and (3) the cooperation modes focus on how the aforementioned inter-city cooperation is assured and grounded. In other words, the cooperation area, including the economic, institutional, and social activities, needs to be well-rounded and inclusive (Perkmann, 2007b). The cooperation mode, regardless of its level of formality, determines the extent of inter-city cooperation (Li & Wu, 2018; Luo & Shen, 2009). Consequently, the spatiotemporal order, cooperation area, and cooperation mode are particularly foregrounded in examining inter-city relations.

4.2.2.3 State–Market–Civil Society Relations

In terms of state–market–civil society relations, Wu (2016) notes that China remains in a state-dominated developmental mode as regards the relationship between state and non-state actors. The extent of development for markets and social actors primarily refers to the extent of involvement of markets and social actors in regional integration. According to the EU, the quasi-or non-state actors are fully engaged in regional integration. This indicates that the degree of involvement of market and social actors should exhibit full or high-level participation in regional integration. To examine the degree of involvement, this book identifies the main stakeholders of the news about inter-city cooperation and further divided them into five categories of cooperation, namely, government-to-government, government-to-market, government-to-society, market-to-market, and society-to-society cooperation. Market-to-society cooperation is excluded because the interactions between them generally are private issues that are irrelevant to regional public issues. By analyzing the proportion of the five types of cooperation in six time periods, this book assesses the involvement degree of the state, market, and civil society in the SDH integration.

Aside from the degree of involvement, this book also focuses on their cooperation areas and organizational forms. Moreover, development status is crucial for analyzing the state and non-state relationships in city-regionalism (Armstrong et al., 2011). The EU experience (see Chap. 2) indicates that state and non-state sectors have been integrated in an embracive partnership by which various services are provided (Chong et al., 2006; Kwak et al., 2009; McKee et al., 2006). Thus, the market and social actors should engage in various types of cooperative areas through privatization. In combination with the division of the above cooperative area, this book further examines the overarching cooperative areas in which these stakeholders engaged through the news about inter-city cooperation.

The organizational form and developmental status primarily denote the forms of market and social actors in regional integration. In the EU context, the organizational forms of market and social actors and public–private partnerships are flexible and various. This book systemically analyzes the inter-city cooperation news to identify the organizational forms of market and social actors in the city-regional integration.

Therefore, the MLG-based analytical framework focuses on (1) the extent to which state, market, and society are involved in regional integration; (2) the types of regional issues in which they are involved; and (3) the organizational forms and development statuses of the market and social actors.

4.2.3 Deciphering the Reshuffling Process of Governance Structure in Response to the Economic Integration of SDH

On the basis of the first two steps in assessing the economic and institutional integration of SDH, how the governance structure reshuffles in response to socioeconomic integration is analyzed from two perspectives. The empirical review of the situation of SDH in Chap. 3 shows that industrial and infrastructure integrations are crucial developmental tasks in SDH's city-regionalism. Thus, this book attempts to examine the reshuffling of governance structure in response to the socioeconomic integration of SDH by investigating the industrial and infrastructure integration processes. Thus, following the case-specific paradigm discussed in Chap. 2, this book selects the cross-boundary economic cooperation zone and inter-city railway project as empirical cases to decipher regional governance and city-regionalism in contemporary China.

After the late 1990s, the industrial integration of PRD was promoted by the phenomenon of industrial transfer, which is a spontaneous or voluntary process for urban industries transferring from core regions, such as Shenzhen and Guangdong, to the peripheral regions of PRD, such as Jiangmen, Qingyuan, Huizhou, and Zhaoqing. In this stage, market-oriented transformation, which is documented in a host of urban and regional agendas, was the key dynamic that enabled industrial transfer. Most transferring industries were labor-intensive and low-value-added industries. In the next stage which started in 2005, industrial transfer was driven by an urban coalition between municipal governments and market actors. In addition to the free-market choice, the government's strong pushing force expanded the scale of industrial transfer and facilitated the emerging trend of clustering and park industrialization. The "Double Transfer" policy reinforced the extent of the industrial integration of PRD with its surrounding areas. In March 2005, the Guangdong Provincial Government promulgated the real-time document "The opinions of promotion of industrial transfer between PRD and mountainous region and eastern and western areas of Guangdong Province," which aimed to establish the mode of industrial transfer parks and formulated a series of regulations on the daily operations of such parks. Based on this document, related provincial policies were promulgated to normalize construction processes; the policies involved the baseline of the industrial transfer park, the tax distribution between municipal and provincial governments, and the preferential policies to create investment-induced industrial parks. From 2005 to the present, 83 industrial transfer parks have been located in PRD; the total investment in these parks is 700 billion Yuan, the accumulated GDP is more than 500 billion Yuan, and the tax collected until 2012 reached 20 billion Yuan.[1] Industrial transfer parks represent a significant motor in the local economic growth and industrial integration of PRD. After 2008, in the wake of the strategy for "revitalization and development of eastern, western, and northern Guangdong," which was proposed by the Guangdong Provincial Government in 2014, the inter-city industrial transfer between cities in PRD and the northern, eastern, and western regions served as a crucial engine in promoting coordinated development at the provincial scale. In this context, industrial transfer parks represent a concrete spatial strategy to enforce balanced-development initiatives.

In response to these empirical contexts, this book uses industrial transfer parks as an example to analyze the reshuffling of governance structure in the industrial integration process of SDH. According to the conceptual review in Chap. 2, the central theme in the construction of cross-boundary economic cooperation zones is to build an integrative governance structure that enables power interactions between cities and inter-scalar governments. Thus, this book primarily examines whether and how the integrative governance mode is constructed between cities and inter-scalar governments in a cross-boundary economic cooperation zone under the city-regionalism of PRD.

---

[1] News. (2012, April 5). The 35 ITP raised more than 500 billion yuan GDP in 3 years in Guangzhou Province. Retrieved from http://news.timedg.com/2012-04/05/9392078.shtml.

In infrastructure integration, Guangdong Province was the first to propose the building of an inter-city railway network, known as the "PRD inter-city railway network," in the early 2000s. The inter-city railway network is a significant component of a regional infrastructure network (see Chap. 3). The planning of the PRD inter-city railway network was approved by the SC in 2005. This well-rounded inter-city network, including SGSIR and the Dongguan–Huizhou inter-city railway, serves as the railway skeleton of SDH. In addition to the inter-city railway network driven by the Guangdong Provincial Government, the "SDH Transportation Integration Plan" was enacted at the end of 2014 in line with the existing transportation conditions of member cities in SDH. The inter-city railway, which was viewed as an appropriate tool to facilitate city-regional integration and improve investment-induced economic growth, has received substantial investments from central and local governments (Xu & Chen, 2014). The construction of the inter-city railway in SDH encapsulates rich institutional designs that aim to enhance economic efficiency and achieve a balanced development among municipalities by restructuring inter-scalar relations. In addition to relations across different levels of government, the Ministry of Railways retains its predominant influence on the entire inter-city railway project. Thus, considering the empirical significance and the complicated inter-scalar interactions under the city-regionalism of PRD, this book selects the inter-city railway project as the case in this book to probe the reshuffling of governance structure, especially in terms of inter-scalar relations.

4.2.3.1 Critical Issues of Cross-Boundary Economic Cooperation Zones and Inter-city Railway Projects

The first step in demonstrating the reshuffling of governance structure is to identify the critical issues in these cases on the basis of the existing literature. The mainstream literature indicates that the critical issues of cross-boundary economic cooperation zones can be categorized into four parts that involve site selection, political legitimatization, daily operation, and industrial development (Liu et al., 2014; Luo & Shen, 2006, 2007; Zhang et al., 2011). These issues are the key domains of the development of the selected cross-boundary economic cooperation zone for investigating the restructuring of regional governance in SDH. Specifically, site selection refers to the negotiation process between actors for the spatial location of a cross-boundary economic cooperation zone. Political legitimatization is based on the site selection process, which refers to the legitimatization of a cross-boundary economic cooperation zone within the existing institutional framework and administrative setting, such as the empowerment of administrative status, provision of construction land quota, and formulation of spatial planning. Daily operation means the administrative and management system of a cross-boundary economic cooperation zone, as well as how it operates between cities and across different levels of government. Industrial development refers to the process of industrial introduction and development in the cross-boundary economic cooperation zone. In summary, this book examines these four correlative processes to demonstrate the governance restructuring process in SDH.

As regards the inter-city railway project, the existing literature indicates that the critical issues of inter-city railways include the making and approval of plans, negotiation of operation model, site selection of inter-city railway stations, and negotiation of local land development (Xu & Chen, 2014; Xu & Yeh, 2013). In particular, in the issue of land development, the central theme is to confirm the governance model used to conduct land development. Generally, local governments are the sole actors in local land development around inter-city railway stations. However, provincial governments also intend to participate in land development by covering operation losses for inter-city railway projects. These issues, which deeply reveal the reshuffling process of the inter-scalar governance structure, have varied due to the divergent local socioeconomic contexts. Therefore, this book critically ascertains these four issues under PRD's city-regionalism.

4.2.3.2 Stakeholders in Cross-Boundary Economic Cooperation Zone and Inter-city Railway Project

The key stakeholders involved in the above issues are further identified by desktop research and semi-structured interviews. As regards the majority of cross-boundary economic cooperation zones, the provincial government is responsible for providing concessionary policy incentives and directly managing the construction and operation processes. These policy incentives include support for the cross-boundary economic cooperation zone with regard to the construction land quota, investment in infrastructure, and reduction of taxes and fees. The incentives effectively reduce the operation cost of enterprises and improve the competitiveness of a cross-boundary economic cooperation zone. The Guangdong Provincial Government has strategically managed construction and operation processes by establishing performance evaluation systems for different zones.

Exporting city governments are responsible for fund collection and the industrial introduction of cross-boundary economic cooperation zones. In construction, exporting city governments can also use their advanced experience to boost cross-boundary economic cooperation zones and coordinate industrial operations. Importing city governments are in charge of the land supply and construction of infrastructure within and outside the zone. Land supply includes the issues of land acquisition, employment, and housing demolition. These tasks can be delegated to different levels of government.

From the perspective of industrial development, to introduce the cluster of industries into a cross-boundary economic cooperation zone and ensure that the different enterprises in the same industry can agglomerate in the zone, guilds can effectively coordinate the relationships between governments and enterprises and assist the government in connecting with enterprises. Enterprises are the main bodies in the industrial development of the zone, are driven by the market economy, and are largely affected by concessionary policy incentives.

The above discussion outlines the basic stakeholders in the construction of cross-boundary economic cooperation zones who are to be identified and then to be examined by desktop research and semi-structured interviews.

For the inter-city railway project, one department of SC, the Ministry of Railways, which oversees 16 railway bureaus and 2 railway group companies, enjoys a monopoly on the construction of inter-city railway projects. The planning process, which includes the approval of plans, investment allocation, and site selection of stations, is largely determined by negotiations among the Ministry of Railways, the provincial government, local governments, and other ministries of SC. To secure sufficient financial support from the central government, the Ministry of Railways is often invited to engage in inter-city railway projects (Xu & Yeh, 2013). The provincial government seeks to promote infrastructure development at the provincial scale and strongly desires to enact inter-city railway plans and cooperate with the Ministry of Railways to facilitate the approval process. The provincial government also needs to coordinate its interests with different local governments in accordance with the route and site selection of stations and revenue distribution of land development. Different levels of government are involved in the implementation and have different political and local demands in regional infrastructure projects.

The above factors outline the framework of stakeholders in inter-city railway projects, and the specific stakeholders vary in specific projects to follow case-specific paradigms and thus had to be explored by observation and interviews.

4.2.3.3 Attributes of Stakeholders and Changing Forms of Relations and Institutional Arrangements

After the identification of stakeholders in every issue, their attributes were explored via interviews to illustrate the changing relationships among these actors. The attributes of stakeholders include aims in negotiation, value orientation, participation level, and relative advantages. Specifically, aims in negotiation refer to the primary objective of these actors in specific cases for different issues. Aims in negotiation are specific and easily identified by interviews. Compared with aims in negotiation, value orientation is more abstract and highly inductive. Following the specific aims in negotiation, value orientation includes public interest (provincial/regional, urban and ministry's interests), private interest, and a combination of distinct interests that share different priorities. Participation level refers to the relationships between stakeholders in specific issues. The content of interviews and desktop research indicates that actors can be categorized into major actors, subordinate actors, and excluded actors according to participation levels. Relative advantages refer to the resources and rights that stakeholders possess, including information, capital, political and administrative powers, and technology. These resources are the bases for interaction between stakeholders and are regulated by formal and informal institutions. On the basis of the stakeholder attributes, the changes in inter-scalar and inter-city relations are illustrated in the current work.

After the investigation of stakeholder attributes through semi-structured interviews, the entire process on how related stakeholders interact with one another to achieve their goals was established with regard to the existing institutional arrangements in inter-scalar and inter-city relations. By probing the entire interaction process in different critical issues, this book summarizes the changing forms of inter-scalar and inter-city, including collaboration, compromise, and resistance (Figs. 4.4 and 4.5). If the aims in negotiation and relative advantages between stakeholders are complementary for certain issues, the relation normally manifests as a collaborative relation. However, when there is tension and struggle of aims in the negotiation among stakeholders, the relative advantages are generally used to give rise to compromises and resistance relations among stakeholders.

Following these three steps, how the governance structure reshuffles itself in response to the industrial and infrastructure integration of SDH can be illustrated with particular reference to inter-scalar and inter-city relations. Based on the assessment and illustration, institutional and policy implications can be formulated to further facilitate SDH integration.

4.3 Research Method

4.3.1 Case Selection

4.3.1.1 Cross-Boundary Economic Cooperation Zone: SSSCZ

To illustrate the governance structure reshuffling in relation to industrial integration, this book selects SSSCZ as the empirical case to elaborate city-regional politics in establishing a cross-boundary economic cooperation zone. SSSCZ is in southern China to the east of PRD. It borders the South China Sea and is connected to Huizhou. It lies in the east of Haifeng County, Shanwei. SSSCZ covers the towns of Ebu, Xiaomo, Chishi, and Houyu in Haifeng County; the total area is 468.3 km^2. The location of SSSCZ is shown in Fig. 4.6.

Following the empirical analysis of the city-regionalism of SDH in Chap. 3, a series of cross-boundary economic cooperation zones were created to facilitate industrial integration and address the different crises of capital accumulation in member cities. Under this empirical background, the following reasons were considered in selecting the cases.

First, SSSCZ which was approved in 2011, is an important policy experiment of the Guangdong Province and has strategically elicited considerable attention from different levels of government, such as the Guangdong Provincial Government, municipal governments, county governments, and town governments (Table 4.5). The SSSCZ initiative aims to treat it as the flagship in cross-boundary cooperation and as a strategy for the revitalization and development of eastern, western, and northern Guangdong. SSSCZ not only serves as a significant policy experiment in

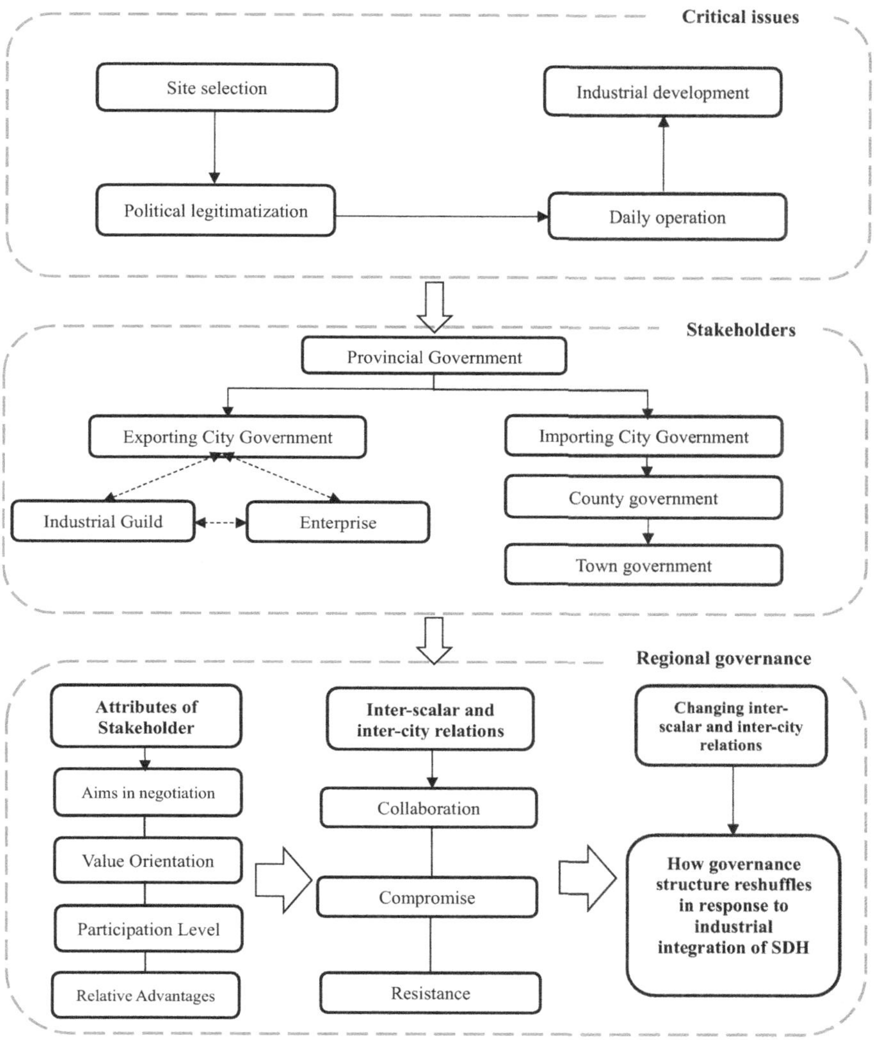

Fig. 4.4 Analytical framework for how governance structure reshuffles to enable the SDH industrial integration. *Source* Author

regional integration but also enjoys substantial concessionary policies from governments. Thus, based on these emerging regional contexts, SSSCZ can be viewed as a testbed for the innovation of integration-first institutions in PRD that can provide some insights for regional governance reshuffling.

Specifically, the Guangdong Provincial Government and Guangdong Provincial Party Committee issued the agenda of "A decision to further promote the development

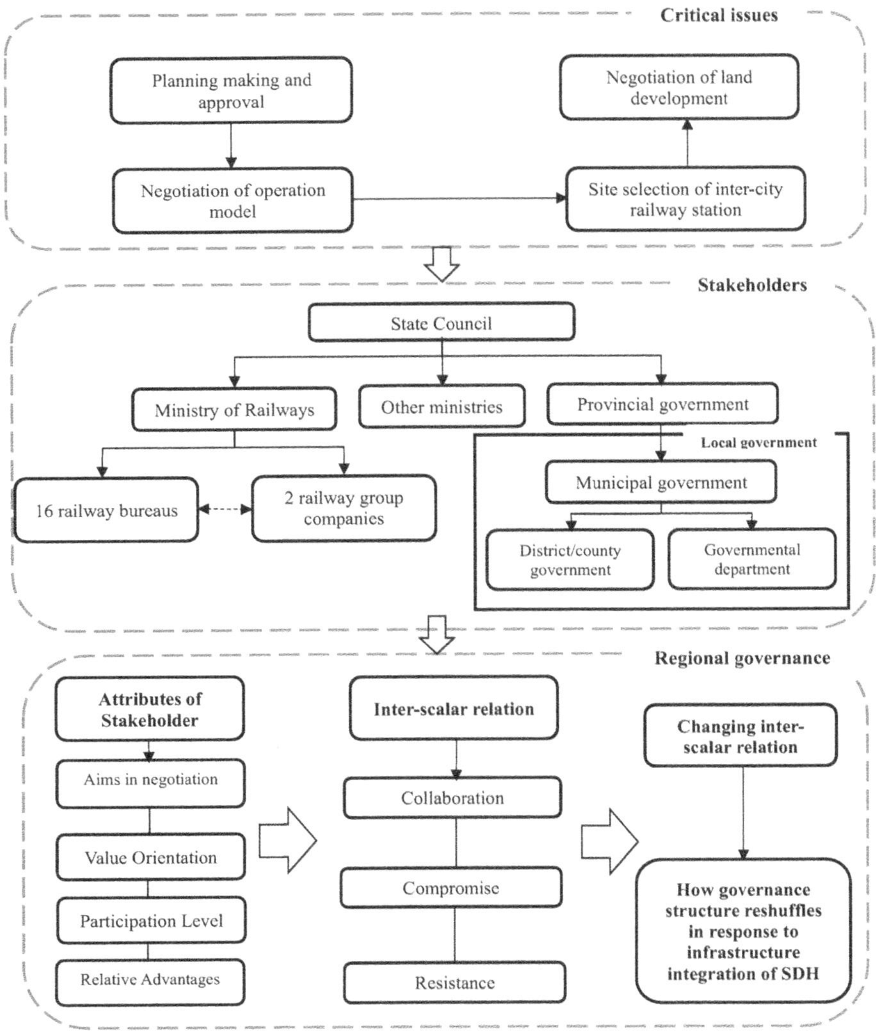

Fig. 4.5 Analytical framework for how governance structure reshuffles to enable the SDH infrastructure integration. *Source* Author

of the northwest region of Guangdong" in 2013. Subsequently, the government spent 13.5 billion Yuan from 2013 to 2017, including 3 billion Yuan as reward money for introducing corporations to investments in SSSCZ. In addition, the Shenzhen Municipal Government issued "A Support Work Plan for Shanwei" in 2014 to offer adequate preferential funds for infrastructure construction and industrial introduction and operation. On the basis of these preferential policies and government endeavors,

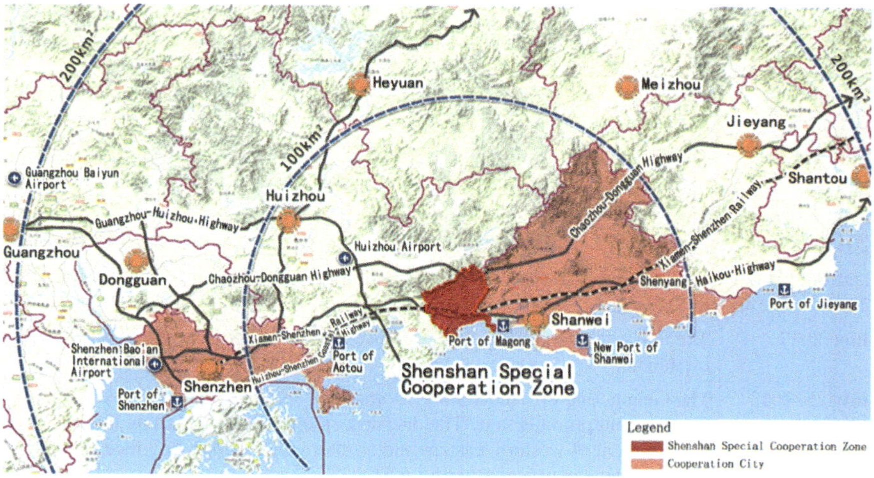

Fig. 4.6 Location of SSSCZ. *Source* The official website of SSSCZ, http://www.shenshan.gov.cn/home/about

institutional innovation in the governance structure for the building of SSSCZ has enabled an industrial collaboration between Shenzhen and Shanwei.

Second, compared with other industrial transfer parks in Guangdong Province, SSSCZ is the only one with prefecture-level administrative power in personnel appointment and removal. Such provincial government-induced administrative empowerment is rare, as shown in the conventional literature on cross-boundary economic cooperation zones. This high administrative level will largely affect the institutional alignment in SSSCZ and is an emerging manifestation of governance structure reshuffling; SSSCZ is therefore a suitable case to investigate governance reshuffling.

Third, SSSCZ developed for several years in terms of industrial development and formed a solid industrial base. These contexts will contribute to the examination of governance reshuffling between the state and market by providing materials and details to document the regionalization process. Specifically, SSSCZ has been established for approximately seven years and has introduced many high-end industrial projects, such as the Tencent cloud-computing data center, the Start-up Square, the Babaihui Biological Science and Technology Park, the Jiulong Bay tourism project, and the Tsinghua Tongfang E-commerce Park. Additionally, SSSCZ has already set up a well-rounded administrative management system and a comprehensive and formal official website. Compared with other industrial transfer parks that have not been in operation for a long time or were developed without a sound management system in SDH, the long daily operation and mature management system of SSSCZ represent a policy laboratory for the examination of the institutional arrangements across different levels of government, between cities, and between state and non-state domains.

Table. 4.5 Major events and timetable of SSSCZ

Time	Major events
February 11, 2011	The Guangdong Provincial Government approved the "Framework of SSSCZ" and established SSSCZ
May 21, 2011	The Guangdong Provincial Government approved "Working Committee of SSSCZ" and "Management Committee of SSSCZ."
September 20, 2011	The first session of "Recommendation Conference on Introduction of Business" was held in Shenzhen
December 26, 2011	The first major project, Cloud computing center of Tencent," was set up in SSSCZ
June 2012	The first group of leaders for SSSCZ selected by Shenzhen and Shanwei was established
July 25, 2013	The Guangdong Provincial Party Committee and the Guangdong Provincial Government promulgated "The decision on further strengthening the development of western, eastern and northern Guangdong Province," thereby representing a new opportunity for SSSCZ
August 8, 2013	The Guangdong Provincial Government decided to optimize the cooperation mechanism of SSSCZ and accelerate its development
April 2014	The adjustment of the new leaders of SSSCZ was completed
April 14, 2014	The joint conference of the Party and Government of Shenzhen and Shanwei proposed "The suggestions on further strengthening the cooperation mechanism of SSSCZ and accelerating the development of it."
May 28, 2014	"SSSCZ Investment Holdings Ltd.," the first registered enterprise, was set up
July 22, 2014	Hu Chunhua, the political bureau chief of the CPC Central Committee and the secretary of the Guangdong Provincial Party Committee, visited SSSCZ
October 31, 2014	The Shanwei municipal government empowered part of the economic and management jurisdiction of the prefecture-level city under the "Management Committee of SSSCZ."
November 13, 2014	"Master planning of SSSCZ (2015–2030)" was adopted in a executive meeting of Guangdong Provincial Government
December 21, 2014	Hu Chunhua, the political bureau chief of the CPC Central Committee and the secretary of the Guangdong Provincial Party Committee, visited SSSCZ and issued important instructions
January 2015	The Guangdong Provincial Government proposed that the financial system of SSSCZ should follow the "Shunde mode," adapting province directly governing SSSCZ
July 23, 2015	"Regulation on Administration of SSSCZ" was adopted by the Guangdong Provincial Government
August 17, 2015	Hu Chunhua, the political bureau chief of the CPC Central Committee and the secretary of the Guangdong Provincial Party Committee, visited SSSCZ

Source Official SSSCZ website, http://www.shenshan.gov.cn/home/about

Fourth, in terms of geographical linkage between Shenzhen and Shanwei, SSSCZ has stood for special inter-city cooperation, spanning the territory of Huizhou which is adjacent to Shenzhen and Shanwei. This phenomenon may indicate that Shenzhen gave up its industrial collaboration with Dongguan and Huizhou and selected Shanwei as its target city for industrial collaboration. Specifically, the outlying city of Shanwei is strategically incorporated into SDH under the initiatives of regional coordinated development from the Guangdong Provincial Government. Shanwei is not directly adjacent to Shenzhen geographically. Why does Shenzhen not select Huizhou or Dongguan as the target city to enable regional industrial cooperation? Does the great geographical distance between Shenzhen and Shanwei trigger great enthusiasm in the development of SSSCZ? Compared with those for the cases of the "Dongguan and Huizhou industrial transfer park" and the "Dongguan and Huidong industrial transfer park," the aforementioned questions continue to be under-discussed. As an emerging phenomenon, the case of SSSCZ can add value to the literature on regional integration.

4.3.1.2 Inter-city Railway Project: SGSIR

SGSIR project was devised to elucidate how the governance structure reshuffles to enable the infrastructure integration of SDH. The project is an inter-city railway that goes through Guangzhou, Shenzhen, and Dongguan and was constructed by the China Railway Second Board and China Railway Port and Channel Engineering Group Co., Ltd., in 2008. SGSIR project is also one of the main lines of the regional transportation network in PRD that has been approved by SC. The overall length of SGSIR is 86.62 km, and the gross investment is 19.69 billion Yuan. The line comprises 29 stations across all three municipalities (Fig. 4.7 and Table 4.6).

Fig. 4.7 SGSIR. *Source* http://vdisk.weibo.com/s/uDU3t5chGKv3z

Table. 4.6 Station list of SGSIR

Inter-city railway	City	Station
SGSIR	Guangzhou	*SGSIR has one station in Guangzhou* Xintang
	Dongguan	*SGSIR has 10 stations in Dongguan* Zhongtang, Wangniudun, Wanghong, Hongmei, Shatian, Houjie, Humen Railway Station, Humen Shopping Mall, Changan Xiabian, Changan Jinsha
	Shenzhen	*SGSIR has four stations in Shenzhen* Maritime Field, Heping, Shenzhen Airport North, Shenzhen Airport

Source Author

On the basis of the territorial practices of SDH in building a regional infrastructure network (see Chap. 3), three reasons are considered in selecting the case. First, the construction of SGSIR involved substantial stakeholders within the state sectors. The construction process involves an intricate interaction that included plan-making and approval, negotiation of the operation model, site selection for railway stations, and local land development. These interlocking issues created a complicated institutional context and served as a tremendous impetus for the reshuffling of the governance model in the inter-city railway project. In PRD, which is a highly fragmented city-region, the governance structure of inter-city railway projects varies enormously across cases. As shown in the conceptual review in Chap. 2, some large-scale transportation projects, such as the Guangzhou–Zhuhai inter-city railway and Guangzhou–Foshan Metro, have been comprehensively examined from the viewpoint of the institution-building process. However, SGSIR remains under-discussed. The investigation into SGSIR will contribute to the mainstream literature on city-regional politics, particularly inter-scalar relations.

Second, SGSIR concatenates the cities of Guangzhou, Dongguan, and Shenzhen, thereby contributing to the examination of how the three cities struggle with upper-level authorities under the prevalent city-regionalism. The existing literature shows that inter-scalar politics, which is central to the politics in the inter-city railway project, can complement the study of SSSCZ as an inter-city cooperation process. These two cases are conducive to developing the whole landscape of the city-regionalism of SDH.

Third, the other main lines of the inter-city railway in SDH remain ongoing, and the scales of these lines are smaller than those of SGSIR, whose construction was completed in 2015. SGSIR has been in operation for more than two years. The development issues, especially for the land development along stations, can be examined in this project.

4.3.2 *Data Collection*

4.3.2.1 Desktop Research

To understand the economic integration and institutional integration of SDH, this book conducted desktop research to collect secondary data. First, the assessment of the economic integration of SDH was based on secondary data from desktop research. These materials comprised the following:

- Guangdong Statistical Yearbook (2005–2016)
- China City Statistical Yearbook (2005–2016)
- China Urban Construction Statistical Yearbook (2005–2016)
- Shenzhen Statistical Yearbook (2005–2016)
- Dongguan Statistical Yearbook (2005–2016)
- Huizhou Statistical Yearbook (2005–2016)
- Shanwei Statistical Yearbook (2005–2016)
- Heyuan Statistical Yearbook (2005–2016)

Second, secondary data are used in the assessment of the institutional integration of SDH as well. To demonstrate the inter-scalar relations of SDH, the administrative setting, and its readjustment after 2000, this book identifies the local innovation in the administrative setting and regional authority on the basis of the secondary data. The materials included the following:

- Shenzhen Statistical Yearbook (2001–2017)
- Dongguan Statistical Yearbook (2001–2017)
- Huizhou Statistical Yearbook (2001–2017)
- Shanwei Statistical Yearbook (2001–2017)
- Heyuan Statistical Yearbook (2001–2017)
- News, publications, planning, and government documents on administrative settings

To elaborate the inter-city and state–market–civil society relations of SDH, this book uses inter-city cooperation news that was collected from professional search engines and official municipal websites (Zhang et al., 2018). Considering the promulgation of the "PRD Urban Cluster Coordinated Development Plan" in 2004, which officially guarantees the formal status of three sub-regions, this book sets the time interval for the news to the period of 2004–2015. Initially, this book used the city's name as the keyword to search for inter-city cooperation news from *Google* and *Baidu*. Two cities from the five SDH cities are selected randomly, and their names were searched on *Google* and *Baidu*; ten combinations of inter-city groups were obtained. Second, this book gathered inter-city cooperation news from the official municipal websites, with the target city's name as the keyword. For instance, to collect news between Dongguan and Shanwei, this research searches the city name of Shanwei in the Dongguan government's official website and the city name of

Dongguan in the Shanwei government's official website. This book obtains 1,015 news items by filtering redundant and irrelevant ones.

Third, this book investigates the processes of governance structure reshuffling under the city-regionalism of SDH by undertaking desktop research in accordance with the analytical framework of the cases of SSSCZ and SGSIR. The collected materials primarily include news, policies, government documents, planning, and agreements that are relevant to the cases of SSSCZ and SGSIR.

4.3.2.2 Semi-structured Interview

The governance structure reshuffling processes are elaborated by semi-structured interviews. In the provision of qualitative data, face-to-face and semi-structured interviews are needed in accordance with the analytical framework. Two rounds of semi-structured interviews are conducted to demonstrate the cases of SSSCZ and SGSIR. Specifically, the first and second rounds of semi-structured interviews were held from March 25, 2016, to April 29, 2016, and from November 2, 2016, to December 26, 2016, respectively, to probe SSSCZ. The third round of semi-structured interviews was held from November 25, 2017, to December 23, 2017, with a primary focus on SGSIR. In total, this book interviews five local officials in SSSCZ and six local officials in SGSIR, four urban planners in SSSCZ and seven urban planners in SGSIR, six staff or managers in enterprises in SSSCZ, and two developers and investors in SSSCZ. The detailed interview questions are listed in the Appendix. Within the semi-structured interview process, this book particularly focuses on the aims in negotiation, participation levels, relative advantages, and value orientations of different stakeholders in different issues, as well as on the interaction and negotiation processes to enable or encumber the socioeconomic integration of SDH.

4.3.3 Data Analysis

To assess the economic integration of SDH, this book collected secondary data on the indicators from the above materials. On the basis of the above analysis, the extent of the market, industrial, and infrastructure integrations at the overall SDH and inter-city group scales are assessed by the selected models in Sect. 4.2.1.

To assess the institutional integration of SDH, this book analyzed the inter-scalar relations on the basis of the data of jurisdictional arrangements that were collected from the yearbooks of five cities. As regards the inter-city and state–market–civil society relations, the data on the inter-city cooperation news are deployed to demonstrate their spatial and temporal order, cooperation area, cooperation mode, involvement degree, and organization and developmental status. Specifically, to analyze these aspects, this book meticulously counts and categorizes the inter-city cooperation news in accordance with different attributes. The attributes of the inter-city cooperation news are shown in Table 4.7 in accordance with the analytical framework of institutional integration (see Sect. 4.2.2).

To probe the governance structure reshuffling process of SDH in the cases of SSSCZ and SGSIR, this book analyzed the qualitative data collected from the semi-structured interviews in accordance with the analytical framework (see Sect. 4.2.3) by demonstrating the interactions across different actors in different issues.

4.4 Conclusions

In this chapter, the analytical framework has been formulated to answer the four research questions in Chap. 1. The economic integration of SDH is assessed at the overall SDH and inter-city group scales from the three dimensions of market, industrial, and infrastructure integration. The institutional integration of SDH is evaluated under the MLG-based analytical framework at overall SDH and inter-city group scales from three dimensions, namely, inter-scalar, inter-city, and state–market–civil society relations. Given the relatively inadequate literature on the governance structure reshuffling process under city-regionalism, the cases of SSSCZ and SGSIR are explored by probing the aims in negotiation, participation levels, value orientations, and relative advantages of different actors and their interactions. Following the above analysis, this book provides a full portrait of the SDH integration and empirically grounded evidence to comprehend the city-regionalism of China.

The chapter elaborates on the research methods as well. SSSCZ and SGSIR projects were selected to demonstrate qualitatively how the governance structure reshuffles itself to enable or encumber industrial and infrastructure integration. Consistent with the analytical framework in this chapter, this book used desktop research and semi-structured interviews to assess economic and institutional integration and decipher the reshuffling process of the governance structure of SDH.

Table. 4.7 Attributes of inter-city cooperation news from social media

Inter-city cooperation group									
Shenzhen–Dongguan	Shenzhen–Huizhou	Shenzhen–Shanwei	Shenzhen–Heyuan	Dongguan–Huizhou	Dongguan–Shanwei	Dongguan–Heyuan	Huizhou–Shanwei	Huizhou–Heyuan	Shanwei–Heyuan
Time group									
2004 and 2005		2006 and 2007	2008 and 2009		2010 and 2011		2012 and 2013		2014 and 2015
State-market-civil society relation									
Government-to-government		Government-to-market		Government-to-society		Market-to-market		Society-to-society	
Cooperation area									
Comprehensive issue				**Economic development**		**Institution design**			
Comprehensive issue				Industrial development Infrastructure construction		Institutional construction Urban management Urban planning			
Secondary function				**Aid activity**		**Others**			
Science and education Entertainment Disaster emergency Resource and environment				Aid activity		Others			

Source Author

References

Armstrong, D., Bello, V., Gilson, J., & Spini, D. (2011). *Civil Society and International Governance: The role of non-state actors in global and regional regulatory frameworks* (Vol. 10, pp. XII-204). Taylor & Francis.

Asche, F., Bremnes, H., & Wessells, C. R. (1999). Product aggregation, market integration, and relationships between prices: An application to world salmon markets. *American Journal of Agricultural Economics, 81*(3), 568–581.

Bai, M. (2011). *Research on industry integration in Pearl River Delta and the other three Peripheral area of Guangdong province* (Master's dissertation). Available from China Master's Theses Full-text Database (*in Chinese*).

Behrman, J. N. (1972). Industrial integration and the multinational enterprise. *The Annals of the American Academy of Political and Social Science, 403*(1), 46–57. https://doi.org/10.1177/000271627240300105

Bekaert, G., & Harvey, C. R. (1995). Time-varying world market integration. *The Journal of Finance, 50*(2), 403–444. https://doi.org/10.2307/2329414

Chen, Y. (1998). The measurement of industrial structure convergence, desired convergence and undesired convergence. *China Industrial Economics, 12*(4), 37–43 (*in Chinese*). https://doi.org/10.19581/j.cnki.ciejournal.1998.04.007

Chen, H., & Li, G. (2009). Study on the level and procedural of Beijing-Tianjin-Hebei metropolitan regional market integration from 1985 to 2007. *Geographical Research, 28*(6), 1476–1483. (in Chinese).

Chong, E., Huet, F., Saussier, S., & Steiner, F. (2006). Public-private partnerships and prices: Evidence from water distribution in France. *Review of Industrial Organization, 29*(1–2), 149.

Douglass, M. (2002). From global intercity competition to cooperation for livable cities and economic resilience in Pacific Asia. *Environment and Urbanization, 14*(1), 53–68. https://doi.org/10.1177/095624780201400105

Fan, F. (2011). The industrial division of labor in Yangtze Delta area: 1998–2008-A comparative analysis at provincial level and at municipal level. *Industrial Economics Research, 10*(4), 8–16 (*in Chinese*). https://doi.org/10.13269/j.cnki.ier.2011.04.009

Fan, F. (2013). An improved method of measuring industrial structure similarity degree between areas. *The Journal of Quantitative & Technical Economics, 30*(7), 98–115 (*in Chinese*). https://doi.org/10.13653/j.cnki.jqte.2013.07.019

Finger, J. M., & Kreinin, M. E. (1979). A measure of export similarity' and its possible uses. *The Economic Journal, 89*(356), 905–912. https://doi.org/10.2307/2231506

Frank, L. K. (1925). The significance of industrial integration. *Journal of Political Economy, 33*(2), 179–195.

Goldberg, P. K., & Verboven, F. (2005). Market integration and convergence to the law of one price: Evidence from the European car market. *Journal of International Economics, 65*(1), 49–73. https://doi.org/10.1016/j.jinteco.2003.12.002

Gui, Y., Chen, M., Lu, M., & Chen, Z. (2006). Does the Chinese commodity market develop towards integration or disjoint: Based on the analysis of relative price. *The Journal of World Economy, 2*, 20–30. (in Chinese).

Hurd, J. (1975). Railways and the expansion of markets in India, 1861–1921. *Explorations in Economic History, 12*(3), 263–288. https://doi.org/10.1016/0014-4983(75)90004-2

Jiang, F., & Luo., X. (2016). Analysis to the cooperative modes of industrial parks: A case study on Jiangsu Province. *Urban Problems, 35*(7), 38–43 (*in Chinese*). https://doi.org/10.13239/j.bjsshkxy.cswt.160706

Jonas, A. E., & Ward, K. (2007). Introduction to a debate on city-regions: New geographies of governance, democracy and social reproduction. *International Journal of Urban and Regional Research, 31*(1), 169–178. https://doi.org/10.1111/j.1468-2427.2007.00711.x

Krugman, P. R. (1991). *Geography and trade*. MIT Press.

Kwak, Y. H., Chih, Y., & Ibbs, C. W. (2009). Towards a comprehensive understanding of public private partnerships for infrastructure development. *California Management Review, 51*(2), 51–78. https://doi.org/10.2307/41166480

Li, F., Ma, D., & Lv, L. (2002). Study on the corn and soybean market integration in China. *Journal of Shanxi Agricultural University (Natural Science Edition), 22*(1), 87–90 (*in Chinese*). https://doi.org/10.13842/j.cnki.issn1671-8151.2002.01.024

Li, Y., & Wu, F. (2018). Understanding city regionalism in China: Regional cooperation in the Yangtze River Delta. *Regional Studies, 52*(3), 313–324. https://doi.org/10.1080/00343404.2017.1307953

Liu, P. (2011). *Research on iso-structuralizing of "Shen-Guan-Hui" industrial development* (Master's thesis). Available from China Master's Theses Full-text Database (*in Chinese*).

Liu, S., Luo, X., Tian, D., & Wang, Y. (2014). Preliminary study on China's cross-boundary new district in forming mechanisms, spatial organizations and governance model. *Economic Geography, 34*(12), 41–47 (*in Chinese*). https://doi.org/10.15957/j.cnki.jjdl.2014.12.007

Liu, Y. (2015). *The research on level measure and counter-measure of economic integration in Nanchang and Jiujiang* (Master's dissertation). Available from China Master's Theses Full-text Database (*in Chinese*).

Luo, X., & Shen, J. (2006). Urban growth between cities: The case of Jiangyin economic development zone in Jingjiang. *Acta Geographica Sinica, 61*(4), 435–445. (in Chinese).

Luo, X., & Shen, J. (2007). Models of inter-city cooperation and its theoretical implications: An empirical study on the Yangtze River Delta. *Acta Geographica Sinica, 62*(2), 115–126. (in Chinese).

Luo, X., & Shen, J. (2008). Why city region planning does not work well in China: The case of Suzhou–Wuxi–Changzhou. *Cities, 25*(4), 207–217. https://doi.org/10.1016/j.cities.2008.04.003

Luo, X., & Shen, J. (2009). A study on inter-city cooperation in the Yangtze river delta region, China. *Habitat International, 33*(1), 52–62. https://doi.org/10.1016/j.habitatint.2008.04.002

Ma, L. J. (2005). Urban administrative restructuring, changing scale relations and local economic development in China. *Political Geography, 24*(4), 477–497. https://doi.org/10.1016/j.polgeo.2004.10.005

McKee, M., Edwards, N., & Atun, R. (2006). Public-private partnerships for hospitals. *Bulletin of the World Health Organization, 84*, 890–896.

Naughton, B. (2003). How much can regional integration do to unify china's market? In N. Hope, D. Yang, & M. Y. Li (Eds.), *How far across the river? Chinese policy reform at the millennium* (pp. 204–232). Stanford University Press

Parsley, D. C., & Wei, S. J. (2001b). Limiting currency volatility to stimulate goods market integration: A price based approach, *Working Paper* (No. w8468). Cambridge, MA, USA: NBER. https://doi.org/10.3386/w8468

Parsley, D. C., & Wei, S. J. (1996). Convergence to the law of one price without trade barriers or currency fluctuations. *The Quarterly Journal of Economics, 111*(4), 1211–1236. https://doi.org/10.2307/2946713

Parsley, D. C., & Wei, S. J. (2001a). Explaining the border effect: The role of exchange rate variability, shipping costs, and geography. *Journal of International Economics, 55*(1), 87–105. https://doi.org/10.1016/S0022-1996(01)00096-4

Perkmann, M. (2007). Construction of new territorial scales: A framework and case study of the EUREGIO cross-border region. *Regional Studies, 41*(2), 253–266. https://doi.org/10.1080/00343400600990517

Ravallion, M. (1986). Testing market integration. *American Journal of Agricultural Economics, 68*(1), 102–109. https://doi.org/10.2307/1241654

Robson, P. (2002). *The economics of international integration.* Routledge.

Rodríguez-Pose, A. (2008). The rise of the "city region" concept and its development policy implications. *European Planning Studies, 16*(8), 1025–1046. https://doi.org/10.1080/09654310802315567

Schrijnen, P. M. (2000). Infrastructure networks and red–green patterns in city regions. *Landscape and Urban Planning, 48*(3), 191–204. https://doi.org/10.1016/S0169-2046(00)00042-6
Scott, A. J. (2001). Globalization and the rise of city regions. *European Planning Studies, 9*(7), 813–826. https://doi.org/10.1080/09654310120079788
Tinbergen, J. (1954). *International economic integration.* Elsevier.
Wang, Z., & Chen, Q. (2006). Methods measuring to the same structure of manufacturing in the Delta area of Changjiang River—Analyzing based on time serial datum. *Industrial Economics Research, 5*(4), 35–41. (in Chinese).
Wu, F. (2016). China's emergent city-region governance: A new form of state spatial selectivity through state-orchestrated rescaling. *International Journal of Urban and Regional Research, 40*(6), 1134–1151. https://doi.org/10.1111/1468-2427.12437
Xu, J. (2008). Governing city-regions in China: Theoretical issues and perspectives for regional strategic planning. *Town Planning Review, 79*(2–3), 157–186.
Xu, J., & Chen, Y. (2014). Planning inter-city railways in China's mega-city regions: Insights from the Pearl River Delta. *China Review, 14*(1), 11–36.
Xu, J., & Yeh, A. G. (2013). Interjurisdictional cooperation through bargaining: The case of the Guangzhou-Zhuhai railway in the Pearl River Delta, China. *The China Quarterly, 213*, 130–151. https://doi.org/10.1017/S0305741013000283
Xu, X. (2002). Have the Chinese provinces become integrated under reform? *China Economic Review, 13*(2–3), 116–133. https://doi.org/10.1016/S1043-951X(02)00058-5
Yang, C. (2005). Multilevel governance in the cross-boundary region of Hong Kong-Pearl River Delta. *China. Environment and Planning A, 37*(12), 2147–2168. https://doi.org/10.1068/a37230
Yu, W., & Huang, J. (1998). Study on China's grain market reform from the perspective of integration degree of rice market. *Economic Research Journal*, pp. 50–57 (*in Chinese*).
Zhang, J., Geng, L., Yin, J., & Luo, X. (2011). The regional cooperative governance from the perspective of regional spatial production—A case study of Jinjiang Park, Jiangyin Economic Development Zone. *Human Geography, 26*(1), 5–9 (*in Chinese*). https://doi.org/10.13959/j.issn.1003-2398.2011.01.016
Zhang, T. (2006). From inter-city competition to collaborative planning: The case of the Yangtze River Delta region of China. *Urban Affairs Review, 42*(1), 26–56.
Zhang, X., Guo, Q., Cheung, D. M. W., & Zhang, T. (2018). Evaluating the institutional performance of the Pearl River Delta integration policy through intercity cooperation network analysis. *Cities, 81*, 131–144. https://doi.org/10.1016/j.cities.2018.04.002
Zheng, Y., & Li, C. (2003). Efficiency loss caused by regional separation. *Social Science in China, 24*(1), 64–72. (in Chinese).

Chapter 5
The Assessment of Economic and Institutional Integrations of SDH

5.1 Introduction

The assessment of economic and institutional integrations of SDH is discussed in this chapter to reach an in-depth understanding of SDH integration within the city-regionalism of the Pearl River Delta (PRD). This chapter attempts to answer two research questions: (1) to what extent does SDH operate as an economically integrated entity as reflected in price similarity, industrial structure similarity, and infrastructure development? and (2) to what extent does SDH operate as an institutionally integrated entity by redefining the governance structure across government hierarchies, between cities, and between the state and non-state domains?

As elaborated in Chap. 4, the assessment of economic integration involves three domains: market, industrial, and infrastructure integrations. Infrastructure integration is further evaluated from the perspectives of equality and connectivity. Aside from these three inter-related dimensions, the overall SDH regional level and inter-city group scales are examined to further understand the economic integration on different spatial scales. The institutional integration is systematically assessed in accordance with three perspectives as well, encompassing inter-scalar, inter-city, and state–market–civil society relations. The inter-scalar relation of SDH is analyzed through ascertaining the current situation of the administrative setting and adjustments of the administrative division. The inter-city relation of SDH examines the time–space distribution feature, cooperation area, and cooperation mode based on the inter-city cooperation news. The state–market–civil society relation of SDH focuses on the degree of involvement; area of cooperation; and organization and development status of state, market, and civil society through probing the inter-city cooperation news as well. These multi-dimensional analyses of the SDH integration will shed light on the related economic and institutional landscapes of SDH under the PRD integration policy.

After the introductory section, this chapter presents the research process and includes three sections. Section 5.1 assesses the economic integration of SDH by dissecting the market, industrial, and infrastructure integrations on the overall SDH

X. Zhang, *Regional Integration and Governance Reshuffling in the Making of China's City-Regions*, Urban Sustainability, https://doi.org/10.1007/978-981-99-2792-0_5

and the inter-city group scales. Section 5.2 assesses the institutional integration of SDH in terms of inter-scalar, inter-city, and state–market–civil society relations. The conclusion systemically reports the extent of the economic and institutional integration of SDH.

5.2 Assessing Economic Integration of SDH

5.2.1 Assessing Market Integration of SDH

To depict the degree of market integration, this book employs relative price variance to measure the market integration on the overall SDH and inter-city group scales. According to the formula presented in Chap. 4, if the value of relative price variance is small, then the price changes in different places converge, and the regional market is integrated, *and* vice versa. An analysis of the relative price variance establishes the values of the relative price variance of SDH and ten inter-city groups (as categorized in Chap. 4), as shown in Table 5.1. Details of the analysis are presented below.

5.2.1.1 Market Integration at the SDH Scale

This book seeks to understand the degree of market integration for the overall SDH from 2004 to 2015 (Fig. 5.1) by calculating the relative price variance (Table 5.1).

Figure 5.1 summarizes two fundamental features of the market integration at the overall SDH scale as follows.

First, the degree of market integration before 2009 was rather unstable with some large fluctuations. The trend of relative price variance before 2009 has experienced some dramatic fluctuations in terms of variation. The peak value was in 2005 with a score of 0.01535. The number experienced a dramatic decrease to 0.00845 from 2005 to 2006. The relative price variance has witnessed a dramatic and continuous increase in the next two years. However, after 2008, the relative price variance declined sharply to 0.00939 in 2009. Overall, the relative price variance of SDH has fluctuated dramatically in the range from 0.00845 to 0.01535 within the six years from 2004 to 2009. The mean of relative price variance was 0.0119167. Thus, the regional market was completely unsettled before 2009, and the degree of economic integration can be described as unstable and burgeoning.

Second, the degree of market integration after 2009 has seen a slight increase with a relatively stable trend. The trend of relative price variance after 2009 has exhibited a smoother fluctuation compared with the previous stage. The valley value was 0.00823 in 2014. However, the peak value was merely 0.01146 in 2011. Overall, the relative price variance of SDH has fluctuated smoothly within the range from 0.00823 to 0.01146 during the six years from 2010 to 2015. In addition, the mean of relative price variance was 0.00985. This was slightly lower than 0.0119167 in the

Table 5.1 Relative price variance of ten inter-city groups and overall SDH

Years	Inter-city groups										SDH
	SS	SD	SH1	HS2	SH2	HS1	DH1	HH	DH2	DS	
2004	0.01364	0.01374	0.00718	0.00868	0.00572	0.00933	0.01073	0.01353	0.01126	0.01075	0.01046
2005	0.01215	0.01906	0.01683	0.01283	0.01753	0.01187	0.01431	0.01204	0.02118	0.01566	0.01535
2006	0.00898	0.00976	0.00568	0.00732	0.00398	0.00487	0.00754	0.01489	0.00927	0.01222	0.00845
2007	0.00402	0.00996	0.00810	0.01219	0.00636	0.00980	0.01032	0.01077	0.00931	0.01236	0.00932
2008	0.00706	0.00685	0.01245	0.01519	0.02263	0.01418	0.01726	0.01432	0.01259	0.01927	0.01418
2009	0.00938	0.01092	0.01066	0.00746	0.00650	0.01193	0.01104	0.00944	0.00836	0.00821	0.00939
2010	0.01015	0.01075	0.01099	0.01383	0.01170	0.00869	0.00498	0.00899	0.01597	0.01363	0.01097
2011	0.00780	0.00949	0.01305	0.01170	0.01179	0.01048	0.01274	0.01563	0.01295	0.00894	0.01146
2012	0.00714	0.00670	0.01054	0.00601	0.01342	0.01606	0.01610	0.00898	0.00881	0.01352	0.01073
2013	0.00791	0.00598	0.01080	0.01102	0.00689	0.00655	0.00657	0.00698	0.01052	0.01164	0.00849
2014	0.00524	0.00612	0.00517	0.00731	0.01205	0.00965	0.01001	0.00585	0.00790	0.01296	0.00823
2015	0.01075	0.00570	0.00675	0.00854	0.00513	0.01214	0.01033	0.01124	0.01279	0.00884	0.00922
Mean	0.00868	0.00958	0.00985	0.01017	0.01031	0.01046	0.01099	0.01105	0.01174	0.01233	—

Note Shenzhen-Dongguan (SD), Shenzhen-Huizhou (SH1), Shenzhen-Shanwei (SS), Shenzhen-Heyuan (SH2), Dongguan-Huizhou (DH1), Dongguan-Shanwei (DS), Dongguan-Heyuan (DH2), Huizhou-Shanwei (HS1), Huizhou-Heyuan (HH) and Heyuan-Shanwei (HS2)
Source Author

Fig. 5.1 Relative price variance of SDH from 2004 to 2015. *Source* Author

previous stage. This result signals that the degree of market integration after 2009 has been strengthened marginally compared with the previous stage, and the regional unified market has become more stabilized.

5.2.1.2 Market Integration at the Inter-City Group Scale

This book seeks to fathom the degree of market integration for inter-city groups from 2004 to 2015 (Fig. 5.2) by calculating the relative price variance (Table 5.1).

In rankings of the mean relative price variances for different inter-city groups based on Table 5.1, Shenzhen–Dongguan, Shenzhen–Huizhou, Shenzhen–Heyuan,

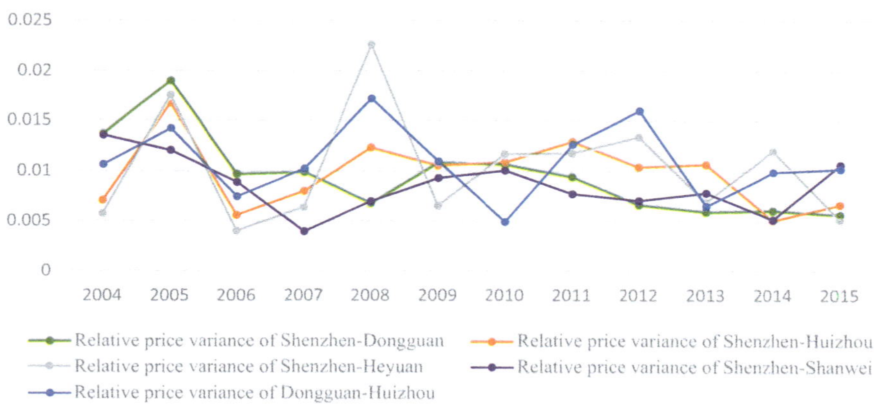

Fig. 5.2 Relative price variance of the inter-city group from 2004 to 2015. *Source* Author

Shenzhen–Shanwei, and Dongguan–Huizhou ranked numbers 2, 3, 5, 1, and 7, respectively, among the ten inter-city groups in SDH. Two fundamental features of the market integration at the inter-city group scale were identified.

First, in terms of Shenzhen with its surrounding cities, Shenzhen with Dongguan, Huizhou, and Shanwei had a high degree of market integration, whereas the market integration between Shenzhen and Heyuan was kept in a medium level. Figure 5.2 shows the tendencies of relative price variance for Shenzhen–Huizhou, Shenzhen–Shanwei, and Shenzhen–Heyuan experienced some dramatic fluctuations from 2004 to 2015. Although its mean of the relative price variance ranked third among the ten inter-city groups in SDH, the market integration of Shenzhen–Huizhou was extremely unstable, manifesting a dramatic fluctuation. However, although Shenzhen–Dongguan ranked second after Shenzhen–Shanwei in first place, the tendency was smoother and overtly downward from 2010 to 2015. The market integration of Shenzhen–Dongguan occurred as a continuous and steady integrating process.

Second, in terms of Shenzhen–Dongguan–Huizhou, the degrees of market integration between Shenzhen and Dongguan as well as Shenzhen and Huizhou was high among all the inter-city groups in SDH. The degree of market integration between Dongguan and Huizhou was equivalent to a medium level within the ten inter-city groups. Compared with the high degree of the market integration of Shenzhen–Dongguan and Shenzhen–Huizhou, the market integration of Dongguan–Huizhou not only experienced a dramatic fluctuation and stayed at a medium level within 12 years, but also exhibited an upward trend after 2013. The market integration of Dongguan–Huizhou generally stayed at a medium level in general and became worse in recent years.

In sum, in terms of the overall SDH, market integration was rather unstable before 2009. However, it was relatively steady and slightly increased after 2009. In terms of inter-city groups, Shenzhen integrated well with Dongguan, Huizhou, and Shanwei and shared a medium degree with Heyuan. However, the market integration between Dongguan and Huizhou stayed at a medium level and became worse after 2013. Thus, the economy of SDH gradually became integrated and stabilized after 2009. Regarding the internal environment, Shenzhen as the regional core city was closely integrated with the member cities in SDH economically, and economic integration between Dongguan and Huizhou remained at a medium level.

5.2.2 Assessing Industrial Integration of SDH

To assess the extent of industrial integration, this book uses the degree of industrial structure similarity to measure the industrial integration at the overall SDH scale and a modified Krugman's index to measure the industrial integration at the inter-city group scale. These two formulas (see details in Chap. 4) measure the degree of industrial isomorphism. If the values of these two formulas are high, then the phenomenon of industrial isomorphism is serious, and, correspondingly, the degree of industrial integration is severely low. Table 5.2 shows the degree of industrial structure similarity of the overall SDH and the modified Krugman's index for the ten inter-city groups, based on the analysis by applying the two formulas. Their results and implications are elaborated in this section.

5.2.2.1 Industrial Integration at the SDH Scale

This book identifies the degree of industrial integration for the overall SDH from 2004 to 2015 (Fig. 5.3) by calculating the degree of industrial structure similarity (Table 5.2).

According to Fig. 5.3, the overall degree of industrial integration of SDH from 2004 to 2015 became increasingly weak and can be divided into three distinct stages as follows:

- In the first stage, from 2004 to 2007, the degree of industrial structure similarity of SDH saw a rapid growth from 0.52915 in 2004 to 0.62372 in 2007. The degree of industrial integration experienced a rapid decline at this stage.
- In the second stage, from 2007 to 2012, the degree of industrial structure similarity of SDH exhibited a smooth and mild change, ranging from 0.62372 in 2007 to 0.62110 in 2012, which affirmed that the industrial integration of SDH remained steady in this stage.
- Nonetheless, in the last stage, from 2012 to 2015, the degree of industrial structure similarity of SDH saw a more significant increase from 0.62110 in 2012 to 0.66471 in 2015. This finding signals that the phenomenon of industrial isomorphism was increasingly severe and that the degree of industrial integration of SDH rapidly and continuously declined.

5.2.2.2 Industrial Integration at the Inter-City Group Scale

This book obtains the degree of industrial integration for inter-city groups from 2004 to 2015 (Fig. 5.4) through calculating the modified Krugman's index (Table 5.2).

Table 5.2 shows the ranking of the mean of the modified Krugman's index in terms of inter-city groups. The positions of Shenzhen–Dongguan, Shenzhen–Huizhou, Shenzhen–Heyuan, Shenzhen–Shanwei, and Dongguan–Huizhou were at the 9th, 10th, 2nd, 1st, and 8th places, respectively, among the ten inter-city groups. The two fundamental features of industrial integration at the inter-city group scale are summarized as follows.

First, in terms of Shenzhen with its surrounding cities, Shenzhen had a low degree of industrial integration with Dongguan and Huizhou, whereas Shenzhen was highly integrated with Heyuan and Shanwei. Figure 5.4 shows that the trend lines of the modified Krugman's index for Shenzhen–Dongguan and Shenzhen–Huizhou were much higher than those of Shenzhen–Heyuan and Shenzhen–Shanwei from 2004 to 2015. This scenario reveals that industrial isomorphism emerged more seriously in

Table 5.2 SDH's degree of industrial structure similarity and modified Krugman's index of ten inter-city groups

Years	Inter-city groups										SDH
	SS	SH2	HS1	HH	DS	HS2	DH2	DH1	SD	SH1	
2004	0.28255	0.16630	0.37993	0.25932	0.36094	0.44921	0.24407	0.71302	0.70352	0.81157	0.52915
2005	0.35104	0.22369	0.44133	0.31138	0.44240	0.45743	0.33032	0.65746	0.66484	0.85087	0.55889
2006	0.38273	0.32356	0.46238	0.39845	0.47482	0.46550	0.43347	0.65455	0.64670	0.76755	0.58680
2007	0.44350	0.39778	0.51152	0.45314	0.54801	0.48400	0.53212	0.64076	0.65235	0.75632	0.62372
2008	0.38885	0.38602	0.46442	0.44489	0.52240	0.54529	0.53568	0.62911	0.64190	0.75412	0.60819
2009	0.36829	0.45135	0.44169	0.47520	0.54428	0.59027	0.59413	0.61054	0.64202	0.72019	0.60965
2010	0.35867	0.49238	0.44110	0.51256	0.53995	0.57303	0.59629	0.63329	0.65362	0.65167	0.61747
2011	0.39911	0.51108	0.42544	0.52093	0.50514	0.55591	0.58967	0.64053	0.66681	0.66136	0.61902
2012	0.39163	0.52880	0.40069	0.50981	0.50256	0.53037	0.60384	0.63956	0.71261	0.65779	0.62110
2013	0.41262	0.55041	0.42692	0.55415	0.51493	0.53791	0.61194	0.68530	0.74545	0.69418	0.64472
2014	0.40044	0.58177	0.44943	0.59010	0.49692	0.50252	0.62284	0.75318	0.75414	0.70118	0.65655
2015	0.34656	0.60300	0.45412	0.62990	0.48034	0.47951	0.64639	0.78500	0.76129	0.72076	0.66471
Mean	0.37716	0.43468	0.44158	0.47165	0.49439	0.51425	0.52840	0.67019	0.68710	0.72896	—

Note Shenzhen–Dongguan (SD), Shenzhen–Huizhou (SH1), Shenzhen–Shanwei (SS), Shenzhen–Heyuan (SH2), Dongguan–Huizhou (DH1), Dongguan–Shanwei (DS), Dongguan–Heyuan (DH2), Huizhou–Shanwei (HS1), Huizhou–Heyuan (HH) and Heyuan–Shanwei (HS2)
Source Author

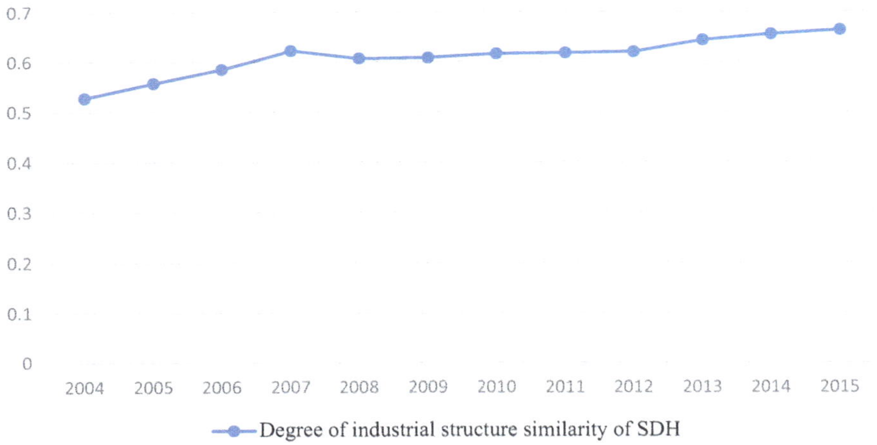

Fig. 5.3 Degree of Industrial structure similarity of SDH from 2004 to 2015. *Source* Author

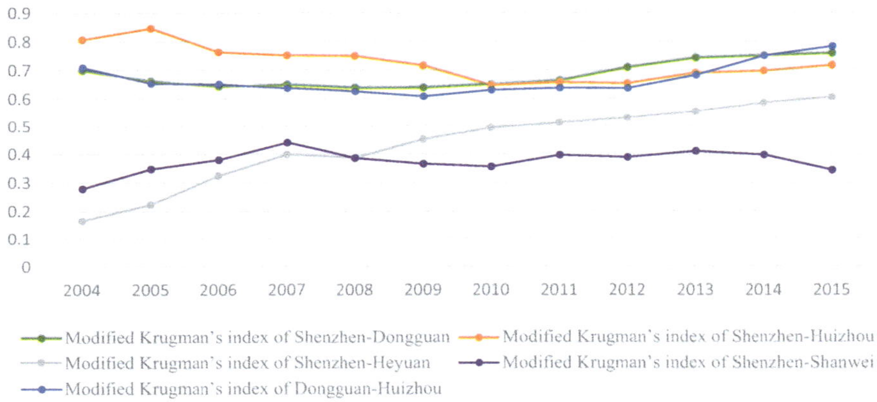

Fig. 5.4 Modified Krugman's index of the inter-city group from 2004 to 2015. *Source* Author

Shenzhen–Dongguan and Shenzhen–Huizhou than in Shenzhen–Heyuan and Shenzhen–Shanwei. In terms of comparison between the industrial integration of Shenzhen–Heyuan and that of Shenzhen–Shanwei, Fig. 5.4 shows that the industrial integration degree of Shenzhen–Heyuan was higher than that of Shenzhen–Shanwei before 2008. However, the degree of industrial integration of Shenzhen–Heyuan after 2008 was lower than that of Shenzhen–Shanwei and experienced a continual downward trend compared with the smooth trend of Shenzhen–Shanwei. Moreover, comparing the industrial integration of Shenzhen–Dongguan and that of Shenzhen–Huizhou, this book corroborates that the modified Krugman's index of Shenzhen–Huizhou was much higher than that of Shenzhen–Dongguan before 2010. The degree of industrial integration of Shenzhen–Huizhou was lower than that of Shenzhen–Dongguan. However, the modified Krugman's index of Shenzhen–Dongguan was more than that of Shenzhen–Huizhou after 2010, which corroborates that the industrial integration of Shenzhen–Dongguan worsened. A higher possibility for industrial cooperation existed between Shenzhen and Heyuan and Shenzhen and Shanwei than with the inter-city groups of Shenzhen–Dongguan and Shenzhen–Huizhou.

Second, in terms of Shenzhen–Dongguan–Huizhou, the degree of industrial integration among them stayed low. Figure 5.4 shows that the industrial integration of Dongguan–Huizhou maintained a steady trend from 2004 to 2012. However, it became worse after 2012, and the modified Krugman's index rapidly rose from 0.63956 in 2012 to 0.78500 in 2015. Hence, severe industrial isomorphism in the three cities existed and became serious in recent years. This phenomenon indicates that underlying industry competition existed among the three cities, thereby challenging the economic integration of SDH.

In sum, in terms of the overall SDH, the degree of industrial integration of SDH from 2004 to 2015 became increasingly weak and manifested in three stages. In terms of inter-city groups, Shenzhen had a high degree of industrial integration with Heyuan and Shanwei and a low degree of industrial integration with Dongguan and Huizhou. In addition, the degrees of industrial integration among Shenzhen–Dongguan–Huizhou remained in a low level from 2004 to 2015. Thus, in terms of production, the homogeneity of the industrial structure of SDH gradually increased from 2004 to 2015, especially after 2012. As for the internal environment, Shenzhen had a high demand for industrial cooperation with Heyuan and Shanwei. Moreover, serious industrial isomorphism emerged among Shenzhen, Dongguan, and Huizhou, and it further worsened in recent years.

5.2.3 Assessing Infrastructure Integration of SDH

To assess infrastructure integration, this book focuses on two essential perspectives, namely, (1) infrastructure equality, and (2) infrastructure connectivity. Infrastructure equality primarily examines the distribution of infrastructure in different jurisdictions based on the initiatives of the construction of regional infrastructure networks. This book has employed the coefficient of variation to measure the degree of infrastructure integration on the overall SDH scale and the degree of infrastructure development similarity to assess the degree of infrastructure integration on the inter-city group scale. According to the formulas presented in Chap. 4, if the values of the coefficient of the variation are high, the equalities of infrastructure integration of overall SDH are low, *and* vice versa. If the degree of infrastructure development similarity is high, then the equality of infrastructure integration for inter-city groups is high, *and* vice versa. Infrastructure connectivity is developed to measure the connectivity of transportation infrastructure in the building process of the regional infrastructure network. The infrastructure connectivity index was formulated to measure the connectivity at the overall SDH and inter-city group scales. Through the computation of the formula (see Chap. 4), if the value of the infrastructure connectivity index is high, then the degree of infrastructure integration is high as well, *and* vice versa.

5.2.3.1 Assessing Infrastructure Integration of SDH in Terms of Equality

This book confirms the specific weight of the indictors through the expert scoring method and analytic hierarchy process (Table 5.3). Table 5.4 shows the coefficient of variation of SDH and degree of infrastructure development similarity among ten inter-city groups according to the analysis of the formulas.

Infrastructure Integration at the SDH Scale

The trend of the coefficient of variation for the SDH scale from 2004 to 2015 (Fig. 5.5) is identified by calculating the coefficient of variation (Table 5.4).

According to Fig. 5.5, the overall degree of infrastructure integration of SDH increasingly weakened from 2004 to 2015. The coefficient of the variation of SDH experienced steady growth from 0.67025 in 2004 to 0.92892 in 2015, which validates that the infrastructure integration increasingly weakened from 2004 to 2015 and infrastructure development was divergent across the city-region. However, the coefficient of the variation of SDH exhibited a slight decrease in 2007 and 2014. The

Table 5.3 Indictors to assess the infrastructure integration in terms of equality

Domain	Weight	Indictor	Weight	Type	Unit
Road and transportation	0.33	Density of Expressways and Class I to IV Highways Network	0.27	Derive	1/km
		Road Lambs Per Kilometer	0.29	Derive	1/km
		Ratio of Road Area	0.54	Derive	%
Telecommunication service	0.32	Internet User Per 100 Inhabitants	0.43	Basic	Per 100 inhabitants
		Subscribers of Mobile Telephones Per 100 Inhabitants	0.38	Basic	Per 100 inhabitants
		Number of Post Offices at Per 10 000 Inhabitants	0.19	Basic	Per 10 000 inhabitants
Ecological environment	0.27	Ratio of Sewage Treatment	0.38	Basic	%
		Ratio of Waste Water and Consumption	0.44	Basic	%
		Ratio of Industrial Solid Wastes Utilized	0.18	Basic	%
Water supply	0.08	Ratio of Urban Water Supply	1	Basic	%

Source Author

Table 5.4 Coefficient of variation of overall SDH and degree of infrastructure development similarity of ten inter-city groups

Years	Inter-city groups										SDH
	HS2	SH1	SD	HH	SH2	DH1	HS1	DH2	SS	DS	
2004	0.65839	0.66897	0.62378	0.59281	0.60841	0.69352	0.48769	0.50823	0.46553	0.39279	0.67025
2005	0.62751	0.59169	0.50735	0.50541	0.57484	0.49414	0.55286	0.44127	0.40077	0.33385	0.77261
2006	0.59719	0.58825	0.61962	0.47681	0.51421	0.51658	0.49546	0.45961	0.34911	0.30298	0.78468
2007	0.61253	0.58694	0.56449	0.59060	0.57269	0.53244	0.52687	0.32255	0.36619	0.32798	0.76454
2008	0.54833	0.53643	0.60019	0.55147	0.59044	0.56684	0.47592	0.37126	0.38157	0.25582	0.79792
2009	0.59678	0.59668	0.64750	0.54272	0.54826	0.57816	0.54573	0.28925	0.31959	0.25559	0.82176
2010	0.60034	0.58418	0.57666	0.62085	0.60995	0.46297	0.43204	0.45585	0.31401	0.29036	0.82996
2011	0.71322	0.56766	0.48325	0.64863	0.53367	0.36061	0.46663	0.35451	0.38719	0.44198	0.84122
2012	0.65410	0.70629	0.50701	0.55074	0.52624	0.30354	0.52908	0.30447	0.37685	0.31774	0.87188
2013	0.65594	0.58323	0.47097	0.52704	0.55672	0.53826	0.36462	0.52222	0.42049	0.37266	0.90067
2014	0.54714	0.59283	0.57629	0.57414	0.52201	0.49018	0.37331	0.50641	0.38582	0.45989	0.83480
2015	0.59054	0.75287	0.54083	0.50799	0.41221	0.54673	0.50263	0.50932	0.35308	0.36896	0.92892
Mean	0.61683	0.61300	0.55983	0.55743	0.54747	0.50700	0.47940	0.42041	0.37668	0.34339	–

Note Shenzhen–Dongguan (SD), Shenzhen–Huizhou (SH1), Shenzhen–Shanwei (SS), Shenzhen–Heyuan (SH2), Dongguan–Huizhou (DH1), Dongguan–Shanwei (DS), Dongguan–Heyuan (DH2), Huizhou–Shanwei (HS1), Huizhou–Heyuan (HH) and Heyuan–Shanwei (HS2)
Source Author

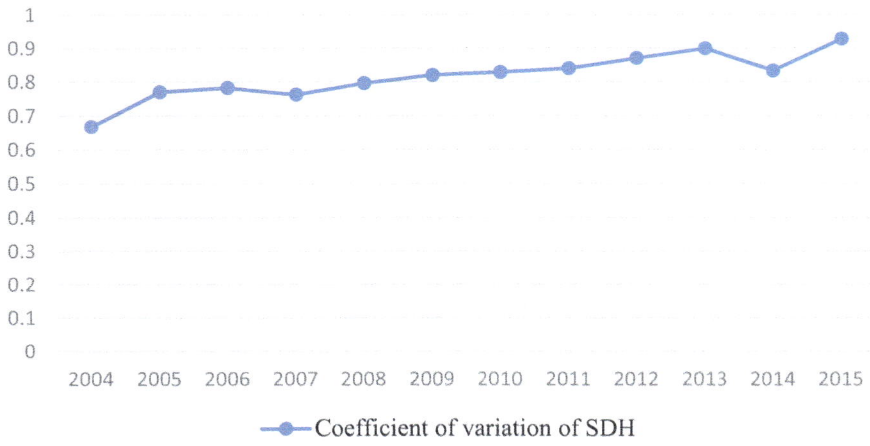

Fig. 5.5 Coefficient of variation of SDH from 2004 to 2015. *Source* Author

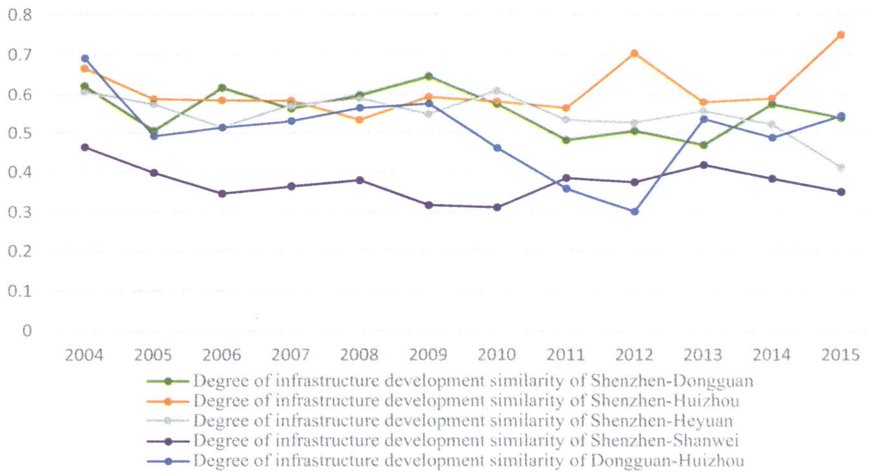

Fig. 5.6 Degree of infrastructure development similarity of inter-city group from 2004 to 2015. *Source* Author

gap of the developmental level of infrastructure in different cities gradually widened and developed toward an unbalanced regional infrastructure network.

Infrastructure Integration at the Inter-City Group Scale

This book formulates the trends in the degree of infrastructure developmental similarity on the inter-city group scale from 2004 to 2015 (Fig. 5.6).

In rankings of the mean of the degree of infrastructure development similarity among ten inter-city groups (Table 5.4), the groups of Shenzhen–Dongguan, Shenzhen–Huizhou, Shenzhen–Heyuan, Shenzhen–Shanwei, and Dongguan–Huizhou stayed at 3rd, 2nd, 5th, 9th, and 6th positions, respectively, within the ten inter-city groups. Thus, there are two fundamental features of infrastructure integration.

First, in terms of Shenzhen with its surrounding cities, Shenzhen with Dongguan and Huizhou had a high degree of infrastructure integration, whereas Shenzhen had a medium degree of infrastructure integration with Heyuan and a low degree of infrastructure integration with Shanwei. Figure 5.6 shows that the infrastructure developmental similarity degree of Shenzhen–Dongguan, Shenzhen–Huizhou, and Shenzhen–Heyuan experienced slight fluctuations from 2004 to 2015. From 2013 to 2015, only the infrastructure integration degree of Shenzhen–Huizhou witnessed an upward tendency, which verified that the infrastructure integration of Shenzhen–Huizhou improved more than Shenzhen–Dongguan and Shenzhen–Heyuan in recent years. The infrastructure integration of Shenzhen–Shanwei has always remained at a low degree with some slight fluctuations. The infrastructure integration of Shenzhen–Shanwei showed a decline from 2013 to 2015. Compared with those of other cities, the developmental levels of the infrastructure of Huizhou and Dongguan were closer to those of Shenzhen. However, Shanwei was behind Shenzhen in terms of the developmental levels of infrastructure.

Second, in terms of Shenzhen–Dongguan–Huizhou, the degrees of the infrastructure integration between Shenzhen and Dongguan and Shenzhen and Huizhou were high. The infrastructure integration degree between Dongguan and Huizhou was of a medium level within the ten inter-city groups. In a ranking of the means of the degree of infrastructure development similarity, the infrastructure integration of Dongguan–Huizhou stayed at a medium stage during the 12 years. The degree of infrastructure developmental similarity of Dongguan–Huizhou rapidly declined from 0.57816 in 2009 to 0.30354 in 2012, which suggests that the infrastructure integration worsened from 2009 to 2012. Shenzhen, Dongguan, and Huizhou had relatively similar developmental levels of infrastructure, providing a solid foundation for local socioeconomic development.

In sum, in terms of the overall SDH, the degree of infrastructure integration increasingly enhanced from 2004 to 2015. In terms of inter-city groups, Shenzhen had a high degree of infrastructure integration with Dongguan and Huizhou, a medium degree with Heyuan, and a low degree with Shanwei. In addition, although Shenzhen integrated well with Dongguan and Huizhou, the infrastructure integration between Dongguan and Huizhou was of a medium level only. Therefore, Shanwei and Heyuan were the backward regions in regional infrastructure development and therefore less integrated into the networks, especially Shanwei. These weak places for infrastructure or facilities posed an obstacle to socioeconomic integration and accelerate social inequality.

5.2.3.2 Assessing Infrastructure Integration of SDH in Terms of Connectivity

On the basis of Chap. 4, this book selects three indicators, namely, the number of inter-city railways, expressways, and national highways, to gauge the connectivity of transportation infrastructure. Subsequently, to systemically assess the connectivity, this book assigns weightings of 1, 2, and 1 to the number of inter-city railways, expressways, and national highways, respectively, through the expert scoring method. Table 5.5 shows the infrastructure connectivity index of SDH and the ten inter-city groups in accordance with the analysis of the formula (see Chap. 4).

Infrastructure Integration at the SDH Scale

Table 5.5 and Fig. 5.7 reveal the trends of infrastructure connectivity in the overall SDH region in the period between 2004 and 2015.

According to Fig. 5.7, the overall degree of infrastructure integration of SDH strengthened from 2004 to 2015. The infrastructure connectivity index of SDH experienced steady growth from 12 in 2004 to 29 in 2015, which confirms that the infrastructure integration increasingly strengthened, enhancing connectivity. After 2009,

Table 5.5 Infrastructure connectivity index of overall SDH and ten inter-city groups

Years	Inter-city groups										SDH
	SD	HH	SH1	HS1	SH2	HS2	DH1	SS	DH2	DS	
2004	5	4	2	1	1	1	0	0	0	0	12
2005	5	4	2	1	1	1	0	0	0	0	12
2006	5	4	2	1	1	1	0	0	0	0	12
2007	8	4	2	1	1	1	0	0	0	0	15
2008	8	4	2	1	1	1	0	0	0	0	15
2009	8	4	2	1	1	1	0	0	0	0	15
2010	8	4	4	1	1	1	0	0	0	0	17
2011	9	4	4	1	1	1	0	0	0	0	18
2012	11	6	4	1	1	1	0	0	0	0	22
2013	15	6	7	2	1	1	2	1	0	0	27
2014	15	6	7	2	1	1	2	1	0	0	27
2015	15	8	9	2	3	1	2	1	0	0	29
Mean	9.33	4.83	3.92	1.25	1.17	1.00	0.50	0.25	0.00	0.00	–

Note Shenzhen–Dongguan (SD), Shenzhen–Huizhou (SH1), Shenzhen–Shanwei (SS), Shenzhen–Heyuan (SH2), Dongguan–Huizhou (DH1), Dongguan–Shanwei (DS), Dongguan–Heyuan (DH2), Huizhou–Shanwei (HS1), Huizhou–Heyuan (HH) and Heyuan–Shanwei (HS2)
Source Author

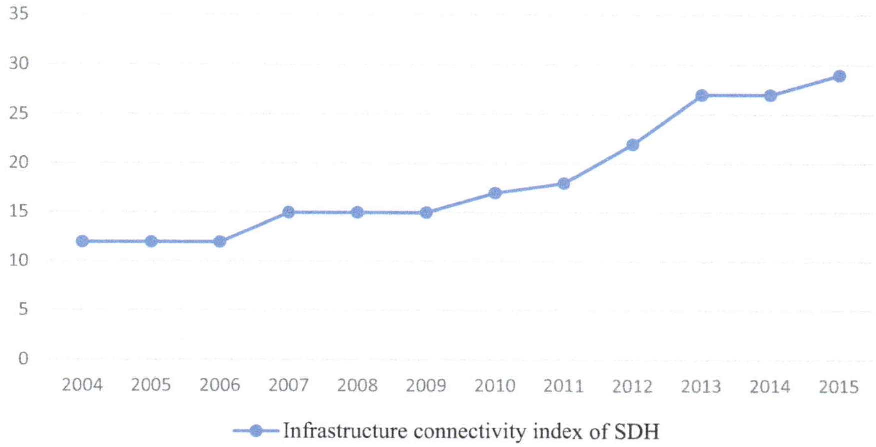

Fig. 5.7 Infrastructure connectivity index of SDH from 2004 to 2015. *Source* Author

the infrastructure connectivity dramatically increased from 15 in 2009 to 29 in 2015, especially during 2011 to 2013, increasing from 18 to 27.

Empirically speaking, the transportation infrastructure offered advantageous preconditions for economic integration and enhancement of the economic competitiveness of SDH. According to the above analysis, the connectivity clearly strengthened after 2009, accelerating the flow of goods and the efficient utilization of factors of production across jurisdictions in SDH.

Infrastructure Integration at the Inter-City Group Scale

Figure 5.8 reveals the historical changes of the infrastructure connectivity index at the inter-city group scale from 2004 to 2015.

In a ranking of the means of the infrastructure connectivity index for different inter-city groups (Table 5.4), Shenzhen–Dongguan, Shenzhen–Huizhou, Shenzhen–Heyuan, Shenzhen–Shanwei, and Dongguan–Huizhou scored positions 1, 3, 5, 8, and 7, respectively, within the ten inter-city groups. Two fundamental features of the infrastructure integration are identified.

First, in terms of Shenzhen with its surrounding cities, Shenzhen with Dongguan and Huizhou had a high degree of infrastructure integration, whereas Shenzhen has a medium degree of infrastructure integration with Heyuan and a low degree of infrastructure integration with Shanwei. The infrastructure integration of Shenzhen–Dongguan ranked first in the ten inter-city groups and stayed at a high level (15) in three recent years. Although the infrastructure integration of Shenzhen–Huizhou ranked third and witnessed rapid growth after 2009, a wide gap remained with Shenzhen–Dongguan. The infrastructure integration of Shenzhen–Heyuan kept the same degree from 2004 to 2014 and increased only to 3 in 2015. The infrastructure integration of

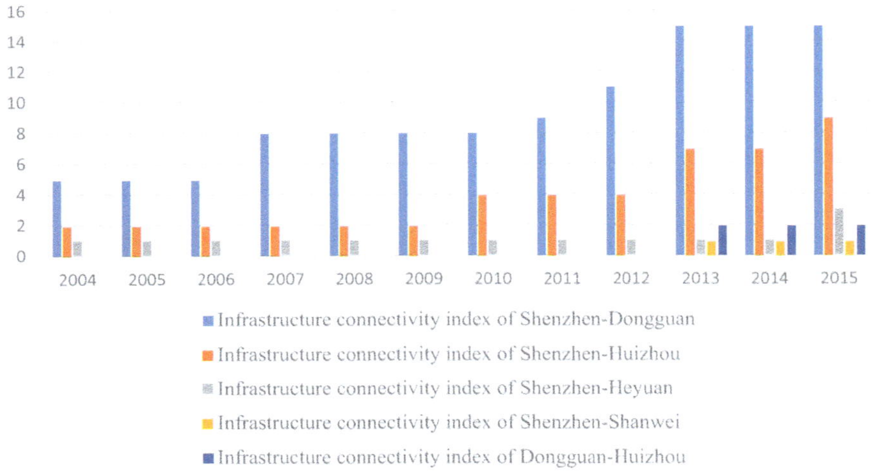

Fig. 5.8 Infrastructure connectivity index of the inter-city group from 2004 to 2015. *Source* Author

Shenzhen–Shanwei stayed at a low stage from 2004 to 2015. Thus, the transportation linkages between Shenzhen and Dongguan and Shenzhen and Huizhou were tight, providing sophisticated facilities to accelerate the flow of goods and factors of production. The outlying cities, especially Shanwei, had poor transportation linkages with Shenzhen, leaving a great challenge for the industrial collaboration between Shenzhen and Shanwei.

Second, in terms of Shenzhen–Dongguan–Huizhou, the degrees of the infrastructure integration between Shenzhen and Dongguan and Shenzhen and Huizhou were high. The infrastructure integration degree between Dongguan and Huizhou was equivalent to a medium level within the ten inter-city groups. According to the rankings of the mean of the infrastructure connectivity index, the infrastructure integration of Dongguan–Huizhou stayed at a medium stage during the 12 years. The infrastructure connectivity index of Dongguan–Huizhou stayed at 0 from 2004 to 2012. However, it increased to 3 after 2012 and remained unchanged from 2013 to 2015. Thus, to build the regional infrastructure network, the transportation linkage between Dongguan and Huizhou must be reinforced to enable market integration and industrial collaboration.

In sum, in terms of the overall SDH, the degree of infrastructure integration of SDH increasingly strengthened from 2004 to 2015, especially after 2009. In terms of inter-city groups, Shenzhen had a high infrastructure integration degree with Dongguan and Huizhou, a medium degree with Heyuan, and a low degree with Shanwei. In addition, although Shenzhen integrated well with Dongguan and Huizhou, the infrastructure integration between Dongguan and Huizhou was only medium. Given that infrastructure connectivity between cities served as the precondition and foundation for productive economic integration, the backward cities should be connected with Shenzhen through enhancing the transportation linkages, which will contribute to industrial collaboration and market integration in coping with the crises of market cycles.

5.2.4 Summary

This section aims to answer the first research question, namely, to what extent does SDH operate as an economically integrated entity as reflected in price similarity, industrial structure similarity, and infrastructure development?

This section argues that the economic integration of SDH should be a well-rounded process that includes three aspects: (1) market, (2) industrial, and (3) infrastructure integration. On the basis of this recognition, this book examines these three aspects from two levels; one is the overall SDH scale, and the other is the inter-city group scale.

5.2.4.1 Economic Integration at the SDH Scale

Figure 5.9 shows the economic integration on the overall SDH scale from 2004 to 2015, in terms of market, industrial, and infrastructure integrations.

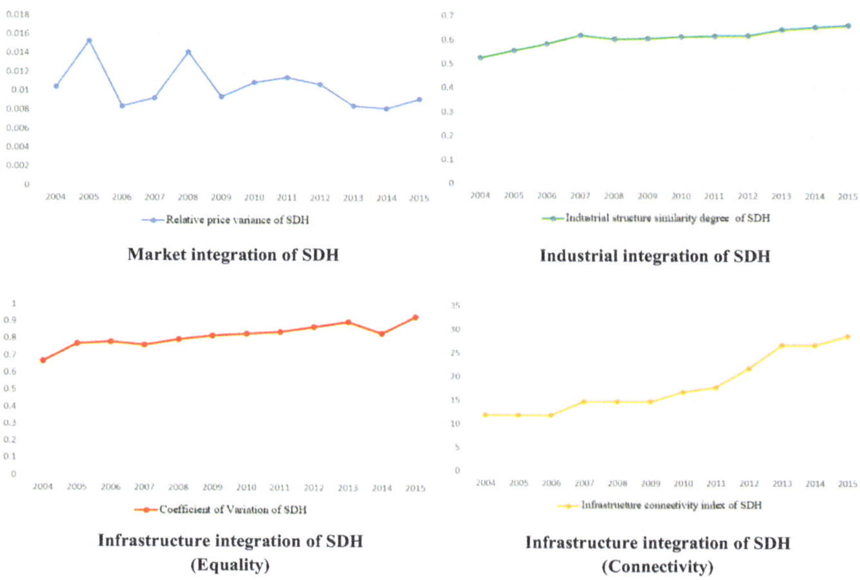

Fig. 5.9 Economic integration at the overall SDH scale from 2004 to 2015. *Source* Author

- *Market integration*—it experienced a dramatic fluctuation before 2009, manifesting as rather unstable market integration. However, it witnessed a slight increase and remained relatively stable after 2009.
- *Industrial integration*—it experienced a steady weakening trend from 2004 to 2015 and is divided into three stages according to its range.
- *Infrastructure integration (equality and connectivity)*—the equality of infrastructure integration weakened from 2004 to 2015, while the connectivity of infrastructure increasingly strengthened from 2004 to 2015, especially after 2009.

According to Pinder (1969), the nature of economic integration is the free movement of commodities and the efficient utilization of factors of production in an integrated region. Market integration can be considered the outcome and final state of economic integration with particular reference to the market of consumer goods. Infrastructure and industrial integrations are the foundation and precondition for market integration. From the perspective of the overall SDH, the development of infrastructure and a well-connected infrastructure system acted as efficacious driving forces for market integration to reduce the transportation cost for the flow of consumer goods, especially after 2009. However, the intensified industrial isomorphism and this worsening industrial integration are underlying obstacles to the economic integration of SDH. After 2012, industrial isomorphism became increasingly severe, which means that the industrial competition between cities accentuates and may further hinder the flow of goods and efficient utilization of factors of production in SDH.

5.2.4.2 Economic Integration at the Inter-City Group Scale

Table 5.6 reveals the economic integration on the inter-city group scale from 2004 to 2015 in terms of market, industrial, and infrastructure integrations.

Table 5.6 Economic integration at the inter-city group scale from 2004 to 2015

	Market integration	Industrial integration	Infrastructure integration	
			Equality	Connectivity
Shenzhen–Dongguan	High	Low	High	High
Shenzhen–Huizhou	High	Low	High	High
Dongguan–Huizhou	Medium	Low	Medium	Medium
Shenzhen–Heyuan	Medium	High	Medium	Medium
Shenzhen–Shanwei	High	High	Low	Low

Source Author

Regarding Shenzhen with its surrounding cities, three majoring findings are identified.

- *Market integration*—although Shenzhen had a high integrative degree with Dongguan, Huizhou, and Shanwei, the level of market integration between Shenzhen and Heyuan was medium.
- *Industrial integration*—Shenzhen had a low degree of industrial integration with Dongguan and Huizhou, whereas Shenzhen had a high degree with Shanwei and Heyuan.
- *Infrastructure integration (equality and connectivity)*—Shenzhen had a high integrative degree with Dongguan and Huizhou, a medium degree with Heyuan, and a low degree with Shanwei.

On the basis of the abovementioned analysis, Shenzhen–Dongguan and Shenzhen–Huizhou had advantageous infrastructure conditions and foundations, and the local government's endeavors to build regional infrastructure networks brought positive effects for market integration. However, the severe industrial isomorphism between Shenzhen–Dongguan and Shenzhen–Huizhou provides uncertainty for creating a regional economic entity, manifesting in vicious industrial competition. As for Shenzhen–Heyuan and Shenzhen–Shanwei, the high level of industrial integration indicates that the differences in the industrial structure were apparent, providing an opportunity for the industrial transfer from Shenzhen and the economic revitalization in Shanwei and Heyuan. The relatively high levels of market integration may have been partially triggered by this industrial complementarity. However, the poor infrastructure conditions, especially between Shenzhen and Shanwei, pose an obstacle for industrial collaboration, the flow of goods, and sharing of factors of production in the SDH region.

Regarding Shenzhen–Dongguan–Huizhou, three conclusions are reported.

- *Market integration*—although Shenzhen had a high degree of integration with Dongguan and Huizhou, the integration between Dongguan and Huizhou was kept at a medium level.
- *Industrial integration*—the degrees among Shenzhen, Dongguan, and Huizhou stayed at a low stage.
- *Infrastructure integration (equality and connectivity)*—Shenzhen had a high degree of integration with Dongguan and Huizhou, whereas the integration between Dongguan and Huizhou was kept at a medium level.

Compared with the economic integration of Shenzhen–Dongguan and Shenzhen–Huizhou, the economic integration of Dongguan still needed to be enhanced, especially by seeking differentiated development of the industrial sector. Moreover, considerable attention should be paid to infrastructure development to maintain connectivity and strengthen equality.

5.3 Assessing Institutional Integration of SDH

5.3.1 Assessing Institutional Integration of SDH from the Inter-scalar Relation

This book examines the inter-scalar relations in the SDH integrating process through investigating the administrative setting and the readjustment of administrative divisions after 2000.

5.3.1.1 Administrative Setting of SDH

In terms of the administrative setting of SDH, this region contains one vice-provincial city (Shenzhen) that is under the jurisdiction of the central government and the Guangdong Provincial Government, four prefecture-level cities (Dongguan, Huizhou, Heyuan, and Shanwei) that are directly under the Guangdong Provincial Government, ten counties, one county-level city, and twelve districts ranked at the same administrative level as counties and county-level cities. Below this administrative level are many towns, townships, and streets, changing rapidly in the wake of urbanization. It needs to be pointed out that there is no formal regional authority to steer and promote the process of SDH integration (Fig. 5.10).

In addition to the formal administrative setting, some informal administrative innovations have also occurred extensively in SDH. First, to facilitate the process of personnel appointment and removal, SSSCZ was endowed with the administrative power of a prefecture-level city from the Guangdong Provincial Government. Hence, although SSSCZ is a cross-boundary economic cooperation zone, it shared equal administrative power with prefecture-level cities when it conducted the activities of personnel appointment and removal. However, such administrative empowerment primarily focused on the field of personnel appointment and removal and other managerial issues. Thus, SSSCZ was not strictly defined as an independent administrative unit. Second, under the approval from SC, Shenzhen set up certain new districts from 2007 and retains two new districts at present, namely, Guangming New District and Dapeng New District. These new districts were also named "functional districts (*Gongnengqu*, 功能区)" and functioned as economic units, not administrative units, but had partial autonomous power. Third, Dongguan reduced the administrative level of the county to establish a tailor-made administrative mode (i.e., municipality-town system). The main purpose of such an institutional configuration was to enhance administrative efficiency and the sustainability of local socio-economic growth.

In sum, formal administrative set-ups and informal institutional innovations co-exist and intertwine in SDH. These two co-constitutive strands exhibit a certain degree of flexibility in jurisdictional arrangement *vis-à-vis* other larger territories.

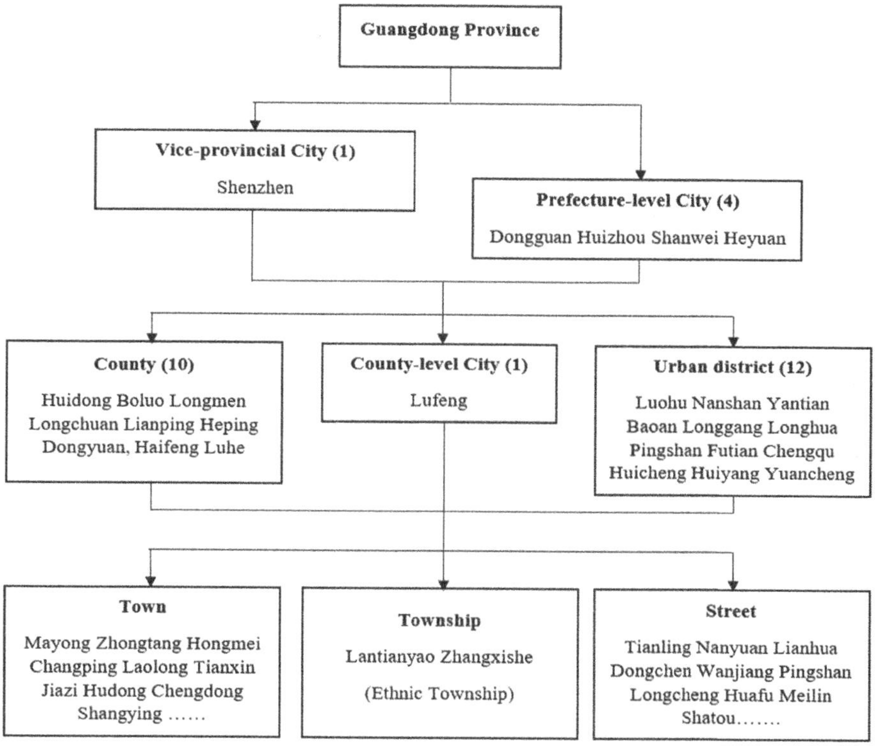

Fig. 5.10 Administrative setting of SDH in 2016. *Source* Author

5.3.1.2 Adjustment of the Administrative Division of SDH After 2000

The primary adjustments of the administrative division are as follows:

- In 2003, Huiyang was converted from a county-level city to an urban district of Huizhou City.
- In 2007, the Guangming New District was established, administered by the Bao'an urban district in Shenzhen.
- In 2009, the Pingshan New District was established, and administered by the Longgang urban district in Shenzhen.
- In 2011, the Longhua and Dapeng New Districts were established and administrated by the Bao'an and Longgang districts, respectively. Moreover, SSSCZ was set up with a prefecture-level administrative status.
- In 2016, the Longhua New District and Pingshan New District were converted from new districts to urban districts,

The administrative division of SDH has remained relatively stable since 2000. Nevertheless, a tumultuous wave of converting a town to a street (*Zhengaijiedao*, 镇改

街道), initiated by the Guangdong Provincial Government, has occurred extensively in the wake of the rapid work to accelerate urbanization and repair the urban–rural fragmentation. The informal institutional tools, such as the setup of new districts and SSSCZ, have been deployed to coordinate fast-tracked economic development and formulaic administrative systems.

In sum, from the abovementioned adjustment of the administrative division in terms of formal and informal institutional settings, this section assesses the inter-scalar relations. Regarding central–local relations, existing formal administrative practices reveal that the central government kept intense control of SDH administratively, and the central government's reaction lagged far behind the local socio-economic development. The restructuring of urban administrative space, i.e., the adjustment of administrative divisions, was a significant institutional strategy in maintaining the power of the central government in localities. According to Ma (2005), within the national state apparatus, the administrative level is bound to urban capabilities for resource acquisition. Ma (2005) affirms: "generally, the higher the level, the more beneficial it would be for its economic development because of the greater administrative power and fiscal resources it will have" (p. 485). However, the central government monopolized power for the majority of the restructuring of the administrative space, except for converting towns to streets, and such a power configuration enabled the state authoritarianism inherited from the Mao era.

However, the delayed action of the central government in response to the local socioeconomic development has been repaired by certain institutional innovations co-produced by the urban and provincial governments. The establishment of SSSCZ, a new district in Shenzhen, and the municipality–town system in Dongguan can be viewed as policy experiments to overcome the crisis of the central government-induced administrative setups under the rapid urban socioeconomic development. Owing to the great economic autonomy of Shenzhen as a "Special Economic Zones" empowered by the central government, the institutional innovations of Shenzhen were widely deployed, manifesting as the decentralization of power to encourage local innovations and compensating for the deficiency of a rigidified administrative system in China. Regarding provincial–municipal relations, the massive converting of towns to streets signifies that the provincial government offered enormous administrative support and development opportunities for local governments to enable socioeconomic development and city-regionalism.

In terms of regional authority, no formal regional authority existed in SDH from 2000 to 2016. The lack of a formal regional authority signals that the tiers of jurisdictional arrangement in SDH were not literally marked as an MLG paradigm.

5.3.2 Assessing Institutional Integration of SDH from the Inter-City Relation

To systemically probe the inter-city relationship in the process of integrating SDH, this book employs the inter-city cooperation news from 2004 to 2015, which is extracted from the governmental official websites and search engines of *Google* and *Baidu* to comprehensively investigate the spatiotemporal layout, inter-city cooperation area, and inter-city cooperation mode (see Chap. 4).

5.3.2.1 Spatiotemporal Layout of Inter-City Cooperation

On the basis of the analysis of inter-city cooperation news, the inter-city relations of SDH are presented in Table 5.7, Figs. 5.11 and 5.12.

Several findings are produced by analyzing the number of inter-city cooperation news items in different time periods. First, from 2004 to 2015, Shenzhen was intensively linked with other cities to serve as a regional core city. The numbers of inter-city cooperation news items for Shenzhen–Heyuan, Shenzhen–Shanwei, Shenzhen–Dongguan, and Shenzhen–Huizhou were 259, 220, 201, and 95, respectively, which were far greater than those for the other inter-city groups in SDH.

Table 5.7 Number of inter-city cooperation of SDH from 2004 to 2015

	2004 and 2005	2006 and 2007	2008 and 2009	2010 and 2011	2012 and 2013	2014–2015	Total
SH2	15	44	57	69	17	57	259
SS	0	2	23	25	30	140	220
SD	2	15	46	50	35	53	201
SH1	1	7	10	24	28	25	95
DH1	0	3	11	23	14	18	69
HH	0	4	9	11	8	10	42
DH2	1	12	12	2	5	10	42
DS	1	2	8	3	3	17	34
HS1	0	1	4	5	11	11	32
HS2	0	1	5	3	2	10	21
Total	20	91	185	215	153	351	1015

Note Shenzhen–Dongguan (SD), Shenzhen–Huizhou (SH1), Shenzhen–Shanwei (SS), Shenzhen–Heyuan (SH2), Dongguan–Huizhou (DH1), Dongguan–Shanwei (DS), Dongguan–Heyuan (DH2), Huizhou–Shanwei (HS1), Huizhou–Heyuan (HH) and Heyuan–Shanwei (HS2)
Source Author

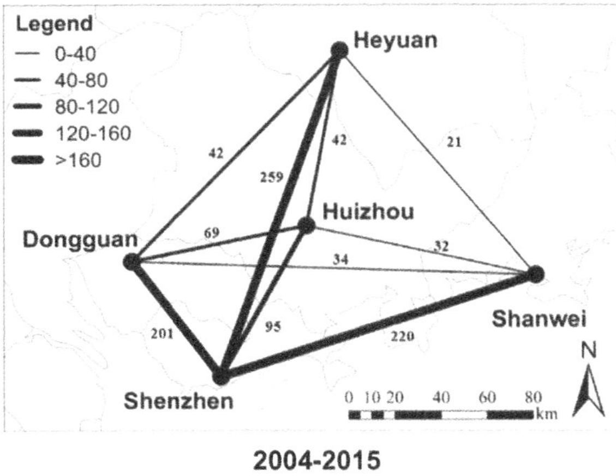

Fig. 5.11 Spatiotemporal evolution of inter-city cooperation (2004 to 2015). *Source* Author

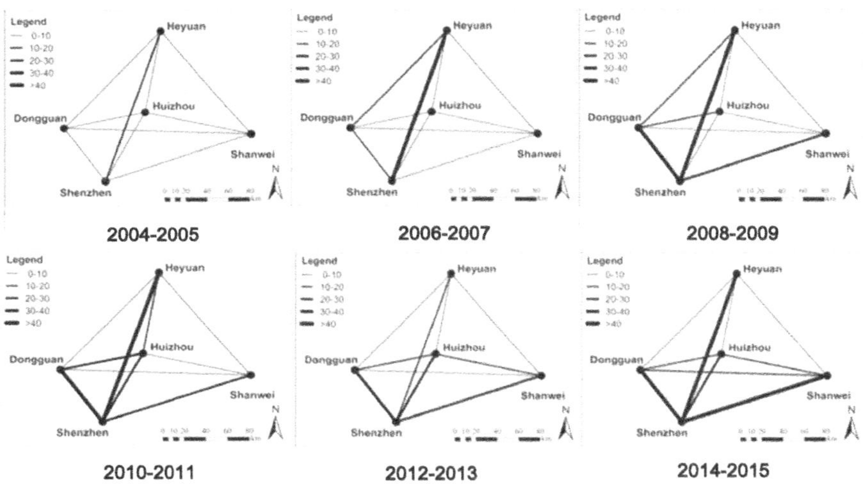

Fig. 5.12 Spatiotemporal evolution of inter-city cooperation in different periods. *Source* Author

Second, there were 59 inter-city cooperation news items between Shenzhen and Heyuan before 2008. Dongguan, Shanwei, and Huizhou kept having merely weakened connections with Shenzhen, reaching 17, 2, and 8 items, respectively *vis-à-vis* the high linkage of Shenzhen–Heyuan.

Third, after 2008, Shenzhen–Heyuan, Shenzhen–Shanwei, Shenzhen–Dongguan, and Shenzhen–Huizhou all experienced rapid growth, especially Shenzhen–Dongguan and Shenzhen–Shanwei.

Fourth, Shenzhen–Shanwei experienced a dramatic increase after 2014 from 30 to 140 items.

5.3.2.2 Inter-City Cooperation Areas

This book categorizes the fields of inter-city cooperation into five themes: comprehensive issues, economic development, institutional design, secondary function, and aid activity. As illustrated in Chap. 4, comprehensive issues refer to the development of comprehensive and multifarious frameworks to steer inter-city cooperation. Economic development in the inter-city cooperation for urban and regional economic growth includes industrial development and infrastructure construction. Institutional design denotes the inter-city cooperation that is designed to formulate integrative institutions, such as urban management, urban planning, and institutional construction. The secondary function is cooperation for facilitating social development, such as science and education, resources, and the environment. Aid activity is a special inter-city assistance activity driven by the strategy of "The Revitalization and Development of Eastern, Western, and Northern Guangdong". The Guangdong Provincial Government experimented with the pair-city strategy by selecting advanced cities to assist the backward cities.

On the basis of the analysis of inter-city cooperation news items, the cooperation areas of SDH and four main inter-city groups are summarized as follows (Figs. 5.13 and 5.14).

First, the primary cooperation area of SDH from 2004 to 2015 focused on economic development, institutional design, and aid activity. Specifically, industrial

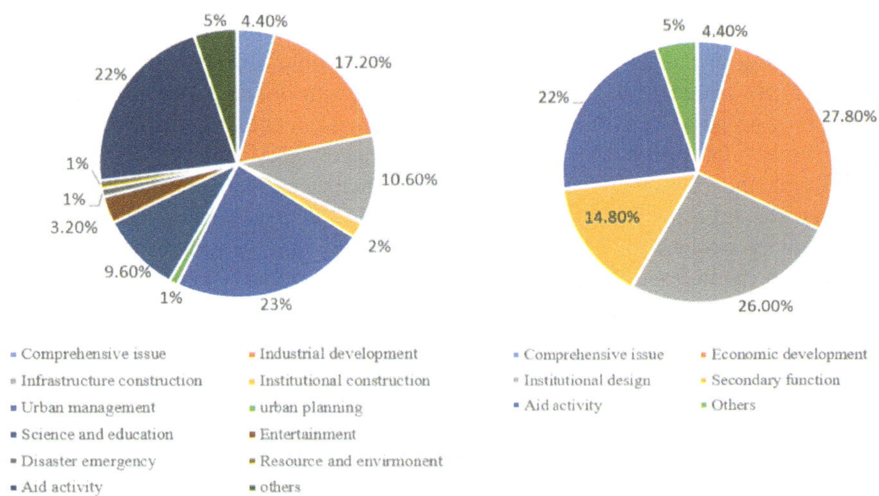

Fig. 5.13 Inter-city cooperation area of SDH from 2004 to 2015. *Source* Author

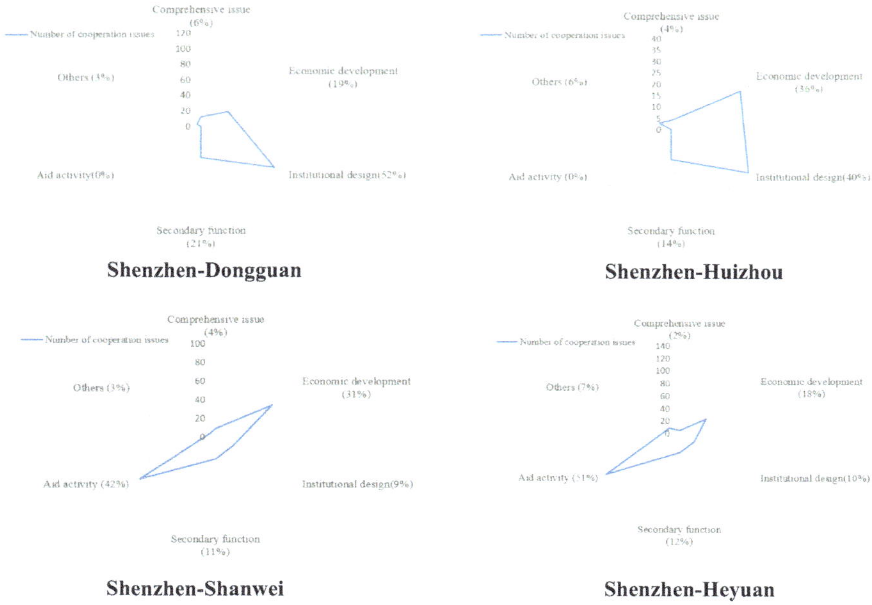

Fig. 5.14 Cooperation area of four inter-city groups from 2004 to 2015. *Source* Author

development, urban management, and aid activity play a dominated role in all the themes of inter-city cooperation, accounting for 17.2%, 23%, and 22%, respectively. Thus, the majority of inter-city cooperation in SDH undertook collaborative activities for economic growth, formulating institutional frameworks to enable integration and inter-city socio-economic assistance. Compared with the voluntary process of city grouping in the EU that is functioning for multiple objectives (see Chap. 2), the inter-city cooperation in SDH remained a growth-oriented and economy-first integrating process. Activities for social, environmental, cultural, and technological developments have been downplayed by policymakers and local officials.

Second, the primary cooperation area of Shenzhen–Shanwei and Shenzhen–Heyuan was aid activity, accounting for 42% and 51% of news items, respectively. The primary cooperation area for Shenzhen–Dongguan and Shenzhen–Huizhou focused on institutional design and economic development. Therefore, Shenzhen acted as a regional economic core, and the city grouping process centered on Shenzhen was driven by a growth-first and economy-oriented ideology that excluded social development.

5.3.2.3 Inter-City Cooperation Modes

The collected data on inter-city cooperation shed light on the cooperation mode, which is a crucial standard to assess the stability of inter-city cooperation. Aside from

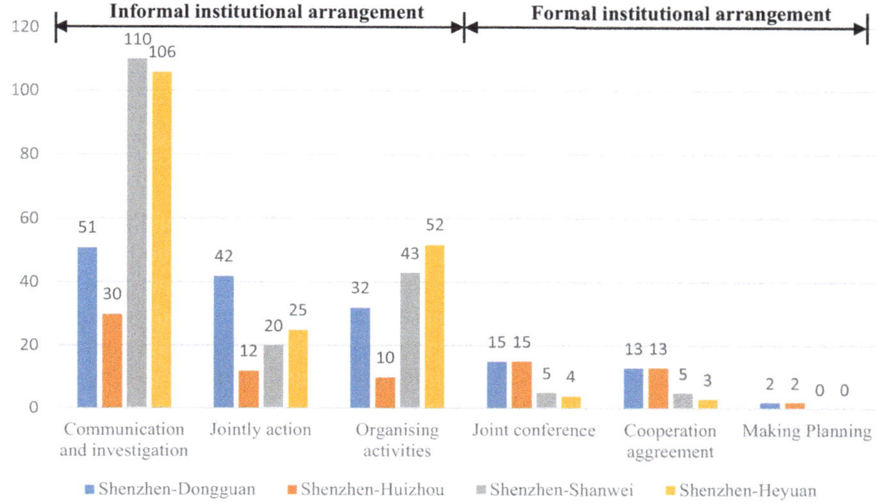

Fig. 5.15 Cooperation mode of four inter-city groups from 2004 to 2015. *Source* Author

the news on the consequences of cooperation, such as the completion of inter-city railways, a significant part of the news uncovered the mode of inter-city cooperation of SDH. The analysis of inter-city cooperation news items reveals the different cooperation modes for the four main inter-city groups, Fig. 5.15 shows.

First, the primary cooperation mode in SDH was the informal institutional arrangement which was far greater than the formal institutional arrangement from 2004 to 2015. Specifically, communication and investigation accounted for the major parts within the three sub-modes of informal institutional arrangement, which was largely owing to institutional flexibility.

Second, although the informal institutional arrangement was far greater than the formal institutional arrangement, the latter played a significant part in inter-city cooperation as well. The majority of the communication and investigation, joint actions, and organizing activities were undertaken under the framework of joint conferences and cooperation agreements. On the basis of the institutional practices in the EU context, the cooperation activities were formalized by inter-city agreements (see Chap. 2). These formalized agreements not only provide guidelines and frameworks for daily activities in the integrating process but also improved the efficiency of inter-city cooperation owing to its institutional stability. From this perspective, we can assert that SDH integration was also articulated and settled by formal institutional arrangements and kept a stabilized and guided institutional integration.

Third, the cooperation modes of Shenzhen–Dongguan and Shenzhen–Huizhou were more formal than those of Shenzhen–Shanwei and Shenzhen–Heyuan. The numbers of joint conferences, cooperation agreements and the amount of planning in Shenzhen–Dongguan and Shenzhen–Huizhou were far higher than those in Shenzhen–Shanwei and Shenzhen–Heyuan, which proves that the integration among

Shenzhen, Dongguan, and Huizhou was more formalized than that between Shenzhen–Shanwei and Shenzhen–Heyuan, and the aid activity, which was widely undertaken in Shenzhen–Shanwei and Shenzhen–Heyuan, was organized by the informal institutional arrangement.

Regarding the formal institutional arrangement, Table 5.8 shows the basic information about major joint conferences in SDH.

- The joint conferences covered various cooperation areas to formalize inter-city cooperation. As shown in Table 5.8, certain agreements under the joint conferences were assigned with reference to the topics of cultural cooperation, policy collaboration, transportation, integration of medical treatment and public health, aid activity, and industrial transfer. The table indicates that the joint conferences were widely acknowledged by local officials in guiding regional integration. Similar to the EU integration, these joint conferences kept the problem-solving and well-rounded.
- Most of the joint conferences had enough continuity in relation to different cooperation areas. Table 5.8 demonstrates that the two longest joint conferences took place 12 times among the five cities. The guiding role of joint conferences sustained certain significant topics of SDH integration. It also proves that having joint conferences as one of the institutional alignments[1] confirmed by the *Outline Plan* in facilitating PRD regional integration was effectively implemented and produced good results.
- Most of the joint conferences drafted inter-city agreements to guide the specific actions between cities. In association with the formalized institutional arrangement in the EU (see Chap. 2), Table 5.8 reveals that, within the framework of joint conferences, the inter-city agreements were easy to align with a technocratic nature, which was similar to the EU. Furthermore, the joint conferences produced considerable achievements through providing a negotiation platform among cities.

In sum, in terms of the overall spatial layout of the inter-city cooperation of SDH, Shenzhen had close connections with Dongguan, Huizhou, Shanwei, and Heyuan, especially for Heyuan, Shanwei, and Dongguan, whereas the linkages between other cities were extremely weak. Before 2008, only Shenzhen–Heyuan had strong inter-city cooperation; after 2008, Shenzhen–Heyuan, Shenzhen–Shanwei, Shenzhen–Dongguan, and Shenzhen–Huizhou experienced a dramatic increase. In addition, Shenzhen–Shanwei showed an apparently increasing tendency under the aid activity policy after 2014. The primary cooperation areas of SDH were economic development, institutional design, and aid activity. The aid activity occurred mainly in the inter-city cooperation between Shenzhen and Shanwei and Shenzhen and Heyuan.

[1] The Outline Plan formulated the three-layer institution to implement PRD integration, specifically, the three-layer institution includes: (1) a leading group, composed of provincial-level officials; (2) the administration of leading group for the evaluation the implementation of the Outline Plan; and (3) joint conferences, composed of local officials for affairs of regional integration.

Table 5.8 Information of major joint conference of SDH from 2004 to 2015

Title	Involving cities	Starting time	Frequency	Agreement under the joint conference
Joint Conference on Cultural Cooperation among Shenzhen, Dongguan, Huizhou, Heyuan and Shanwei	Shenzhen, Dongguan, Huizhou Heyuan, Shanwei	2011	6 (2016)	Cultural Cooperation among Five Cities Framework Agreement
Joint Conference on Policing Collaboration among Shenzhen, Dongguan, Huizhou, Heyuan and Shanwei	Shenzhen, Dongguan, Huizhou Heyuan, Shanwei	2006	12 (2017)	The Public Security Bureau Police Collaboration among Five Cities Framework Agreement
Joint Conference on Transportation Integration among Shenzhen, Dongguan and Huizhou	Shenzhen, Dongguan, Huizhou	2013	3 (2015)	SDH Transportation Integration Plan
Joint Conference for Party and Government Main Leaders among Shenzhen, Dongguan, Huizhou, Heyuan and Shanwei	Shenzhen, Dongguan, Huizhou Heyuan, Shanwei	2007	10 (2016)	A Series of Agreement related Large Infrastructure Projects
Forum on Emergency Management among Shenzhen, Dongguan and Huizhou	Shenzhen, Dongguan, Huizhou	2013	12 (2016)	Emergency Management Collaborative Agreement
Joint Conference on Consumer Council Cooperation among Shenzhen, Dongguan, Huizhou	Shenzhen, Dongguan, Huizhou	2006	11 (2016)	Shenzhen, Huizhou, Dongguan Consumer Council Cooperation Agreement
Joint Conference on Anti-counterfeiting Cooperation among Shenzhen, Dongguan, Huizhou	Shenzhen, Dongguan, Huizhou	2010	7 (2016)	Anti-counterfeiting Cooperation among Shenzhen, Dongguan, Huizhou Memorandum
Joint Conference on Medical Treatment and Public Health among Shenzhen, Dongguan, Huizhou	Shenzhen, Dongguan, Huizhou	2009	8 (2016)	Medical Treatment and Public Health among Shenzhen, Dongguan, Huizhou Agreement
Joint Conference on Aid activity between Shenzhen and Shanwei and Shenzhen and Heyuan	Shenzhen, Shanwei, Heyuan	2015	2 (2016)	–
Shenzhen and Shanwei Investment Promotion	Shenzhen, Shanwei	2014	–	A series of Agreement on Commercial Cooperation

(continued)

Table 5.8 (continued)

Title	Involving cities	Starting time	Frequency	Agreement under the joint conference
Joint Conference on industrial transfer among Shenzhen, Shanwei and Chaozhou	Shenzhen, Shanwei, Chaozhou	2010	–	–

Source Based on the database constructed by author

Shenzhen also had great cooperation with Shanwei in terms of economic development. Regarding the adjacent cities, Shenzhen had strong cooperation on institutional design and economic development with Dongguan and Huizhou as well. The cooperation mode in SDH was mainly the informal institutional arrangement, which referred to communication and investigation. However, the formal institutional arrangement also played a significant role in inter-city cooperation through building the framework for inter-jurisdictional behaviors. Regarding the inter-city groups, Shenzhen–Dongguan and Shenzhen–Huizhou were more formalized than Shenzhen–Shanwei and Shenzhen–Heyuan, which also affirmed that the aid activity of Shenzhen–Shanwei and Shenzhen–Heyuan occurred through informal institutional arrangements. Regarding the formal institutional arrangement, the joint conferences covered various cooperation areas; most of them had good continuity and drafted agreements to guide the inter-jurisdictional cooperative actions.

5.3.3 Assessing Institutional Integration of SDH from the State–Market–Civil Society Relation

Chapter 4 proposes that this book searches the inter-city cooperation news items about major actors in different cities to provide information to examine the relationship of state–market–civil society in the SDH integration. These major actors are divided into five groups: government-to-government, government-to-market, government-to-society, market-to-market, and society-to-society. By counting the number of the different types of inter-city cooperation news items and analyzing the contents of such news items, such as time, organization form, and types of activities, this book demonstrates the relationship of state–market–civil society from three perspectives: degree of involvement, cooperation area, and organization and development status.

5.3.3.1 Involvement Degree of State, Market, and Civil Society

Figure 5.16 shows certain integrating features of the involvement of state, market, and civil society.

Generally speaking, the government had a high degree of involvement and played a dominant role in the institutional integration of SDH. The market and civil society

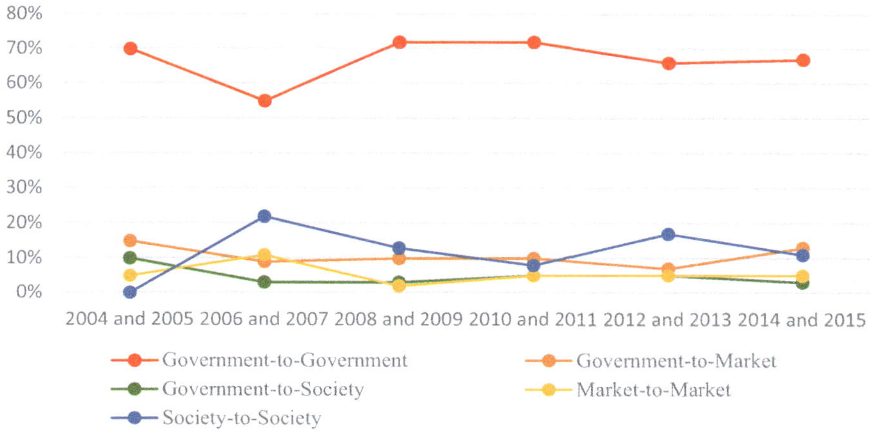

Fig. 5.16 Ratio of five types of cooperation from 2004 to 2015. *Source* Author

kept a roughly equal low degree of involvement and stayed on the margins of the SDH institutional integration. Figure 5.16 reveals the changing ratio of the five types of cooperation from 2004 to 2015. The ratio of government-to-government cooperation was much higher than that of the other four groups, signifying the prominent role that the government played in the integration process. Almost all ratios for the remaining four groups were lower than 20% from 2004 to 2015. The groups of government to market and society to society were somewhat higher than those of the other two groups. However, the low ratios of the four groups prove that the market and society maintained roughly equal development and stayed on the margins. Therefore, the integration of SDH remained a government-led process in which the non-state actors were weakly developed, and it was clearly distinguished from the wide cooperation between the public and private sectors in the EU (see Chap. 2).

5.3.3.2 Cooperation Area of State, Market, and Civil Society

Figure 5.17 summarizes the cooperation areas where the state, market, and civil society engaged as follows.

Generally speaking, the government played a prominent role in all well-rounded cooperation areas. The markets participated mainly in aid activity and economic development through cooperation with each other or the government. Society participated primarily in the activities of the secondary function and aid activity. Government-to-government cooperation focused on economic development, institutional design, secondary function, and aid activity. Moreover, the government cooperated with the market to conduct economic development and with society to enable aid activity, which indicates that the government was the most significant stakeholder

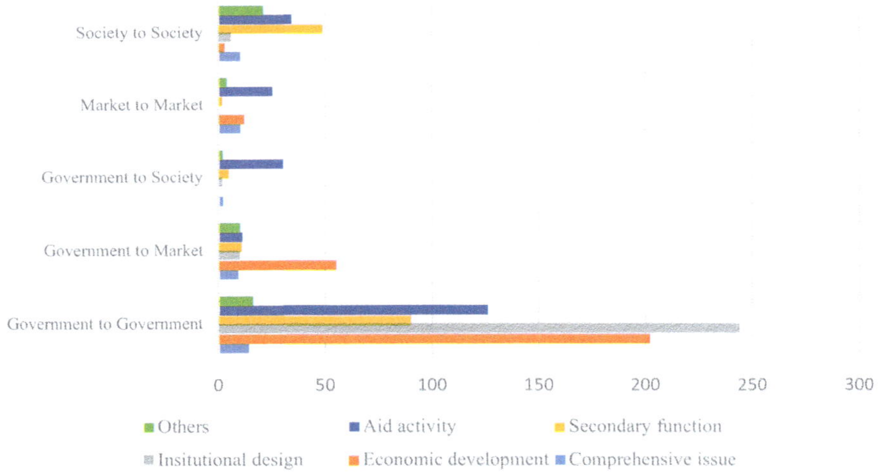

Fig. 5.17 Cooperation area of state, market and civil society in the institutional integration of SDH from 2004 to 2015. *Source* Author

in the SDH integration and actively engaged in the everyday practices of all the cooperation areas. Market-to-market cooperation concentrated primarily on the field of aid activity and was largely involved in economic development through cooperation with the government. Society also actively engaged in aid activity through cooperation with the government and other social actors. Society-to-society cooperation also focused on the entertainment, education, and scientific and technological activities that were the core of the secondary function.

5.3.3.3 Organization and Development Status of Market and Civil Society

Table 5.9 summarizes the organization and development statuses of the market based on analysis of the inter-city cooperation news items and additional information from websites as follows.

- Part of the cooperation in which the market was involved was between the government and the enterprise, and the enterprise was led or guided by the government to strengthen economic linkages across jurisdictions.
- Aside from the enterprise, the main organizational form of the market in the SDH integration was the chambers of commerce, such as the Shenzhen Fuyong Chamber of Commerce, Dongguan Luhe Chamber of Commerce, Shenzhen Songgang Chamber of Commerce, and Dongguan Industry Chamber of Commerce.
- Regarding the spatial distribution of these organizations, the majority of the market actors were in Shenzhen and Dongguan, especially Shenzhen. Moreover, their

Table 5.9 Information on major chambers of commerce in the inter-city cooperation

Title	Location	Time	Issue	Cooperation city
Shenzhen Hi-tech Industry Association	Shenzhen	2011	Shenzhen Hi-tech Industry Association went to Huizhou for investigation the Hi-tech industry development	Huizhou
Heyuan Chamber of Commerce in Shenzhen	Shenzhen	2012	The establishment of Heyuan Chamber of Commerce in Shenzhen	Heyuan
Shenzhen Fuyong Chamber of Commerce	Shenzhen	2010	Shenzhen Fuyong Chamber of Commerce donates 850,000 Yuan for Heyuan to construct the primary school	Heyuan
Shenzhen Songgang Chamber of Commerce	Shenzhen	2009	Shenzhen Songgang Chamber of Commerce donates 100,000 Yuan for Heyuan to construct the primary school	Heyuan
Timepiece Industry Joint Chamber of Commerce between Shenzhen and Hong Kong	Shenzhen	2007	Timepiece Industry Joint Chamber of Commerce between Shenzhen and Hong Kong recommend Heping in Shenzhen	Heyuan
Shenzhen Chamber of Commerce	Shenzhen	2015	SSSCZ signs agreement with Shenzhen Chamber of Commerce to build Innovative business incubators	Shanwei
Shenzhen–Shanwei Chamber of Commerce	Shenzhen	2012	Shenzhen-Shanwei Chamber of Commerce is established to deep the economic cooperation between two municipalities	Shanwei
Shenzhen, SSSCZ Chamber of Commerce	SSSCZ	2015	The establishment of Shenzhen, SSSCZ Chamber of Commerce, aiming to facilitate economic cooperation between two municipalities	Shenzhen Shanwei
Dongguan Luhe Chamber of Commerce	Dongguan	2015	Director went to Dongguan Luhe Chamber of Commerce for learning and investigation	Shanwei
Dongguan Taiwan Investment Enterprise Association	Dongguan	2008	Dongguan Taiwan Investment Enterprise Association aims to promote commercial development with Shanwei	Shanwei
Dongguan Industry and chamber of commerce	Dongguan	2011	Dongguan Industry and chamber of commerce investigates Huizhou for industrial cooperation	Huizhou

(continued)

Table 5.9 (continued)

Title	Location	Time	Issue	Cooperation city
Dongguan Dongkeng chamber of commerce	Dongguan	2009	The directors of Dongguan Dongkeng chamber of commerce went to Hi-tech industrial park in Heyuan for investigation	Heyuan

Source Based on the database constructed by author

main responsibility was to strengthen economic development across jurisdictions, such as undertaking the investigation of Hi-tech industry development and building innovative business incubators.

According to Table 5.9 and the abovementioned findings, while market actors had a relatively low degree of involvement, the different types of chambers of commerce in the member cities positively affected SDH integration, notably in the field of economic cooperation. This finding corroborates that SDH integration had substantial room for the participation of market actors. Future institutional design should also focus on fulfilling the role of the chambers of commerce through endowing them with additional authority in regional public issues and enabling them to serve as the underlying driving forces of city-regionalism.

The organization and development statuses of civil society are summarized as follows (Table 5.10).

- The civil society engaged primarily in public benefit activities across jurisdictions that focused on the themes of entertainment, environmental protection, education and culture, and aid activity activities, such as donating books, providing medical help, taking action on waste treatment for cross-boundary rivers, and donating goods to disaster-affected areas.
- The organizations that led these public benefit activities were public institutions, such as public hospitals and public schools, non-governmental organizations (NGOs), and communities.
- The majority of NGOs were in Shenzhen, and these public benefit activities were minor incidents with minimal effects on the SDH integration.

The inter-city cooperation news showed that the involved NGOs included the Shenzhen Advertisement Association, Shenzhen Literature and Arts Association, Shenzhen Lions Clubs, Shenzhen Overseas Chinese Federation, Shenzhen Chaoshan Culture Research Association, The Nature Conservancy, The Paradise International Foundation, and Guangdong Council for the Promotion of Philanthropy-Shanwei Branch. The majority of these NGOs were in Shenzhen, and the public benefit activities they led were minor incidents, such as tree planting and river protection. These activities were local practices under the action framework of the SDH integration. These specific actions were regulated and driven by intergovernmental cooperation;

Table 5.10 Major cooperation between social organizations in SDH integration

Organization	Cooperation theme	Issue	Time
Shenzhen Advertisement Association	Culture communication	Participating in advertising exhibition	2010
Public school	Aid activity	Donating books	2011
Shenzhen Lions Clubs	Aid activity	Donating goods to disaster-affected area	2013
Shenzhen Literature and Arts Association	Culture communication	Discussing the channels to strengthen culture communication	2015
Shenzhen Chaoshan Culture Research Association	Aid activity	Medical helping	2015
Public school	Education	Exchanging experience on college entrance examination	2009
Public school	Entertainment	Holding variety show jointly	2007
The Paradise International Foundation	Environmental protection	Taking activity on waste treatment for cross-boundary river	2009
Community	Environmental protection	Taking activity on waste treatment for cross-boundary river	2011

Source Based on the database constructed by author

and these NGOs, public institutions, and communities served merely as executors for intergovernmental decisions. Social actors were excluded from regional governance while they could undertake certain activities for social development. Compared with public–private partnerships in the EU context, especially referring to the privatization of public infrastructure that manifested as the full engagement of quasi- and non-state actors (see Chap. 2), the involvement of social actors in China was extremely poor and strongly regulated by governments through limiting the actors to marginal roles. Therefore, empowering social actors with additional powers and responsibilities, such as the provision of public services and the construction of regional infrastructure, was one of the key tasks to enable complete institutional integration in SDH.

In sum, in terms of the developmental extent of state, market, and civil society in the SDH integration, the government actively engaged in the SDH integration and played a dominant role in the process. The degrees of involvement of the market and civil society kept roughly equal and stayed on the margins in the SDH institutional integration. In terms of the cooperation areas of state, market, and civil society in the SDH integration, the government was the main body in all cooperation areas of SDH integration. Markets cooperated with one another and the government to participate in economic development and aid activity. Society engaged primarily in the activities of the secondary function and aid activity. In terms of the organization and development status of the market and civil society in the SDH integration, the role of the market was weak and focused primarily on economic development and aid activity. Although the chambers of commerce actively engaged in the SDH integration, the distribution and cooperation areas were extremely asymmetric. The role of civil society in the SDH integration was weak, focusing primarily on public benefit activities with the themes of public service and aid activity, and had minimal effects on the SDH integration. The organizations of civil society in the SDH integration were public institutions, NGOs, and communities. Compared with neighboring cities, the development status of NGOs in Shenzhen was relatively flourishing.

5.3.4 Summary

This section aims to answer the second research question, namely, to what extent does SDH operate as an institutionally integrated entity by redefining the governance structure across government hierarchies, between cities, and between the state and non-state domains?

This section argues that the evaluation of the institutional integration of SDH is equal to the examination of the MLG in SDH and includes three relations: (1) inter-scalar, (2) inter-city, and (3) state-market-civil society. On the basis of this recognition, this book evaluates these three aspects to provide an insightful grasp of institutional integration.

Inter-scalar relation

- *Central–local relations*—the relatively stable administrative setting from 2000 to 2016 signaled that the central government continued vigorous administrative interventions into and control over localities, which created barriers to the adjustment of local economic development. Additionally, the phenomena of political decentralization occurred in SDH and enabled Shenzhen to undertake institutional innovations and policy experimentation to compensate for deficiencies and enhance its economic growth.
- *Provincial–local relations*—the provincial government provided enormous administrative support and development opportunities for local governments to enable socioeconomic development and city-regionalism through institutional innovation and policy experimentation.
- *Regional authority*—the lack of a formal regional authority signaled that the tiers of jurisdictional arrangement in SDH were not literally marked as an MLG paradigm.

In sum, the theoretical mode of MLG in the EU was not formed devoid of the formal regional authority in the inter-scalar relation of SDH. Regarding the hierarchical government relations, although the existing rigid administrative setting limited the local development, the decentralization process and administrative support from the Guangdong Provincial Government stimulated the local institutional innovations to create new governance platforms and governing entities in response to the socioeconomic restructuring, such as the establishment of the new district and SSSCZ.

Inter-city relation

- *Spatiotemporal layout*—Shenzhen played a dominant role in inter-city cooperation and had intimate intergovernmental linkages with Dongguan, Huizhou, Shanwei, and Heyuan.
- *Cooperation areas*—the overarching cooperation areas were economic cooperation, institutional design, and aid activity. Shenzhen cooperated with Dongguan and Huizhou on the themes of economic development and institutional design primarily. However, the themes in Shenzhen–Shanwei and Shenzhen–Heyuan focused on aid activity and economic development.
- *Cooperation mode*—the informal institutional arrangement primarily referred to communication and investigation and played a prominent role in inter-city cooperation. The formal institutional arrangement offered a high-level action framework to guide inter-city cooperation.

In sum, regarding the inter-city relation of SDH, the spatial layout centered in Shenzhen was consistent with the city-region building initiatives, manifesting as the voluntary process of city grouping in the EU (see Chap. 3). The inter-city cooperation on the secondary functions, such as social development, environmental protection, and cultural communication, was inadequate. This phenomenon indicates that the SDH integration remained a growth-oriented and economy-first process of city-regionalism, whereas the institutional framework had already been articulated in the inter-city cooperation. Formal institutional arrangements, i.e., joint conferences and agreements, played a guiding role in inter-city cooperation through providing an action framework for the informal institutional arrangements. Compared with the MLG practice in the EU context (see Chap. 2), although the inter-city cooperation in the SDH integration was growth-oriented and economy-first rather than multiple objectives, it was assured by a formal agreement. The cooperation indicates that the institutional integration in SDH was highly formalized and operated within a mature framework.

State–market–civil society relation

- *Involvement degree*—intergovernmental cooperation was the overarching form to promote SDH integration, whereas market and civil society stayed on the edges of the SDH institutional integration.
- *Cooperation area*—inter-governmental cooperation focused primarily on all cooperation themes. Inter-market cooperation and government–market cooperation concentrated on issues for economic development and aid activity. However, social actors actively participated in the secondary functions and aid activity.
- *Organization and development status*—the chamber of commerce was the primary organizational form of market actors in the SDH integration. However, the spatial

distribution and cooperation area were extremely asymmetric, triggering the weak role of the market in the SDH integration. Regarding society, public institutions, NGOs, and communities were the primary forms of social actors that facilitated SDH integration. Although Shenzhen had a relatively flourishing NGO development, the role of society was weak in the SDH integration and primarily manifested as public benefit activities with the themes of public service and aid activity.

In sum, regarding the state–market–civil society relations of SDH, intergovernmental cooperation was the primary form that enabled the SDH integration, manifesting as active engagement in all areas of cooperation. Market and society were excluded from the SDH integration and participated merely in some limited fields with the specific forms of chambers of commerce, NGOs, public institutions, and communities. Revisiting the MLG practice in the EU, this book affirms that a complete involvement of quasi- and non-state sectors existed in regional governance (see Chap. 2). The engagement of market and social actors in the SDH integration was less developed and merely limited in certain marginal fields with minor influence on regional integration.

5.4 Conclusions

This chapter presents the assessment of SDH integration in relation to the economy and institutions. This book treats economic and institutional integrations as multifaceted inclusive concepts and undertakes the assessments using different dimensions.

- Economic integration

From the overall city-region perspective, market integration, as the dimension to measure the final state of the city-region integration (i.e., forming a regional economic unit), increasingly improved and remained relatively stable. The infrastructure integration of SDH also experienced a strengthening trend from the perspective of infrastructure connectivity. However, the equality of infrastructure gradually weakened from 2004 to 2015. In addition, the industrial integration of SDH experienced a steady weakening trend from the perspective of the industrial structure degree of similarity. The favorable degree of infrastructure integration contributed to the flow of goods and efficient utilization of factors of production and further accelerated the market integration process of SDH. The mounting degree of industrial isomorphism will give rise to inter-city competition, leaving the underlying crisis for the economic integration of SDH. From the perspective of an inter-city group, Shenzhen had high degrees of market and infrastructure integration and a low degree of industrial integration with Dongguan and Huizhou. It had a low degree of infrastructure integration and high degrees of market and industrial integration with Shanwei. It also shared medium degrees of market and infrastructure integration and a high degree of industrial integration with Shanwei. Thus, the good condition of infrastructure accelerated the market integration between Shenzhen and Dongguan and Shenzhen and Huizhou. The severe industrial competition among them will also sabotage productive economic integration. While the infrastructure conditions between Shenzhen and Shanwei and Shenzhen and Heyuan will undisputedly limit the construction of regional economic units, the enormous differences in the industrial structures will provide opportunities for regional economic integration and industrial collaboration. Regarding the economic integration between Dongguan and Huizhou, the market and infrastructure integrations stayed at a medium stage, and the industrial integration remained at a low stage. This finding signified that all the dimensions of economic integration between Dongguan and Huizhou must be enhanced to enable the city-regionalism of SDH.

- Institutional integration

Connected to the territorial practices of MLG under the context of the EU's integration (see Chap. 2), Table 5.11 shows a comprehensive comparison of the typologies of institutional integration in the EU and SDH.

Regarding the inter-scalar relations across hierarchal governments, the powerful central government control restrained the local socioeconomic development through a rigidified administrative setting. Essentially, such administrative control was the path-dependence of state authoritarianism, maintaining an accumulation regime where the central state played the predominant role in the Mao era in terms of resource allocation and administration. However, the process of decentralizing power from the central government downward to the local government gave rise to opportunities for local statehood to accelerate institutional innovations and political reshuffling to restructure the economy. The provincial government also offered administrative support to localities as a means of creating city-regionalism. However, given the lack of formal regional authority, no MLG has formed in SDH in terms of governing daily activities across jurisdictions compared with the region-based authorities for the provision of public service in the EU context.

In terms of inter-city relations, Shenzhen had an intimate inter-city connection with the other member cities of SDH, and such a spatial layout is consistent with SDH city-region-building initiatives. The inter-city cooperation on the secondary functions lagged far behind economic development, institutional design, and aid activity. Moreover, Shenzhen, the regional core city, took the primary responsibility for aid activity to revitalize Shanwei and Heyuan. The overarching inter-city cooperation mode was the informal institutional arrangement that occurred under the action framework from the formal institutional arrangement. Although the inter-city relationship of SDH resembled the MLG praxis in the EU, manifesting as a voluntary process of city grouping that was formalized by agreements, this relationship remained a growth-oriented and economy-first integrating process rather than a multi-objective process.

In terms of the state–market–civil society relations, the government played the dominant role in the SDH integration, whereas other non-state actors stayed at the edges. The government proactively engaged in all cooperation themes in the SDH

Table 5.11 Typologies of institutional integration in EU and SDH

	EU	SDH
Typologies	MLG paradigm	Government-led integration process
Inter-scalar relation	1. Upscaling of state space to super-national organization; 2. Downscaling of state space to sub-national organization; 3. Establishing region-based authorities for the provision of public service	1. Path dependence of state authoritarianism, remained the predominant role of central state through administrative setting; 2. Flourishing local institutional innovations and policy experimentations, stemming from power decentralization; 3. Lack of region-based authorities; provincial government serves as sub-regional fabric virtually
Inter-city relation	1. A voluntary process of city grouping; 2. Problem-solving cooperation mode which has functioned with multi-objective; 3. Inter-city cooperation network, crystalized by the formalized agreement	1. A voluntary process of city grouping, centered on Shenzhen (regional core city); 2. Growth-oriented and economy-first inter-city cooperation instead of the multi-objective integrating process; 3. Abundant informal institutions guided by formal institutions, e.g., joint conference and agreement
State-market-civil society relation	1. Full engagement of quasi- and non-state actors in regional integration; 2. Multiple fields of public services (Privatization of public infrastructure); 3. A variety of institutional forms of Public–Private Partnerships	1. Government-led regional integration where market and social actors stay at the edge position; 2. Certain issues that are relatively marginal; 3. Limited institutional forms, such as chamber of commerce, public institutions, NGOs, and community

Source Author

integration. Market actors participated primarily in the issues of economic development and aid activity. Social actors engaged mainly in the secondary function and aid activity. The primary organizational form of market actors was the chamber of commerce, which had an extremely asymmetric spatial distribution and cooperation area. The primary organizational forms of social actors were the public institutions, NGOs, and communities with active engagement in public benefits activities. Compared with the MLG in the EU, the complete engagement of quasi- and non-state actors was not created in SDH. The field of the engagement of market and social actors in SDH was restricted to certain issues that were relatively marginal instead of the specific key issues in the EU, such as the provision of public services and the construction of infrastructure.

According to the above-mentioned features, institutional integration in SDH was a salient government-led integration process and remained a departure from the

MLG of the EU at this stage. Chapters 6 and 7 will systemically present how such a government-led integration process reshuffles itself in response to the socioeconomic integration of SDH, especially from the perspectives of inter-scalar and inter-city relations.

References

Ma, L. J. (2005). Urban administrative restructuring, changing scale relations and local economic development in China. *Political Geography, 24*(4), 477–497. https://doi.org/10.1016/j.polgeo.2004.10.005

Pinder, J. (1969). Problems of European integration. In G. Denton (Ed.) *Economic integration in Europe*. Weidenfeld and Nicolson.

Chapter 6
Exploring the Reshuffling Process of Governance Structure to Enable the Industrial Integration of SDH: A Case Study of SSSCZ

6.1 Introduction

This chapter addresses the third research question by demonstrating how the governance structure enables the industrial integration of SDH. Industrial transfer parks are actualized by the integration-first ideology of the Guangdong Provincial Government, aiming to reduce economic-spatial inequality by promoting industrial collaboration between cities in PRD. SSSCZ was thereby devised as an experimental case by the Guangdong Provincial Government and probed in this book to elaborate on how the governance system can be restructured in terms of inter-scalar and inter-city relations in response to the crisis of economic development deadlock. SSSCZ is a cross-boundary industrial project involving several key stakeholders, such as the Guangdong Provincial Government, Shenzhen and Shanwei Municipal Governments, industrial guilds, and related enterprises. On the basis of the meticulous examination of four critical issues in SSSCZ, power configurations and institutional arrangements across territorial organizations are identified in the context of PRD city-regionalism.

Apart from the introduction, this chapter comprises seven sections based on the analytical framework in Chap. 4. The second section systematically depicts the empirical background of SSSCZ by embedding it in the context of PRD coordinated development because SSSCZ serves as a crucial spatial strategy to eliminate the economic–spatial inequality. The next four sections primarily foreground four issues (i.e., negotiation of the construction of SSSCZ, legitimization of SSSCZ, daily operation, and industrial development of SSSCZ) to decipher how inter-scalar, inter-city, and state-market relations were restructured in the search for the efficacious making of industrial transfer parks. The third section reveals how Shenzhen preferred Shanwei to Dongguan as its cooperation target and primary arena for capital accumulation under the political mobilization of the Guangdong Provincial Government. The fourth section probes how territorial states legitimatized SSSCZ through the provision of land quotas and established the legal planning system (i.e., land and planning

X. Zhang, *Regional Integration and Governance Reshuffling in the Making of China's City-Regions*, Urban Sustainability, https://doi.org/10.1007/978-981-99-2792-0_6

politics). The fifth section investigates the administrative system in SSSCZ to elaborate on the relations between Shenzhen and Shanwei based on the characteristics of the administrative system. In the industrial development of SSSCZ, the sixth section focuses on how the Shenzhen Municipal Government drove the industrial development of SSSCZ and how enterprises responded to the re-territorialization strategy. The seventh section summarizes the major finding of the case study of SSSCZ and its theoretical implications. The final section proposes major conclusions to address the third research question.

6.2 Empirical Background: Integration-Oriented Policy and Industrial Transfer Parks Under PRD City-Regionalism

PRD has undergone socioeconomic restructuring driven by a series of integration-oriented policies, reorienting from the ideology of "vacating the cage to change the bird" (*Tenglonghuanniao*, 腾笼换鸟) to "expanding the cage to boost the bird" (*Kuolongzhuangniao*, 扩笼壮鸟). Such integration-oriented policies reveal a new orientation of the Guangdong Provincial Government, moving from geographically specialized development to geographically integrated development, fostering a reduction in economic–spatial inequality.

In the early 2000s, the socioeconomic development of the core cities in PRD, such as Guangzhou, Shenzhen, and Foshan, approached saturation. These cities not only needed urban space to accommodate socio-economic activities but also needed to upgrade their industrial structures to eliminate backward industries. Considering such a development deadlock and the huge shocks from the financial crisis in 2008, the Guangdong Provincial Government proposed the strategy of "vacating the cage to change the bird" in 2008, specifically referring to relocating labor-intensive industries in PRD to other places, such as the inland provinces and Southeast Asia region, and leaving adequate room for high-end industries. In May 2008, the Guangdong Provincial Government formulated the crucial agenda, "the decision to promote industrial transfer and labor transfer" (*Guanyu tuijin chanye zhuanyi yu laodongli zhuanyi de jueding*, 关于推进产业转移与劳动力转移的决定) and a series of related policies with varied institutional tools (e.g., providing financial investment in industrial upgrading as well as readjusting industrial structures and improving the threshold for labor-intensive industries).[1] However, during the test period from 2008 to 2010, the extent of this strategy did not match the intention of the Guangdong Provincial Government. On the one hand, the absence of a benefit-sharing mechanism for industrial upgrading led to tensions and conflicts within enterprises and across multiple levels of government. On the other hand, this round of industrial upgrading failed

[1] Guangdong Provincial Party Committee and Guangdong Provincial Government (2008). The decision to promote industrial and labor transfers (*Guanyu tuijin chanye zhuanyi yu laodongli zhuanyi de jueding*).

to provide the geographical guidelines for relocation as well as policy and financial support for displaced industries. Such a culpable institutional arrangement not only triggered discontent from low-end enterprises but also caused a potential crisis for the industrial upgrading of PRD (Interview with a local official, 03/2016).

To overcome such defects, the Guangdong Provincial Government preferred "expanding the cage to boost the bird" over others as the overarching socioeconomic strategy by relocating low-end industries to backward regions, such as western, eastern, and northern areas of Guangdong Province, and by promoting labor in these backward regions to move to areas in PRD where secondary and tertiary industrial development was taking place. In June 2010, the Guangdong Provincial Government issued the government agenda of "some suggestions on accelerating transformation of the mode of economic development" (*Guanyu jiakuai jingji fazhan fangshi zhuanbian de ruogan yijian*, 关于加快经济发展方式转变的若干意见). Based on this governmental document, the so-called low-end industries settled within the territory of Guangdong Province through co-producing a host of industrial transfer parks in backward regions. Four types of institutional arrangements were undertaken to support the construction of industrial transfer parks in Guangdong Province. First, the Guangdong Provincial Government invested 0.5 billion Yuan from its fiscal revenue as the financial base for constructing industrial transfer parks in the next five years. Second, the Guangdong Provincial Government encouraged institutional innovation in inter-city cooperation for the co-construction of industrial transfer parks, especially for establishing joint-stock companies between cities in the development process. Third, the Guangdong Provincial Government allows local governments to enact an allocation mechanism of benefits for industrial transfer parks independently. The provincial government also made an overall arrangement for the construction land quota. Finally, fees of pollution discharge were rationalized, and electricity prices were differentiated for low-end industries.[2] The related industrial policies driven by "vacating the cage to change the bird" and "expanding the cage to boost the bird", are called strategies of "double transfer" (*Shuangzhuanyi*, 双转移).

By 2015, the Guangdong Provincial Government constructed 83 industrial transfer parks with various forms. These parks could be categorized into three types, namely, city- and country-level industrial transfer parks and industrial agglomeration districts (Fig. 6.1). These industrial transfer parks introduced more than 4900 industrial enterprises, and the added value of industries above a designated scale for these parks has reached 197.7 billion Yuan. The total tax reached 29.9 billion Yuan in 2015.[3]

Such institutional arrangements lay a solid foundation for SSSCZ and profoundly mirrored inter-scalar, inter-city, and state-market relations in SSSCZ. The predecessor of SSSCZ was one of the industrial transfer parks, which was co-constructed

[2] Guangdong Provincial Party Committee and Guangdong Provincial Government (2010). Some suggestions on accelerating transformation of mode of economic development (*Guanyu jiakuai jingji fazhan fangshi zhuanbian de ruogan yijian*).

[3] Guangdong Provincial Government (2017). The 13th Five-Year Plan for industrial transfer parks in western, eastern, and northern areas of Guangdong province (*Yue dongxibei shengchanyezhuanyiyuan fazhan shisanwu guihua*).

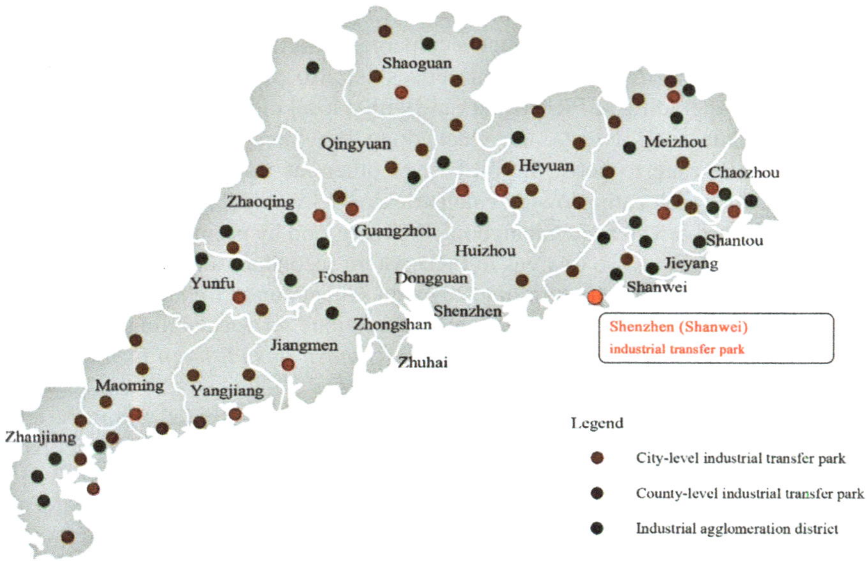

Fig. 6.1 Distribution of industrial transfer parks in PRD. *Source* Author

by Shenzhen and Shanwei and called "Shenzhen (Shanwei) industrial transfer park" (*Shenzhen Shanwei chanye zhuanyi yuan*, 深圳 (汕尾) 产业转移园).

6.3 Negotiating the Construction of SSSCZ: Stable Allocation Mechanism of Benefits Across Inter-scalar and Inter-city Domains

The initiatives of building SSSCZ were driven by an ad hoc and intricate negotiation, involving inter-city and inter-scalar politics. In inter-city relations, Shenzhen socioeconomically repositioned its cooperation target from Dongguan to Shanwei, and such a geographical reconfiguration was chiefly determined by the reciprocal advantages between Shenzhen and Shanwei under the local crises of economic development. In inter-scalar relations, the construction of SSSCZ was the result of the political mobilization of the Guangdong Provincial Government, driven by reducing economic–spatial inequality, combined with positive responses from the Shenzhen and Shanwei Municipal Governments. These governments aimed to break the local developmental deadlock. Territorial states also assembled as an integrative governance apparatus by creating a stable allocation mechanism for benefits between Shenzhen and Shanwei and across provincial and local jurisdictions.

6.3.1 Crises of Economic Development in Shenzhen and Shanwei

6.3.1.1 Shenzhen: Industrial Upgrading constrained by Insufficient Urban Space

During the last two decades of the twentieth century, PRD's economy was strategically oriented toward Hong Kong and served as a hinterland to accommodate the manufacturing relocated from Hong Kong. Such a geographic industrial division known as "front shop, rear factory" has been restructured in recent decades, especially appearing in the repositioning of Shenzhen with its rapid industrial upgrading.

In the recent decades, Shenzhen has embarked on tremendous industrial upgrading to develop well-rounded industrial structures, involving four pillar industries (i.e., cultural and creative industries, high and new technological industries, modern logistics, and finance), six emerging strategic industries (i.e., bio-industry, new energy industries, the Internet, advanced material production, information technology, and energy conservation and environmental protection industries), and five future industries (i.e., life health, marine, aerospace, robotic, and intelligent equipment industries). Based on governmental statistics, the added value of emerging strategic industries increased from 276 billion Yuan in 2010 to more than 700 billion Yuan in 2015, and the average annual growth rate reached 17.4%. The GDP of the strategic emerging industry accounted for 40% of the total GDP from 2010 to 2015.[4] Certain well-known enterprises, such as Huawei (华为), Tencent (腾讯), and BYD (比亚迪), emerged in Shenzhen and embodied its reorientation from low-end industry and manufacturing to high-tech industry and modern services.

However, this high-tech-oriented economic development mode in Shenzhen faced persistent land constraints. The shortage of construction land not only gave rise to a high access threshold for low-end and high-tech industries but also triggered spillover effects, such as the relocation of supply chain of Huawei from Shenzhen to Dongguan. Figure 6.2 shows that the planned supply of construction land in Shenzhen experienced a steady decline from 1.75 thousand ha in 2014 to 1.35 thousand ha in 2016, whereas the actual supply of construction land experienced a relative fluctuation from 1.06 thousand ha in 2011 to 1.17 thousand ha in 2016. Generally, the clear decline demonstrates a serious land constraint for the industrial development of Shenzhen despite the fact that Shenzhen had a strong predominant regulatory power over land resources. A government official in Shenzhen noted:

> Shenzhen now faces a severe constraint for land resources and the supply of construction land is extremely limited. Besides, a considerable portion of land will be used for real estate development and the provision of public facilities. The land leasing for industrial development is no more than 1 km^2 annually. (Interview with a local official, 04/2016)

---

[4] Development and Reform Commission of Shenzhen Municipality (2016). The 13th Five-Year Plan for the development of strategic emerging industry in Shenzhen (*Shenzhenshi zhanluexing xinxing chanye fazhan shisanwu guihua*).

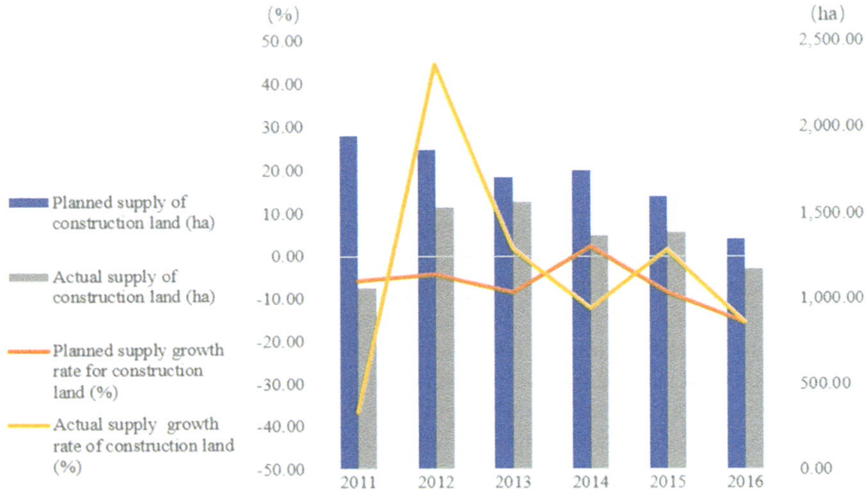

	2011	2012	2013	2014	2015	2016
Planned supply of construction land (ha)	1952	1868	1710	1750	1600	1350
Actual supply of construction land (ha)	1062	1536	1563	1369	1390	1173
Planned supply growth rate for construction land (%)	-5.70	-4.30	-8.46	2.34	-8.57	-15.63
Actual supply growth rate of construction land (%)	-36.75	44.63	1.76	-12.42	1.50	-15.58

Fig. 6.2 Planned and actual supply of construction land and their growth rates from 2011 to 2016. *Source* Shenzhen China, Retrieved from http://www.sz.gov.cn/cn/xxgk/zfxxgj/ghjh/csgh/jq/

Resources and other input needed from Shenzhen in the development of SSSCZ included its high-tech industrial production factors such as industrial connections, equipment and human resources, capital, and technology for other industrial development. Confronted by keen economic competition in China's post-reform era, the primary task to tackle was acquiring adequate land resources to accommodate economic activities and their associated social and infrastructure development such as housing and transportation.

6.3.1.2 Shanwei: Revitalizing a Local Economy constrained by Weak Local Socioeconomic Conditions

Shanwei is a municipality in the eastern region of Guangdong Province. Figure 6.3 reveals the GDP of certain primary cities in Guangdong Province from 2014 to 2016, and their degrees of economic development during the three years remained at low levels in terms of their GDP position in Guangdong Province. The growth momentum of Shanwei was also smooth, and the increment was slight. Thus, Shanwei was labeled "collapsed Shanwei" for its floundering economic situation.

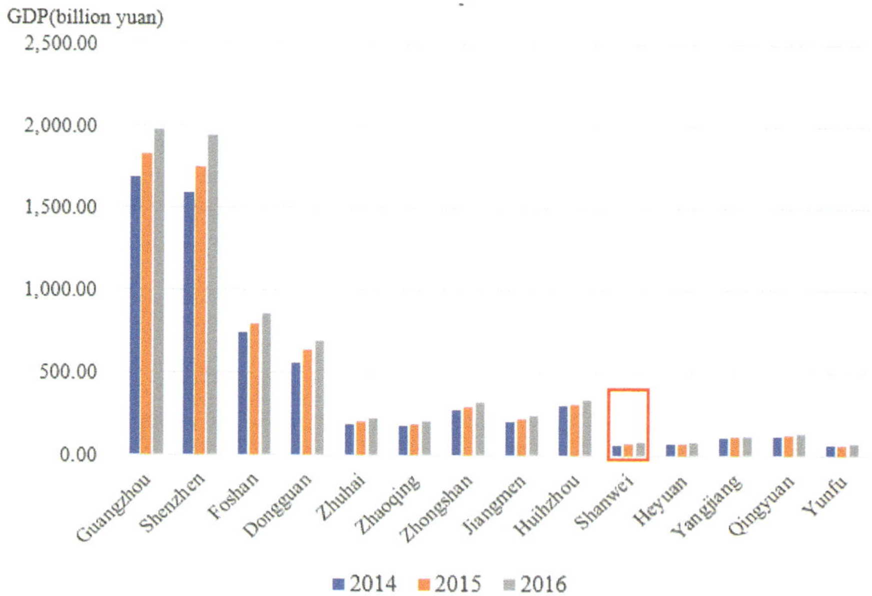

Fig. 6.3 GDP of 14 primary cities in Guangdong Province from 2014 to 2016. *Source* Guangdong Statistical Yearbook (2015–2017)

However, as for its land resources, Shanwei's administrative area reached 5.27 km^2, which was 2.6 times Shenzhen's administrative area. Shanwei's construction land had low-density and was smaller than Shenzhen's. In terms of the construction of SSSCZ, Shanwei had adequate land resources to exchange with Shenzhen for co-development. The primary aim for Shanwei was to achieve rapid economic growth through industrial transformation and upgrading. To match both cities' demands, the key issue for an agreement on the construction of SSSCZ was to directly build a stable allocation mechanism of benefits between Shenzhen and Shanwei for the income of SSSCZ.

6.3.2 *Industrial Transfer Parks as Crisis-Management Strategy: From Shenzhen–Dongguan Cooperation to Shenzhen–Shanwei Cooperation*

6.3.2.1 Establishing Industrial Transfer Parks—Political Mobilization by the Guangdong Provincial Government

Based on the "double transfer" strategies, the Guangdong Provincial Government vigorously pushed forward the construction of industrial transfer parks to reduce the economic–spatial inequality. To provide financial incentives and concession policies for backward cities, the Guangdong Provincial Government further proposed the strategy of "The Revitalization and Development of Eastern, Western, and Northern Guangdong" in 2014 to expand the three sub-regions spatially. These integration-oriented regional policies proffered mobilization and assistance to localities. The Guangdong Provincial Government specifically offered fund subsidies, land quota provision, industrial guidance, and technological support for the construction of industrial transfer parks (Table 6.1).

Table 6.1 Concessionary policies from the Guangdong Provincial Government for the construction of SSSCZ

Category	Content
Fund subsidy	• Providing 1.5 billion Yuan of transfer payment for the construction of industrial transfer parks in backward regions from 2008 to 2012 • Providing 0.5 billion Yuan of funding awards and more than 1 billion Yuan of special funds for the training of surplus labor annually from 2008 to 2012 • Providing 2.5 billion Yuan of special funds for the construction of sewage disposal facilities in eastern, western, and northern Guangdong Province; these special funds can be used to improve facilities in industrial transfer parks • Providing 0.9 billion Yuan to establish competitive support funds for the construction of industrial transfer parks and 0.3 billion Yuan of funding awards to encourage the industrial transfer between cities from 2010 to 2013
Land quota provision	• Providing certain land quotas for the transfer from agricultural land to construction land in industrial transfer parks • If land transferring is beyond the jurisdiction of the Guangdong Provincial Government, then the Department of Land Resource of Guangdong Province applies for land quotas to the Ministry of Land and Resources • Providing additional land quotas for industrial transfer parks with good performance and special arrangement of such quotas for large projects with more than 1-billion-Yuan investment
Industrial guidance	• Assisting local governments to conduct industrial chain-based investment attraction (*Chanyelian zhaoshang*) and focusing on the introduction of large-scale enterprises, Hong Kong, Macao, Taiwan, and high-tech enterprises • Encouraging local governments to conduct differential introduction policy and improve the threshold for industrial introduction and pollutant emission • Encouraging local governments to construct one to two large-scale industrial transfer parks with high-tech industries
Technological support	• Strengthening the vocational and technical education in western, eastern, and northern Guangdong Province and helping local governments train certain professional technological personnel • Training the rural labor forces with special conditions for free and providing skill upgrading training for more than 1 million peasant workers

Source Summary from certain related government agendas by author

6.3.2.2 Changing Inter-city Cooperation from Shenzhen–Dongguan to Shenzhen–Shanwei

To overcome the problem of insufficient space, the Shenzhen Municipal Government attempted to seek a solution through the strategy of city-regionalism. Based on existing socioeconomic linkages, the Shenzhen Municipal Government initially expressed a strong willingness to cooperate with Dongguan to co-produce an integrated industrial system.

However, the mismatch of aims in negotiation failed to enable the Shenzhen–Dongguan cooperation in terms of industrial collaboration. On the one hand, Dongguan faced a severe burden from industrial upgrading to reposition itself from a manufacturing-based city to a high-tech and modern service city in PRD. On the other hand, Dongguan strongly desired to retain the regulatory power for its jurisdiction and still attempted to play a dominant role in developing its urban space (Interview with a local official, 11/2016). Such conflicting interests between Shenzhen and Dongguan triggered the failure of inter-city industrial collaboration through the establishment of industrial parks.

This situation also applied to the inter-city cooperation between Shenzhen and Huizhou. Confronting a dilemma for site selection, the Shenzhen Municipal Government strategically reoriented to Shanwei for industrial collaboration, propelled by the political mobilization from the Guangdong Provincial Government.

6.3.2.3 Constructing a Stable Allocation Mechanism of Benefit for SSSCZ

The negotiation for the construction of SSSCZ between Shenzhen and Shanwei started in 2009 and underwent several rounds to formulate a stable allocation mechanism for benefits among the Shenzhen and Shanwei Municipal Governments and the Guangdong Provincial Government. Based on the above analysis, Table 6.2[5] summarizes the allocation mechanism of benefits among the three governments.

It shows that the Guangdong Provincial Government and the Shenzhen and Shanwei Municipal Governments could acquire potential benefits and benefits from daily operations in SSSCZ. Regarding the Guangdong Provincial Government, SSSCZ served as a demonstration plot to effectively promote a coordinated development strategy at the overall provincial scale. The Guangdong Provincial Government also faced political competition from other provinces, and the performance assessment for officials was based on the degree of socioeconomic development. Thus, the construction of SSSCZ not only promoted socioeconomic development for Guangdong Province but also brought political performance and bonuses to the officials of the Guangdong Provincial Government. As for the daily operation of SSSCZ, the

[5] Shenzhen Municipal Government, Shanwei Municipal Government, (2011). The fundamental framework for Shenzhen-Shanwei Special Cooperation Zone (*Shenshan(wei) tebie hezuoqu jiben kuangjia fangan*).

Table 6.2 Allocation mechanism of benefit among the Guangdong Provincial Government and the Shenzhen and Shanwei Municipal Governments

Actor	Potential benefits	Benefits from daily operation
Guangdong Provincial Government	1. Obtaining a grasp of the regional coordination strategy 2. Obtaining the socioeconomic and political revenue from the regional coordination strategy;	Obtaining a certain proportion of local taxes
Shenzhen Municipal Government	Obtaining the socioeconomic revenue through relocating surplus industry to SSSCZ	1. Obtaining 12.5% of local taxes from SSSCZ from 2015 to 2020 2. Obtaining 25% of local taxes from SSSCZ after 2020
Shanwei Municipal Government	Obtaining investments and industries from SSSCZ and enhancing employment, especially for Haifeng County	1. Obtaining 12.5% of local taxes from the SSSCZ from 2015 to 2020 2. Obtaining 25% of local taxes from SSSCZ after 2020 3. Obtaining 2% of land-transferring fees from SSSCZ

Source Author

Guangdong Provincial Government obtained a certain part of the local taxes from the fiscal revenue of SSSCZ.

For the Shenzhen Municipal Government, the relocation of surplus industries from Shenzhen to SSSCZ provided additional urban space to accommodate certain high-tech and emerging strategic industries. Such industrial upgrading brought potential benefits for the Shenzhen Municipal Government and alleviated the crisis for local socio-economic development. The agreement between Shenzhen and Shanwei[6] confirmed that Shenzhen would share 12.5% of local taxes from SSSCZ from 2015 to 2020 and 25% of local taxes from SSSCZ after 2020.

With respect to the Shanwei Municipal Government, the construction of SSSCZ was predicted to promote the economic revitalization of four towns (i.e., Xiaomo, Houmen, E-bu, and Chishi) under Haifeng County. SSSCZ also attracted some relatively high-tech industries to Shanwei, which had an exemplary effect on the laggard region of Shanwei. For example, these relocated industries enhanced local employment and promote industrial upgrading. In the daily operation of SSSCZ, Shanwei resembled Shenzhen in sharing 12.5% of local taxes from SSSCZ from 2015 to 2020 and 25% of local taxes from SSSCZ after 2020. Aside from the shared

---

[6] Shenzhen and Shanwei municipal governments (2011). The basic framework for Shenzhen–Shanwei Special Cooperation Zone (*Shenshan(wei) tebie hezuoqu jiben kuangjia fangan*).

Table 6.3 Relative advantages, participation levels, aims in negotiation, and value orientations of different actors

Actor	Relative advantages	Participation levels	Aims in negotiation	Value orientations
Guangdong Provincial Government	• Administrative power at the regional scale • Resources to promote regional development (e.g., land quota and technology)	Major actor	• Promoting coordinated regional development • Enhancing economic growth of the whole province	Public interest (provincial interest)
Shenzhen Municipal Government	• Advanced industries • Advanced management experience and technology	Major actor	• Obtaining urban space to accommodate high-tech industries • Retaining the regulatory power for surplus industries	Public interest (urban interest)
Shanwei Municipal Government	• Substantial available space and land resources	Major actor	• Obtaining relatively high-end industries to promote their industrial transformation and economic growth	Public interest (urban interest)
Dongguan Municipal Government	• Relative solid socioeconomic foundation with Shenzhen • Substantial available space and land resources	Excluded	• Retaining the regulatory power for jurisdiction • Obtaining high-end industries to promote their industrial transformation	Public interest (urban interest)

Source Author

revenue from local taxes, the Shanwei Municipal Government could obtain 12% of land-transferring fees from SSSCZ.

To summarize, Table 6.3 shows the relative advantages, participation levels, aims in negotiation, and value orientations of different actors in the negotiation and the construction of SSSCZ.

6.4 Legitimatizing SSSCZ: Land and Planning Politics Across Inter-scalar and Inter-city Domains

Considering the recognition of the necessity to build SSSCZ, two critical issues had to be addressed to legitimatize SSSCZ. The first issue was to obtain an adequate construction land quota to enable the construction of SSSCZ. The second issue was to build a well-rounded planning system to arrange the various functions of SSSCZ comprehensively (Interview with an urban planner, 12/2016).

6.4.1 Land Politics in SSSCZ

6.4.1.1 Planned Construction Land Quota

SSSCZ was formally established in 2011, which was later than the formulation and approval of land use planning of the four towns. The land-use plans for the towns of Xiaomo, Houmen, E-bu, and Chishi were approved in 2010, and the predicted area of "Shenzhen (Shanwei) industrial transfer parks" in 2020 was merely 2760 ha (Interview with an urban planner, 03/2016). Compared with the area of SSSCZ in 2009, which was 1824 hectares, merely 936 ha of construction land quota has been endowed to SSSCZ.

The Shenzhen and Shanwei Municipal Governments invited the "Planning and Design Institute of Peking University (Shenzhen)" (*Beijingdaxue Shenzhen guihua sheji yanjiu zhongxin*, 北京大学 (深圳) 规划设计研究中心) to formulate the land use planning for SSSCZ. However, the predicted area of construction land for SSSCZ in 2020 reaches 7934 ha in this plan and increases to 5174 ha compared with the predictive volume in the land use planning of four towns.[7]

To fill the gap of the construction land quota (i.e., 5174 ha), three approaches were proposed through the regional collaborative provision. The first approach was to seek a concessionary policy and quota subsidy from the central government and the Guangdong Provincial Government to provide 2000 ha of construction land. The second approach was to obtain the 2900 ha construction land quota by optimizing the existing construction land layout among different levels of government in Shanwei. However, the coordination across different levels of government was extremely difficult in the sense of the divergent local socioeconomic environments and urbanization interests. The last approach was to collect around 200 ha of the construction land quota through intensifying land-use, the linkage between urban land taking and rural land giving (*Chengxiang jianshe yongdi zengjian guagou*, 城乡建设用地增减挂钩), and the reclamation project in SSSCZ (Interview with an urban planner, 12/2016).

6.4.1.2 Shanwei's Dilemma in Gaining Construction Land Quota

Allocating the construction land quota of 2900 hectares across different levels of government in Shanwei by the planning mechanism was a real dilemma. The primary

---

[7] Planning and Design Institute of Peking University (Shenzhen) (2010). Land use planning for Shenzhen–Shanwei Special Cooperation Zone (2010–2020) (*Shenzhen Shanwei tebie hezuoqu tudi liyong zongti guihua*).

negotiating actors were the Shanwei Municipal Government and lower-level governments (i.e., related county, district, and town governments). The Shanwei Municipal Government was a proactive actor in supplying the construction land quota and offered around 200–300 ha of construction land to SSSCZ by revising the land-use planning of its central urban area. The Shanwei Municipal Government also actively persuaded the Haifeng County Government to proffer 100 ha of the construction land quota to SSSCZ. However, lower-tier governments, specifically the town governments, boycotted the mobilization from the municipal governments. These governments had a strong local interest in retaining the construction land quota to benefit from the rapid urbanization process, as the land revenue was the primary source of finance to support urbanization. Thus, these lower-tier governments argued that most construction land quotas had already been proposed in land-use plans, and they had difficulties in readjusting such plans (Interview with a local official in SSSCZ, 04/2016).

6.4.1.3 The Active Role of the Provincial Government in Providing Construction Land Quota

To fill the huge gap between the construction land quota and the actual need for SSSCZ to take off, the Guangdong Provincial Government coordinated with the Shanwei Municipal Government and provided land quotas annually from its spare quotas for the construction of SSSCZ. In 2014 and 2015, it provided around 70 ha and 370 ha of construction land quotas for SSSCZ, respectively (Interview with a local official in SSSCZ, 04/2016). Based on the agreed supply of construction land quotas across different levels of government, the central government offered additional construction land quotas to provincial governments to enable the construction of large-scale infrastructure and public facilities. To summarize, land politics and conflicts across different levels of government in Shanwei hindered the collection of adequate construction land quotas for the operation of SSSCZ. The provincial government had to play an active role to plug the gap to enable SSSCZ to develop as envisioned.

6.4.2 *Planning Politics in SSSCZ*

SSSCZ is an industrial transfer park, which was empowered by prefectural-level administrative power in terms of personnel appointment and removal. However, SSSCZ was not an independent administrative unit and continued to play the role of an inter-jurisdictional economic space. As an economic space without an independent administrative status, land-use, urban, and development planning cannot be approved separately and should be integrated with the related plans of Shanwei.

To efficaciously enable the various functions in SSSCZ to materialize, such as industrial development, efficient use of land, and infrastructure construction, the

Shanwei Municipal Government selected an integrative approach by making a development master plan. First, the Shanwei Municipal Government invited different planning institutions to formulate related special plans (Table 6.4). These plans provided only specific guidelines on different subjects but without legal effects in terms of implementation. Second, to establish authority over the plans, the Shanwei Municipal Government collaborated with one of the planning institutions (i.e., New Land Tool) to formulate the "SSSCZ Development Master Plan (2015–2030)" (*Shenshan tebie hezuoqu fazhan zongti guihua*, 深汕特别合作区发展总体规划). This comprehensive planning integrated the main and compulsory contents of different special plans and was approved in the executive meeting of the Guangdong Provincial Government (Interview with an urban planner, 04/2016). Such an institutional design offered a sound solution to the planning issues of SSSCZ.

Table 6.4 Information of different types of plans for SSSCZ

Planning	Time	Formulating institution	Approval	Approval department
SSSCZ Urban Master Plan (2010–2030) (*Shenshan tebie hezuoqu chengshi zongti guihua*, 深汕特别合作区城市总体规划)	2012	New Land Tool	No	–
SSSCZ Land-use Master Plan (2010–2020) (*Shenshan tebie hezuoqu tudi liyong zongti guihua*, 深汕特别合作区土地利用总体规划)	2012	Planning and Design Institute of Peking University (Shenzhen)	No	–
The 12th Five-year Plan for SSSCZ (*Shenshan tebie hezuoqu dishierge wunian guihua gangyao*, 深汕特别合作区第十二个五年规划纲要)	2012	Administrative Committee of SSSCZ	No	——
SSSCZ Industrial Development Plan (2012–2015) (*Shenshan tebie hezuoqu Chanye fazhan guihua*, 深汕特别合作区产业发展规划)	2012	China Development Institute	No	–
SSSCZ Development Master Plan (2015–2030) (*Shenshan tebie hezuoqu fazhan zongti guihua*, 深汕特别合作区发展总体规划)	2014	New Land Tool	Yes	Executive meeting of Guangdong provincial government

Source Interview with an urban planner, 04/2016

6.5 Daily Operation of SSSCZ: Co-producing an Integrated Administrative System Between Shenzhen and Shanwei

Due to the legitimization issues of SSSCZ, building an integrated administrative system to govern and enable daily operation was not easy, but quite essential. It is largely a matter of rationalizing and allocating responsibilities equitably and effectively between Shenzhen and Shanwei. After several rounds of negotiation among the Guangdong Provincial Government and the Shenzhen and Shanwei Municipal Governments, an integrated administrative system was established, creating a well-defined distribution of responsibility and profound functional integration among the cities.

The administrative system of SSSCZ was constructed in 2011 through collaboration among the Guangdong Provincial Government and the Shenzhen and Shanwei Municipal Governments. The system was a "three-tier system". The first tier was the leading group, which comprised local officials of Shenzhen and Shanwei who were department-level cadres and were responsible for making decisions. The second tier was the Party Working Committee and Management Committee of SSSCZ, which took responsibility for the management of the Zone. The last tier was the "SSSCZ Development and Construction Co., Ltd." (*Shenshan tebie hezuoqu kaifa jianshe youxian gongsi*, 深汕特别合作区开发建设有限公司), which took direct charge of daily operations.

Under such a three-tier administrative system, the leaders of the Party Working Committee and Management Committee of SSSCZ were empowered as deputy departmental-level cadres. The membership of these committees was approved by the Guangdong Provincial Government and established in 2011. The Party Working Committee was co-located with the Management Committee, and they shared the prefecture-level administrative status in terms of personnel appointment and removal. As standing agencies of the Guangdong Provincial Government, each committee was served by nine bureaus. Moreover, SSSCZ set up separate national tax, local tax, and public security bureaus (Fig. 6.4). Two major characteristics of inter-scalar and inter-city relations can be identified in the administrative system of SSSCZ: (1) the devolution of power from provincial to municipality government, and (2) the integration between Shenzhen and Shanwei, as elaborated in the next section.

6.5.1 Devolution from Province to Municipality

To construct an integrated administrative system in SSSCZ, the Guangdong Provincial Government decentralized certain powers to local levels, such as the Shenzhen and Shanwei Municipal Governments and SSSCZ, to enable inter-jurisdictional cooperation. Apart from the land quota provision, the Guangdong Provincial Government devolved and deregulated itself in two primary aspects.

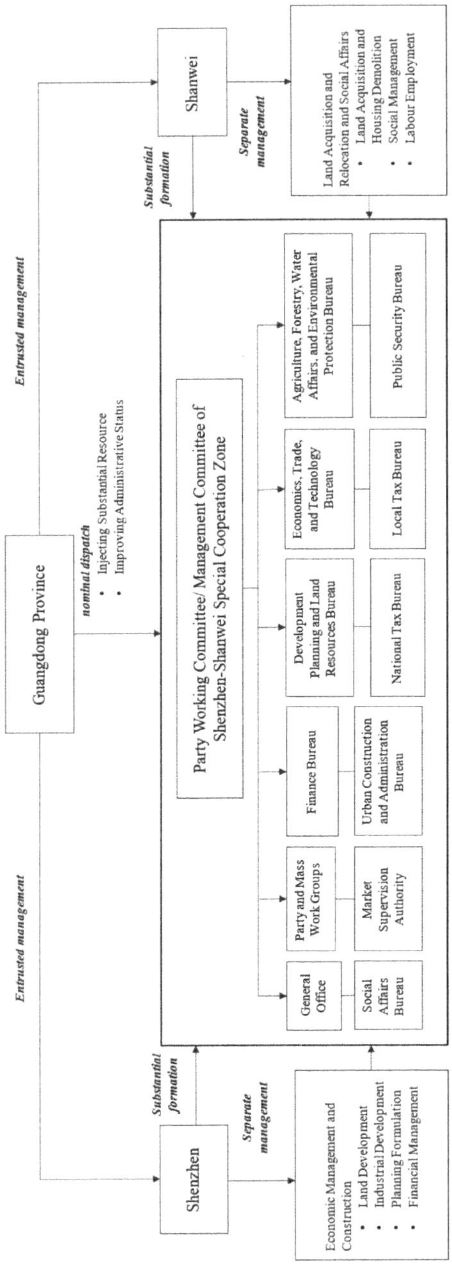

Fig. 6.4 Administrative system of SSSCZ. *Source* Author

First, to facilitate the daily operation of SSSCZ in the organization of staffing, the Guangdong Provincial Government empowered SSSCZ to share prefectural-level administrative power in terms of personnel appointment and removal. However, such empowerment did not mean that SSSCZ was a prefecture-level administrative unit. SSSCZ shared only certain important priorities *vis-à-vis* general industrial transfer parks.

Second, in building the administrative system of SSSCZ, the Party Working Committee and Management Committee of SSSCZ were the agencies of the Guangdong Provincial Government in terms of administrative relations. Such relations were merely nominal, and the Guangdong Provincial Government rarely guided and supervised the daily operation of SSSCZ directly. Moreover, in personnel arrangement, the Guangdong Provincial Government assigned hardly any officials to be stationed in SSSCZ for supervision and coordination.

6.5.2 Integration Between Shenzhen and Shanwei

The administrative system of SSSCZ revealed high integration between Shenzhen and Shanwei. Considering the negotiation between the two cities, Shenzhen was primarily responsible for economic development, such as industrial introduction and land development. Shanwei took direct charge of social and spatial development, such as the management for social development and land acquisition. Regardless of such division of responsibility in the administrative system, the officials of Shenzhen and Shanwei were integrated as an institutional entity through personnel arrangement.

First, one of the local officials in Shenzhen served as the director of the Management Committee of SSSCZ concurrently. Correspondingly, one of the local officials in Shanwei served as the director of the Party Working Committee of SSSCZ concurrently. Second, two of the three deputy directors of the Management Committee of SSSCZ, who took responsibility for industrial introduction and planning, came from the Shenzhen Municipal Government. The other deputy director, who took charge of social issues, was selected from the local officials of Shanwei. Third, for low-level departments in the administrative system, the principals of departments, who took responsibility for economic development, were selected by Shenzhen from its local officials. Moreover, principals for certain social issue-oriented departments were appointed from local officials of Shanwei. However, the deputies of these departments practiced alternate appointments. For example, the director of the social affairs department was selected from the local officials of Shanwei, whereas its deputy director was appointed by the Shenzhen Municipal Government. The specific personnel arrangements were negotiated between Shenzhen and Shanwei.

The alternate appointments of personnel efficaciously repaired the governmental fragmentation triggered by the differentiated division of responsibility in the administrative system. This integrative institutional design enabled high-efficiency management and sustained the negotiation mechanism in the daily operation of SSSCZ.

6.6 Industrial Development of SSSCZ: Shenzhen-Led Re-territorization Strategy

6.6.1 Industrial System and Economic Performance of SSSCZ

The ideology of industrial development in SSSCZ was "headquarters in Shenzhen, bases in SSSCZ." Based on this recognition, the primarily imported enterprises were the branches of enterprises in Shenzhen, and approximately 90% of enterprises were relocated from Shenzhen. Most remaining enterprises were the local enterprises of Shanwei (Interview with a local official, 04/2016). In terms of individual enterprises, 65 were contracted, 59 constructing, and 11 other enterprises were completed and put into operation (Interview with a local official, 04/2016).

However, these enterprises were strictly introduced in accordance with the arrangement of the industrial system in SSSCZ. According to the "SSSCZ Development Master Plan (2015–2030)," SSSCZ was defined as the demonstration plot for the regional coordination and new industrial pillar in Guangdong Province. Five types of industries were acknowledged as the leading industries for SSSCZ: advanced manufacturing, environmentally friendly agriculture, modern services, modern tourism, and advantageous traditional industries. Moreover, large data centers (*Dashuju zhongxin*) remained a preeminent part in the industrial structure. China Resources (*Huarun*), Huawei, Print-Rite (*Tianwei*), and Tencent relocated their large data centers to SSSCZ (Interview with a local official, 04/2016).

Table 6.5 shows the GDP for SSSCZ, Shanwei, and Shenzhen from 2011 to 2017. The economic performance of SSSCZ experienced steady growth, increasing from 2.1 billion Yuan in 2011 to 4.7 billion Yuan in 2017. Compared with the GDP of Shanwei, that of SSSCZ had a considerably larger economic volume. However, SSSCZ was in the start-up period, and most enterprises remained at the construction stage. Thus, economic volume was relatively low in terms of GDP.

6.6.2 Re-territorization Strategy Initiated by the Shenzhen Municipal Government

Considering the division of responsibility between Shenzhen and Shanwei, the industrial development of SSSCZ was the core task for the Shenzhen Municipal Government. Existing practices of Shenzhen manifested as two dimensions to re-territorize the economic flow: (1) designating the concessionary policy of industrial introduction and (2) constructing the regional infrastructure (Interview with a manager of the local enterprise in SSSCZ, 12/2016).

Table 6.5 GDP of SSSCZ, Shanwei, and Shenzhen from 2011 to 2017

Year'	SSSCZ	Shanwei	Shenzhen
2011	2.1	53.0	1151.6
2012	2.5	58.5	1297.1
2013	2.7	64.5	1457.2
2014	3.1	68.1	1600.0
2015	3.5	72.7	1750.3
2016	4.1	79.3	1949.3
2017	4.7	85.5	2239.1

Units billion Yuan
Source Shenzhen National Economy and Society Developed Statistical Bulletin 2011–2017, Shanwei National Economy and Society Developed Statistical Bulletin 2011–2017, "SSSCZ Master Plan (2017–2035) (draft)"

The concessionary policy of industrial introduction involved three dimensions (Interview with a local official, 04/2016).

- Shenzhen and Shanwei will set up more than 1 billion Yuan funds for industrial experimentation.
- Shenzhen will provide an 88-million-Yuan subsidy for large projects in industrial transfer parks annually and 0.4 billion Yuan in five years in total.
- Enterprises in SSSCZ will share similar concessionary policies (e.g., industrial and social policies) with the local enterprises in Shenzhen.

Moreover, the Shenzhen Municipal Government was devoted to facilitating industrial collaboration between its counties/districts with SSSCZ, such as the industrial collaboration between Guangming New District and SSSCZ as well as Longhua New District and SSSCZ.

The Shenzhen Municipal Government also built an extensive regional infrastructure to improve the attractiveness of SSSCZ, especially in the field of regional transportation. For example, Shenzhen Urban Transportation Association and Shenzhen Bus Group Co., Ltd. collaborated with SSSCZ to set up two inter-city bus lines between Shenzhen and SSSCZ. Moreover, the Shenzhen–Shanwei inter-city railway is being constructed to enhance the connectivity of SSSCZ (Interview with a local official, 04/2016).

6.6.3 Hesitation of Enterprises for the Future of SSSCZ

Industrial development was behind the blueprint for the economic development of local governments. According to the "SSSCZ Development Master Plan (2015–2030)," the target GDP in 2020 is 22.5 billion Yuan, which is an extremely big jump from the GDP in 2017 (i.e., around 4.7 billion Yuan). The Shenzhen Municipal

Government provided substantial concessionary policies and optimized the use of regional infrastructure in SSSCZ, but enterprises still hesitated to relocate to SSSCZ.

Enterprises in SSSCZ suffered from three primary misgivings: locational disadvantages, a weak economic base, and the expected high level of public services. One of the business executives in SSSCZ noted:

> For us, we need great courage to make the decision. First, SSSCZ is far away from Shenzhen; even there is a high-speed train to Shenzhen, we still need around 1 hour. The regional transportation is not as sound as Shenzhen. This poor locational condition is the primary impediment for us. In addition, the economic base of SSSCZ is weak. It is a great challenge to promote the growth of enterprises without a solid economic base. Finally, we are very concerned about the level of public services provision. Shenzhen has a high degree of public service provisions, such as public transportation and related social welfare. But SSSCZ is a new district, and it may be extremely difficult to provide a considerably high degree of public service. Generally speaking, we are not optimistic for the future of SSSCZ, and maybe Qianhai is better than SSSCZ. (Interview with a business executive in SSSCZ, 12/2016)

6.7 Findings and Discussions

6.7.1 Major Findings

This section aims to systemically illustrate the governance reshuffling process to enable the industrial integration of SDH through deciphering the inter-scalar and inter-city relations in the case of SSSCZ. Based on the above analysis, this book reports the following empirical findings with reference to SSSCZ.

First, in the negotiation for the site selection of SSSCZ, the agreement between Shenzhen and Shanwei to collaborate was due to the recognition that Shenzhen's crisis, a deadlock in economic development, could be best addressed by Shanwei. Specifically, Shenzhen needed additional urban space to accommodate its socioeconomic activities, especially for its industries. For Shanwei, it was imperative to revitalize its backward economy through industrial transformation and upgrading. Thus, these crises triggered the two cities to set up SSSCZ. Moreover, the Guangdong Provincial Government aimed to address the economic–spatial inequality in the entire Guangdong Province by adopting a series of integration-oriented regional policies. Such concessionary policies and political mobilization from the Guangdong Provincial Government accelerated the establishment of SSSCZ, especially by providing firm institutional guarantees and political incentives to localities. To enable efficient construction of SSSCZ, a steady mechanism for allocating benefits among the Guangdong Provincial Government and the Shenzhen and Shanwei Municipal Governments was formulated; the governments could obtain potential benefits and benefits from daily operation simultaneously.

Second, to legitimatize SSSCZ, the Guangdong Provincial Government allocated construction land quotas to SSSCZ to mitigate the conflicts among and between different levels of government in Shanwei, incurred by this development constraint. This new land incentive of the provincial government reflected that it had become a proactive co-sponsor and collaborator with local governments in the city-regionalization process. In addition, SSSCZ confronted difficulties in obtaining planning approval due to its independent economic status and thus insufficient administrative power. The hurdle was nonetheless solved by the Shanwei Municipal Government by employing a consolidative approach to integrate the fragmented planning system. Making the "SSSCZ Development Master Plan (2015–2030)" through political deregulation signified that the Guangdong Provincial Government left adequate room for the Shanwei Municipal Government to enable institutional innovation.

Third, an administrative system was established, forming two integrative relations across multiple actors to manage the daily operation of SSSCZ. On one hand, the Guangdong Provincial Government empowered SSSCZ to have prefecture-level administrative power in terms of personnel appointment and removal. The deregulation of the Guangdong Provincial Government also contributed to accelerating the construction of SSSCZ. Specifically, the Guangdong Provincial Government merely defined the Party Working Committee and Management Committee of SSSCZ as its nominal subordinate body and virtually dropped out of the administrative system of SSSCZ. On the other hand, while an explicit responsibility division existed between Shenzhen and Shanwei, the alternate appointment of personnel arrangement ensured mutual exchange and supervision simultaneously by the two cities and gave rise to high levels of inter-city integration in the administrative system.

Fourth, to accelerate the industrial development of SSSCZ, Shenzhen provided a series of re-territorization strategies to introduce industries, such as concessions given to the new industries and the construction of regional infrastructure. These Shenzhen-led strategies attracted more than 60 enterprises to relocate to SSSCZ, and the GDP of SSSCZ increased from 2.1 billion Yuan in 2011 to 4.7 billion Yuan in 2017. However, although provincial and local governments had great enthusiasm to advertise and promote the construction and development of SSSCZ, market participation was behind the pace of government in the sense of the scruples about locational disadvantages, a weak local socioeconomic base, and inadequate public services.

6.7.2 Discussion

This book analyzes the relative advantages, participation levels, aims in negotiation, and value orientations of different stakeholders of SSSCZ to decipher the evolution of SSSCZ's governance structure.

6.7.2.1 Rationale of the Collaborative Governance: Inter-scalar and Inter-city Relations

From the perspective of the Guangdong Provincial Government, in the process of city-regionalism, its administrative power on local governments, especially the Shanwei Municipal Government, remained ascendant. The backward economy of

Shanwei was a component of the political performance of the Guangdong Provincial Government. These two integrative strands gave rise to the political mobilization of the Guangdong Provincial Government in forming a pair-city between Shenzhen and Shanwei through constructing an industrial transfer park. Thus, the Guangdong Provincial Government was actively engaged in the development of SSSCZ to promote coordinated regional development and enhance the economic growth of the province. In addition to its active role in political mobilization, the Guangdong Provincial Government also increased the provision of developmental resources and policy incentives based on its relative advantages in administrative power and resources (see Table 6.3), such as the quota of construction land and concessionary policies in the industrial introduction. Despite the active role of the Guangdong Provincial Government in political mobilization and provision of developmental resources and policy incentives, political deregulation was strategically employed to remove obstacles in the development of SSSCZ. Thus, the Guangdong Provincial Government shifted its role from that of regulator and coordinator to a proactive animator and collaborator to promote pro-growth initiatives and coordinated development initiatives through active engagement and political deregulation.

In terms of cooperation between Shenzhen and Shanwei, the complementing of the relative advantages of Shenzhen and Shanwei was the primary driving force that facilitated inter-city cooperation. Table 6.3 shows that Shenzhen confronted constraints from a limited urban space. However, Shenzhen was adept in urban socioeconomic growth by occupying advanced industries and bringing management experience and technology. By contrast, Shanwei retained its massive land resource but was subjected to backward urban economy and industrial structure. These city-based accumulation crises generated collaborative initiatives that were further catalyzed and consolidated by political mobilization from the Guangdong Provincial Government. The example of the site selection of SSSCZ suggests that the expanded territory of Shenzhen revealed the rise of collaborative initiatives between governments. Moreover, the convergence of action frameworks across actors was intensified by the stable mechanism for allocating benefits in SSSCZ (see Table 6.2). The strategy of different local governments such as the Shanwei Municipal Government assumed the responsibility of managing social development and land acquisition due to its jurisdiction over SSSCZ. This finding applies, especially if the strategy manifests as institutional innovation in integrating the fragmented and illegal planning system in SSSCZ. The Shenzhen Municipal Government played a guiding role in the development of SSSCZ, including industrial development, building infrastructure projects, and formulating concessionary policies to attract outside capital. Thus, SSSCZ was efficiently constructed in terms of land acquisition, making plans, and industrial introduction. The rise of collaborative governance in SSSCZ was settled and enabled by the complementary demands between Shenzhen and Shanwei; however, the higher level of socioeconomic development in Shenzhen determined the uneven power relations between them, which generated the dominant role of Shenzhen in the industrial development of SSSCZ *vis-a-vis* the subordinate role of Shanwei in its focus on social affairs. This uneven power relation planted the seeds of failure for collaborative governance of intercity cooperation.

6.7.2.2 Positioning the Collaborative Governance Mode: Conventional Typologies of Governance Modes

The ongoing theoretical discussion on the reshuffling process of governance structures in establishing industrial collaboration identifies four modes of governance modes: multilevel governance, sub-regional differential governance, local government-dominated governance, and hosting governance (see Chap. 2). The governance mode identified in SSSCZ shares some distinctive characteristics of the conventional modes, but it forms a mode of its own.

The provincial government was not directly engaged in the administrative management of SSSCZ. By contrast, MLG manifests as the existence of an upper committee established by the provincial government and sub-regional differential governance, which designates a district administrated by the provincial government. The high level of deregulation from the provincial government in SSSCZ mobilized the enthusiasm of local developments for institutional innovation. However, the retreat of provincial governments resembled the local government-dominated governance mode. The integrative administrative setting in SSSCZ was more collaborative than the fragmented spatial division between cities in the local government-dominated governance mode. In hosting governance, the two cities organized a unified administrative system where the clear duty division was forged, but this system was merely a collage within the administrative system without interference. The collaborative governance mode in SSSCZ manifested as a high degree of integration between cities. This mode of governance had a clear division of duties and mutual supervision and help through alternate personnel appointments. Thus, the case of SSSCZ reveals the rise of a high degree of collaborative governance in SDH city-regionalism.

6.7.2.3 Dynamics of Collaborative Governance: Contingency and Causality

Connected with the conclusion in Chap. 5, in the case of SSSCZ, the government-led integration process in SDH has manifested as a rise of collaborative governance in the state sector. However, the rise of collaborative governance was deeply embedded in the specific local socioeconomic environment, and the contextualization of this typology of governance calls attention to the contingency and causality between the broader institutional context and the socioeconomic crises of localities.

According to Xu and Yeh (2010), PRD is a highly fragmented and growing polycentric city-region. The concessionary policies from the Guangdong Provincial Government have lasted over a long-term period, acting as a regional strategic tool to stimulate local economic growth and coordinate uneven regional development. However, the actual effect of these concessionary policies remained ambiguous for the Guangdong Provincial Government. The decentralizing feature of PRD gave rise to the weak regulatory capability of the Guangdong Provincial Government to localities (Interview with an urban planner, 04/2016). The political mobilization was entangled with the concessionary policies, promoting the regional integration between core cities in PRD and cities in the peripheral region of Guangdong Province. In this case, such an institutional context provided the institutional pathway for Shenzhen and Shanwei to break the local development deadlock through path generation. The reason that macro-level political mobilization received a positive response from Shenzhen and Shanwei was two urgent requests: to expand the territory of Shenzhen to accommodate superfluous industries and to boost the backward economy of Shanwei. The convergence of these three correlative strands triggered the steady power matrix and allocation mechanism of benefits among the Guangdong Provincial Government and the Shenzhen and Shanwei Municipal Governments in the case of SSSCZ. Similarly, the steady mechanism to allocate benefits was formalized through an agreement (see Table 6.2). Thus, the rise of collaborative governance in the case of SSSCZ could be viewed as one of the successful policy experimentations by the Guangdong Provincial Government in localities. This macro-level political mobilization catered to the request of local government to address crises of over-accumulation. In a sense, the such institutional coupling is casual, emerging as a random result of policy experimentation in PRD. For example, for the city of Guangzhou, in the sense of its larger territory and substantial undeveloped lands, political mobilization and concessionary policies may be failures given the indifferent local response for expanding the territory.

However, the above-mentioned contingent match necessarily generated collaborative governance in SSSCZ. Some institutional pathways in making the collaborative governance resembled the territorial practices of MLG in the EU (see Chap. 2). First, the benefit distribution between actors was formalized and crystalized by the agreement in the negotiation process of SSSCZ. This formal institutional arrangement was the precondition and institutional assurance for the operation of collaborative governance. Second, political decentralization of administrative power and deregulation were widely employed in the construction of SSSCZ. The decentralizing and deregulating processes empowered greater economic autonomy for cooperation between Shenzhen and Shanwei, packed tight with a specialized division of duties (see the administrative system of SSSCZ), resulting in the emergence of effectual integration between Shenzhen and Shanwei. Third, aside from the political mobilization from the Guangdong Provincial Government, cooperation between Shenzhen and Shanwei was a voluntary and inter-city type (see Sect. 6.3.2.2) that largely avoided the risk of institutional failure. Thus, the rise of collaborative governance in SSSCZ appears inevitable since the local demand to break the development deadlock was closely linked to and matched by political mobilization from the Guangdong Provincial Government.

6.8 Conclusions

This chapter attempts to conceptualize how the governance structure can be reassembled to enable industrial integration through the case of SSSCZ. Inter-scalar, intercity, and preliminary state-market relations are meticulously elaborated in the four critical issues of SSSCZ, involving a series of actors, such as the Guangdong Provincial Government, the Shenzhen and Shanwei Municipal Governments, relevant county, and town governments. Certain enterprises were also introduced into SSSCZ.

This case reveals that collaborative governance, as an emerging and prevalent approach, has occurred in SSSCZ across different levels of government and between cities. Driven by such a collaborative mode within the state sector, SSSCZ has been established and constructed rapidly in terms of site selection, planning, and land and infrastructure development. The governmental system was operated highly efficiently and collaboratively, but market participation fell behind the pace of government due to divergent considerations for SSSCZ. In terms of inter-scalar, inter-city and state-market relations, this book draws three conclusions.

First, the inter-scalar relation was significantly restructured toward a collaborative mode. In the traditional paradigm, the provincial government serves as the regulator and mediator of local development through regulation and an institutional coordinating mechanism. The provincial government acts as a sub-regional fabric actually in regional governance. However, it has two tasks: both achieving coordinated regional development in the single city-region and promoting pro-growth initiatives at the overall provincial scale vis-à-vis pure regional institutions. The provincial government is perpetually confronted with considerable tough and ad hoc tasks emanating from city-regionalism. In this book, the provincial government reduced the political and institutional intervention for inter-city cooperation. Political deregulation was entangled with the provision of developmental resources and political mobilization to facilitate the construction of SSSCZ. These two emerging trends triggered the shift in the role of the Guangdong Provincial Government from a regulator and mediator to a proactive animator and collaborator.

Second, inter-city relations showed a high degree of integration that manifested as an explicit division of duties, collaboration, and mutual supervision. The closely complementary advantages and aims of negotiation between cities and political mobilization were preconditions to the rise of collaborative governance. Inter-city integration not only manifested in site selection and political legitimization processes, but it also created a highly unified administrative system in SSSCZ that is distinct from the conventional mode of governance.

Finally, collaborative governance originated from and operated within the governmental system. Departure from the market and social involvement is explicitly demonstrated by this book. City-regional governance in China is a state-centric mode of governance despite the context of market-oriented transformation and political decentralization. Furthermore, Li and Chiu (2018) have exposed the germinating state capitalism in the new town development of Shanghai to reveal that local states strategically operate state-affiliated agencies as market vehicles to participate in local capital accumulation. This book shows that place-making through the improvement of infrastructures, social facilities, and concessionary policies in industrial development still remains an essential role in state-market relations; in the short term, this strategy can produce certain effects in industrial introduction. Thus, while this book shows that the government-led integration process reshuffles itself toward the collaborative governance mode in response to industrial collaboration, such collaborative governance still remains the key power reconfiguration within the internal state sector and the government-led re-orchestration in the state apparatus.

In sum, the case study of SSSCZ demonstrates that provincial and local governments reassembled as a collaborative approach to enable an effectual industrial collaboration in SDH. This emerging collaborative governance was crystallized and settled by the steady mechanism for allocating benefits, which was a win–win situation for multiple actors. Facts prove that such collaborative governance was steady and high-efficient in the construction of SSSCZ. However, due to the inherent defects of SSSCZ, such as poor geographic location and local socioeconomic environment as well as the high requirement of public service degree for SSSCZ, enterprises lacked high enthusiasm in response to the call of governments and remained hesitant in terms of the development of SSSCZ.

References

Li, J., & Chiu, L. R. (2018). Urban investment and development corporations, new town development and China's local state restructuring—The case of Songjiang new town, Shanghai. *Urban Geography, 39*(5), 687–705. https://doi.org/10.1080/02723638.2017.1382308

Xu, J., & Yeh, A. G. (2010). Coordinating the fragmented mega-city regions in China: State reconstruction and regional strategic planning. In J. Xu & A. G. Yet (Eds.), Governance and planning of mega-city regions: An internatinal comparative perspective (pp. 213–235). Routledge.

Chapter 7
Exploring the Reshuffling of the Governance Structure to Enable the Infrastructure Integration of SDH: A Case Study of SGSIR

7.1 Introduction

This chapter aims to answer the third research question: how does the reshuffling of the governance structure enable the infrastructure integration of SDH? SGSIR is selected as a case to reveal how inter-scalar relations, a key dimension of governance structure, restructure to enable the planning, construction, and land development of an inter-city railway project. However, as the construction of SGSIR is an indivisible part of the PRD inter-city railway network, this chapter analyzes SGSIR in the broader context of the PRD inter-city railway network. Meticulous examination of such a large-scale trans-jurisdictional regional project identifies the institutional reconfiguration across different levels of authority with particular reference to the relations between the Guangdong Provincial Government and local governments.

Apart from the introduction, this chapter comprises six sections, in line with the analytical framework. The second section briefly discusses the empirical background of the PRD inter-city railway network. The next four sections systematically reveal the inter-scalar relations and their restructuring on the basis of four critical issues: planning approval and negotiation of operating models, site selection for the railway station, financial crises, and local land development. In every section, the relative advantages, participation levels, aims in negotiation, value orientations, and the interaction between the Ministry of Railways and the Guangdong Provincial Government and local governments are systemically analyzed according to semi-structured interviews, with a particular focus on power configurations and specific institutional arrangements. The third section covers the issue of planning approval and the operating model, the reconfiguration of inter-scalar relations from ministerial–provincial–local cooperation to the provincial–local mode of cooperation. However, information asymmetry among different levels of authority helps confer on the local government an advantageous position in negotiating for site selection of the inter-city railway stations under both modes, thereby triggering station peripheralization. The fourth section reveals that the Guangdong Provincial Government encountered a serious financial crisis due to station peripheralization and the hollowing-out of

X. Zhang, *Regional Integration and Governance Reshuffling in the Making of China's City-Regions*, Urban Sustainability, https://doi.org/10.1007/978-981-99-2792-0_7

financial power under a large operation loss from SGSIR. On the issue of local land development, to cover operation losses, the Guangdong Provincial Government reconfigured its power horizontally and vertically, aiming to engage in local land development. The final section offers major findings and answers the research question accordingly.

7.2 Empirical Background: The Project of an Inter-city Railway Network Under PRD City-Regionalism

In the early 2000s, the Guangdong Provincial Government initiated the project of the PRD inter-city railway network as a regional strategy to facilitate regional integration. In June 2000, the Guangdong Provincial Railway Construction Investment Group (hereafter the *Investment Group*), on behalf of the Guangdong Provincial Government, cooperated with China Railway Siyuan Survey and Design Group and Guangdong Provincial Department of Housing and Urban Construction to formulate a feasibility study for the planning and construction of the PRD inter-city railway network. The Guangdong Provincial Government then incorporated the suggestions from different provincial departments to prepare the "Rapid Rail Transit Network Plan for the Pearl River Delta Economic Zone". This version of the plan was later submitted to the Planning Department of the Ministry of Railways for confirmation in August 2003. Subsequently, on the basis of the "Mid–Long Term Railway Network Plan (2004–2020)" (*zhongchangqi tieluwang guihua*) created by the Ministry of Railways, the *Investment Group* submitted a modified plan, namely, "Rapid Rail Transit Network Plan for the Pearl River Delta (2005–2020)", to the NDRC for review. In April 2004, NDRC finished the final review of the "Rapid Rail Transit Network Plan for the Pearl River Delta (2005–2020)" and integrated this plan as part of the "Mid–Long Term Railway Network Plan (2004–2020)" to submit to SC for approval. SC confirmed these endeavors in March 2005, thereby approving the "Mid–Long Term Railway Network Plan (2004–2020)". Thus, the project of the PRD inter-city railway network received national approval and administrative support in 2005.

In general, to facilitate the planning approval process of the PRD inter-city railway network from SC, the Guangdong Provincial Government actively collaborated with the Ministry of Railways, aiming to integrate this plan as part of the "Mid-Long Term Railway Network Plan (2004–2020)". The "Mid-Long Term Railway Network Plan (2004–2020)" as a national-level transportation plan was approved much more quickly by the central government than by provincial-level and local-level plans.

According to the PRD inter-city railway network plan, the overall network will be accomplished in 2020, and the spatial structure of this network is summarized as "Two Principal Axes, Three Auxiliary Axes" (Fig. 7.1). Importantly, SGSIR and the "Guangzhou–Zhuhai Inter-city Railway" were the two principal

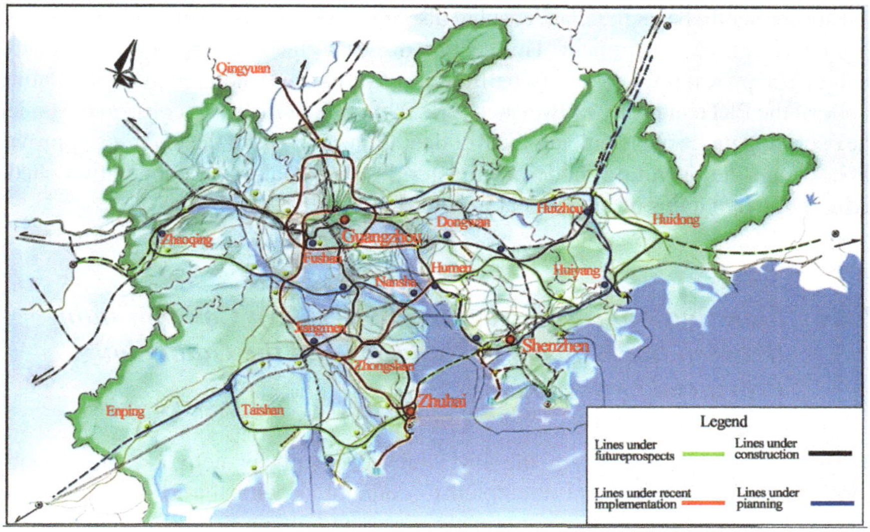

Fig. 7.1 PRD inter-city railway network. *Source* PRD infrastructure integration plan (2009–2020)

axes in this network. The three auxiliary axes included the "Guangzhou–Dongguan–Huizhou Inter-city Railway," "Guangzhou–Xiaolan–Jiangmen–Xinhui Inter-city Railway," and "Guangzhou–Nanhai–Foshan–Sanshui–Zhaoqin Inter-city Railway". The overall length of the PRD inter-city railway network reached approximately 590 km. SGSIR and Guangzhou-Zhuhai Inter-city Railway were the key lines in the network, reflecting the re-articulation by the Guangdong Provincial Government of the infrastructure integration among large cities in PRD.

SGSIR was one of the two principal axes in the PRD inter-city railway network, and its construction will strengthen the geographical accessibility among Shenzhen, Guangzhou, and Dongguan. In addition, the Guangdong Provincial Government was deeply involved in the construction process of SGSIR, which started in 2008 and was undertaken by the China Railway Second Board and the China Railway Port Channel Engineering Group Co., Ltd.

7.3 Planning Approval Process and Negotiation of Operation Mode: Changing Inter-scalar Politics

Planning approval and the mode of operation are central issues in the implementation of SGSIR, and the operating mode in particular has witnessed a transformation from ministerial–provincial–local to provincial–local cooperation. Investigating such a transformation provides insight into the inter-scalar politics underlying SGSIR. In addition, the operating mode not only refers to the agent who pays the invited costs and subsequently bears the main cost but also involves responsibility-sharing among different levels of government. This section discusses the inter-scalar politics of the planning approval process and two different governance structures in the operating mode of the PRD inter-city railway network. Although SGSIR was constructed under the operating mode of provincial–local cooperation, analysis of the planning approval process and the operating mode of ministerial–provincial–local cooperation sheds light on the dynamics of inter-scalar politics for SGSIR as well.

7.3.1 Planning Approval Process (2000–2005): Collaboration Between the Ministry of Railways and the Guangdong Provincial Government

Before the implementation of the PRD inter-city railway network, the Guangdong Provincial Government attempted to collaborate with the Ministry of Railways to accelerate the planning approval process and obtain sufficient financial and technological support. As a local official puts it:

> To facilitate the construction process of the PRD inter-city railway network, the Guangdong Provincial Government foresees that if they can put their inter-city railway network plan into the "Mid–Long Term Railway Network Plan" and obtain support from the Ministry of Railways, the approval from central government will be faster than if they do it separately (Interview with a local official, 12/2017).

From the perspective of the Guangdong Provincial Government, to promote coordinated regional development in the PRD and optimize the regional spatial structure, the primary aim in negotiations during the planning approval stage was to accelerate the planning approval process. Essentially, such a demand for promoting city-regionalism in China was argued to represent the public interest. As the major actor in the PRD inter-city railway network, the Guangdong Provincial Government played the dominant role in the planning process, selecting the Ministry of Railways as its primary collaborator and excluding the local governments. By inviting the Ministry of Railways as a controlling shareholder in the operating process of the PRD inter-city railway network, the Guangdong Provincial Government successfully integrated the "Rapid Rail Transit Network Plan for the Pearl River Delta Economic Zone" into a chapter of the national scheme "Mid–Long Term Railway Network Plan (2004–2020)," and the project of PRD inter-city railway network was approved by SC in March 2005.

The collaboration model for obtaining planning approval as discussed is summarized in Table 7.1. In terms of the Ministry of Railways, the primary aim in the negotiations was to strengthen its administrative regulation and control of the inter-city railway development in the PRD (Interview with a local official, 12/2017), which also facilitated collaboration with the Guangdong Provincial Government. However, although the Ministry of Railways as the national authority represented the public interest, this primary aim in the negotiation indicates that the Ministry was also driven by its own interest, manifested in struggles with other departments and localities over-regulating and controlling local socioeconomic development. In addition, the Ministry possessed abundant capital, technologies, and administrative power, as recounted by a local official.

> The plan made by the Ministry of Railways is easier to get approved from the central state, compared with a local plan. Besides, the Ministry of Railways has abundant capital from the central state to help PRD implement its inter-city railway network strategy, and its technology has a great attraction for Guangdong Provincial Government as well (Interview with a local official, 12/ 2017).

Consequently, due to the collaboration between the Ministry of Railways and the Guangdong Provincial Government, the plan for a PRD inter-city railway network was approved by SC in 2005. In addition, such a collaborative model in the planning approval process also directly determined the operating models of the PRD inter-city railway network: the ministerial–provincial–local cooperation and the provincial–local cooperation as elaborated in the next section.

Table 7.1 Collaboration model in the planning approval process between the Ministry of Railways and the Guangdong Provincial Government

Actor	Relative advantages	Participation level	Aims in negotiation	Value orientations
Ministry of Railways	• Administrative control of the railway system (regulatory capacity) • Closed power relation to the central government for railway development • Technology • Capital	Major actor	• Strengthen administrative regulation and control for local inter-city railway development	Public interest Ministry's interest
Guangdong Provincial Government	• Administrative power on PRD	Major actor	• Accelerate planning approval • Obtain technological and capital support from the Ministry of Railways • Promote coordinated regional development • Optimize the regional spatial structure	Public interest Provincial/ regional interest
Local government	–	Excluded	–	–

Source Author

7.3.2 *Negotiation of Operation Model: From Ministerial–Provincial–Local Cooperation to Provincial–Local Cooperation*

Given the strategy of gaining the planning approval presented above, the negotiation and implementation of the operating model of the PRD inter-city railway network in the period between 2000 and 2010 experienced a transformation from ministerial–provincial–local cooperation to provincial–local cooperation. The operating model involves not only investment-sharing but also responsibility-sharing in the processes of construction, operation, and management across different government levels. Specifically, the stage of ministerial–provincial–local cooperation from 2000 to 2007 was deeply affected by the collaborative model in the application for the planning approval process completed in 2005. Since then, the Guangzhou-Zhuhai inter-city railway was constructed under the same model. The subsequent operating model of provincial-local cooperation started from 2008 to 2010, and SGSIR was constructed under the same operating model. In order to comprehend the formation of this model of operation, this study systemically analyzes the rise of a ministerial-provincial-local cooperative model and how this mode was broken and translated into the operating model of provincial-local cooperation where SGSIR remained.

7.3.2.1 Stage One (2000–2007): Ministerial–Provincial–Local Cooperation

Based on the agreed model of collaboration between the Ministry of Railways and the Guangdong Provincial Government in the planning approval process from 2000 to 2005, the operation of the PRD inter-city railway network also used the ministerial-provincial-local cooperative model and lasted from 2000 to 2007.

The Guangdong Provincial Government signed an agreement with the Ministry of Railways in the "Pan-Pearl River Delta Economic and Trade Cooperation Forum" (*Fanzhusanjiao jingmaohezuo qiatanhui,* 泛珠三角经贸合作洽谈会) in 2004. Hereafter, the operating model of ministerial–provincial–local cooperation was confirmed with reference to the investment- and responsibility-sharing across different government levels. According to their negotiations, the division for investment and responsibility was such that the Ministry of Railways and the Guangdong Provincial Government each bore fifty per cent of the responsibility for investing in the railway construction and the related equipment, and they jointly took responsibility for the construction, daily operation, and management of the inter-city railway. The local governments merely took charge of land acquisition and related housing demolition. Note that the local governments did not bear the operating loss of the railway line. Hence, the Ministry of Railways and the Guangdong Provincial Government played overwhelming roles in investment allocation and responsibility, whereas the local government played a lesser role (Table 7.2). Under the operating model of ministerial–provincial–local cooperation, the construction of one principal axis in the

PRD inter-city railway network, namely, the Guangzhou–Zhuhai inter-city railway, started soon, in December 2005.

In general, this operating model was the by-product of the collaboration model between the Ministry of Railways and the Guangdong Provincial Government in the planning approval process. It is important to note that the local government was a subordinate actor in this operating model, differing from the collaborative model in the planning approval process, where the local government was excluded. In terms of the aims in negotiation, the Ministry of Railways aimed to strengthen its regulatory capacities in local development through investing in the PRD inter-city railway network and bearing related responsibilities. Aside from facilitating the planning approval process, the Guangdong Provincial Government intended to acquire the

Table 7.2 Comparison of the two collaboration modes for investment allocation

Operation model	Actor	Responsibility	Investment allocation	Operating loss
Ministerial–provincial–local cooperation	Ministry of Railways (major actor)	Construction, operation, management, and technology consulting	Fifty percent of investment for the railway construction and the related equipment	Fifty percent of operating loss
	Guangdong Provincial Government (major actor)	Construction, operation, and management	Fifty percent of investment for the railway construction and the related equipment	Fifty percent of operating loss
	Local government (subordinate actor)	Land acquisition and housing demolition	All of the investment for land acquisition and housing demolition	–
Provincial–local cooperation	Ministry of Railways (subordinate actor)	Technology consulting	–	–
	Guangdong Provincial Government (major actor)	Construction, operation, and management	All of investment for the railway construction and the related equipment	All of operating loss
	Local government (subordinate actor)	Land acquisition and housing demolition	All of the investment for land acquisition and housing demolition	–

Source Author

capital and technologies from the Ministry of Railways. Regarding local governments, they could gain socioeconomic benefits from the PRD inter-city railway that was freely parachuted in from upper-level governments (Interview with a local official, 11/2017). Thus, local governments actively participated in the operating model and paid merely for the cost of land acquisition and related housing demolition due to their administrative powers over the territories. Within the operating model of ministerial–provincial–local cooperation, the Ministry of Railways and the Guangdong Provincial Government remained the major actors standing for the Ministry's interest and provincial/regional interests, respectively, while local governments turned into subordinate actors.

7.3.2.2 Stage Two (2008–2010): Provincial–Local Cooperation

However, after 2008, the interests of the Guangdong Provincial Government changed from acquiring technological and capital support from the Ministry of Railways to acquiring the leadership in the projected PRD inter-city railway network. Thus, the operating model of ministerial–provincial–local cooperation was translated into the model of provincial-local cooperation, which lasted from 2008 to 2010 for constructing SGSIR.

The shifting aim of the Guangdong Provincial Government was based on the emerging context, as explained by an urban planner.

> The Guangdong Provincial Government changed its cooperation partner largely due to the construction process of the PRD inter-city railway network. Because the Ministry of Railways injected the major capital into the Guangzhou–Zhuhai inter-city railway, the other railways were delayed, which triggered the Guangdong Provincial Government's dissatisfaction. In addition, the strong regulatory power from the Ministry of Railways destroyed the voice of the Guangdong Provincial Government in the construction process (Interview with an urban planner, 11/2017).

Aside from the dissatisfaction with the unbalanced investment and strong voice of the Ministry of Railways in the project, SC had already approved the PRD inter-city railway network in 2005. Thus, given its rich fiscal revenue from local industries, the Guangdong Provincial Government believed that it could invest in the PRD inter-city railway and bear its operating losses independently (Interview with an urban planner, 12/2017). In addition, the Ministry of Railways confronted huge fiscal deficits in the construction of railway networks at the national scale. To cope with this plight, the Ministry of Railways attempted to concentrate its investments on the main national railways (Interview with an urban planner, 12/2017). Thus, after negotiations, the Guangdong Provincial Government became the sole actor in the construction, operation, and management of the PRD inter-city railway network in 2008. On behalf of the Guangdong Provincial Government, the *Investment Group* oversaw the construction, operation, and management of the said project. Essentially, the shift of the operating mode from ministerial–provincial–local cooperation to provincial–local cooperation (Table 7.2) was due to the provincial government's objection to the regulation from the Ministry of Railways.

According to the operating model of the provincial–local cooperation, the Guangdong Provincial Government made all the investments for the railway construction and the related equipment and assumed the sole responsibility for the construction, daily operation, and management of the inter-city railway. Correspondingly, the provincial government needed to bear all the operating losses from the PRD inter-city railway network. The local governments continued to oversee the land acquisition and housing demolition. However, the Ministry of Railways did not totally retreat from the PRD inter-city railway project and provided the function of technological consultation for the Guangdong Provincial Government (Table 7.2). The Guangdong Provincial Government revised the plan for the PRD inter-city railway network and formulated a new version in 2009. On the basis of the new version, SGSIR became one of the major inter-city railways in the future plan. Moreover, the major actor for its construction, operation, and management was the *Investment Group*, on behalf of the Guangdong Provincial Government.

With reference to the analytical framework on inter-scalar governance presented in Chap. 4, the negotiation aim of the Guangdong Provincial Government shifted from facilitating the planning approval to acquiring leadership of the PRD inter-city railway network, whereas the Ministry of Railways retreated from the PRD inter-city railway network due to its limited funds and main national-railway-first strategy. Through merely providing technological consultation, the Ministry of Railways resorted to a subordinate role, akin to that of the local governments, while the Guangdong Provincial Government remained the sole major actor in this operating model. The setup of the operating model of provincial-local cooperation gave rise to the intricate inter-scalar interactions in the site selection and land development processes of SGSIR as well.

7.3.3 Summary

In the planning approval process and negotiations of the operating models of the PRD inter-city railway network, the changing aims in negotiation for different actors in the state sector triggered a restructuring of inter-scalar relations (Fig. 7.2).

Three major findings can be reported based on the investigation above.

First, the closed power relation to the central government, sufficient funds and technologies, and the regulatory capacity for inter-city railway development of the Ministry of Railways led to a collaborative model between it and the Guangdong Provincial Government in the planning approval process under the latter's intention to facilitate the approval of the plan for the PRD inter-city railway network. Such a collaborative model excluded the involvement of the local governments and gave rise to the operating model of ministerial–provincial–local cooperation from 2000 to 2007.

Second, the collaborative model between the Ministry of Railways and the Guangdong Provincial Government further stipulated the equal division of responsibility-

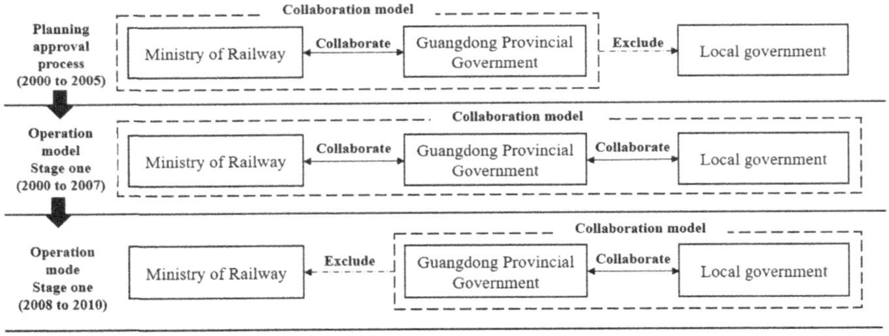

Fig. 7.2 Changing inter-scalar relations in the planning approval process and negotiation of operation models in different stages. *Source* Author

and investment-sharing in the operating model of ministerial–provincial–local cooperation. As the major actors, the Ministry of Railways and the Guangdong Provincial Government were jointly responsible for the construction, operation, and management of the inter-city railway network. Local governments merely oversaw the land acquisition and related housing demolition.

Third, as a result of the provincial opposition to the regulation and administrative control of the Ministry of Railways, the operating model shifted to provincial–local cooperation from 2008 to 2010, centering on the Guangdong Provincial Government. The Guangdong Provincial Government as the sole major actor subsequently took charge of the construction, operation, and management of the PRD inter-city railway network independently. The Ministry of Railways was retained merely as a technological consultant for the PRD inter-city railway network. The local government continued to take responsibility for land acquisition and related housing demolition.

To summarize, the Guangdong Provincial Government reshuffled the forms of collaboration across different levels of authorities to ensure its dominant role in the PRD inter-city railway network. For instance, it collaborated with the Ministry of Railways and excluded local governments from the planning approval process, collaborated with the Ministry of Railways and the local governments during stage one (2000–2007), and both excluded the Ministry of Railways and collaborated with the local governments in stage two (2008–2010). These changing inter-scalar relations further gave rise to peri-urban locations for the SGSIR stations, as elaborated in the next section.

7.4 Site Selection for the Railway Station: Peri-Urban Locations for the SGSIR Stations Under Inter-scalar Politics

SGSIR was constructed in 2008 under the operating model of provincial–local cooperation. There were complicated inter-scalar and inter-city politics between the Guangdong Provincial Government and local governments underlying the site selection process of the SGSIR stations. This book investigates the site selection process for railway stations under the operating model of ministerial–provincial–local cooperation initially, contributing to an understanding of the site selection of the SGSIR stations under the operating model of provincial–local cooperation.

7.4.1 Inter-scalar Politics for Site Selection Under Ministerial–Provincial–Local Cooperation

7.4.1.1 First Phase of Inter-scalar Bargaining for Site Selection

The first phase of inter-scalar bargaining for site selection occurred from 2000 to 2004, which corresponds to the administrative approval of the "Mid–Long Term Railway Network Plan (2004–2020)". As indicated in the prior analysis, the Guangdong Provincial Government attempted to raise the PRD inter-city railway network to the national level in the planning approval process. Hence, the Ministry of Railways retained a strong decision-making power in the proposed PRD inter-city railway network until 2008 (see Sect. 7.3.2.2), and the primary objective in the first phase of site selection was to confirm the railway alignment and the cities or districts on the route rather than determining the specific railway site. As such, inter-scalar bargaining was conducted in two phases. The inter-scalar politics in this phase is shown in Fig. 7.3.

The first phase was the negotiation between the Ministry of Railways and the Guangdong Provincial Government. Although the outline of the PRD inter-city railway network at this stage was determined by both entities jointly, given the strong regulatory capacity of the Ministry of Railways, the Guangdong Provincial Government was subordinate to the Ministry of Railways (Interview with a local official, 12/2017). According to the division of responsibility in the ministerial–provincial–local cooperation, the Ministry of Railways bore fifty per cent of the operation losses of the PRD inter-city railway network. Therefore, its primary aim in negotiation was to maximize the operation's income to cover the financial deficit, standing in for the Ministry's interest. Thus, the Ministry maintained that the PRD inter-city railway should pass through a highly populated area or urban center to absorb substantial passenger flow for compensating for the fiscal losses. The Guangdong Provincial Government, however, aimed to enhance coordinated regional development. Their aim indirectly aligned with that of the Ministry, as more populated cities, had greater

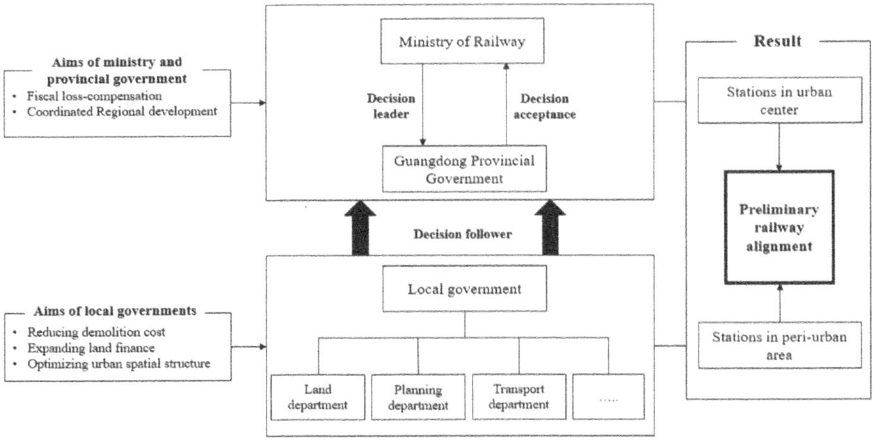

Fig. 7.3 First phase of inter-scalar bargaining (2000–2004). *Source* Author

regional importance, and the government supported the proposition of the Ministry of Railways for site selection (Interview with an urban planner, 12/2017).

The second phase refers to the negotiations between the Ministry of Railways and local governments. The PRD inter-city railway network, as a large-scale and trans-jurisdictional infrastructure project, would lead to a large socioeconomic benefit for urban development. In addition, the local governments needed merely to bear the cost of land acquisition and housing demolition. Thus, at this phase, in terms of aims in negotiation, the local governments attempted to become one of the member cities in the PRD inter-city railway network and hoped that the railway could pass through their urban areas. Thus, they rarely opposed the intentions of the Ministry of Railways on site selection (Interview with an urban planner, 12/2017).

To sum up, the strong decision-making power of the Ministry of Railways restrained the common interest of the provincial and local governments in the selected railway station sites. Thus, the first phase of site selection went rather smoothly.

7.4.1.2 Second Phase of Inter-scalar Bargaining for Site Selection

After the preliminary railway alignment in the first phase of inter-scalar bargaining, the primary objective of the second phase of inter-scalar bargaining was to confirm the specific location of every station, which occurred from 2005 to 2008. The Guangdong Provincial Government continued to remain a subordinate actor in the inter-scalar bargaining. However, unlike in the first phase of inter-scalar bargaining, the local government had a different aim in the negotiations to confirm specific site selections. The inter-scalar politics in this phase are summarized in Fig. 7.4.

According to the aims of the local governments, the inter-city railway station should have been located in the suburban areas, new town areas, and undeveloped

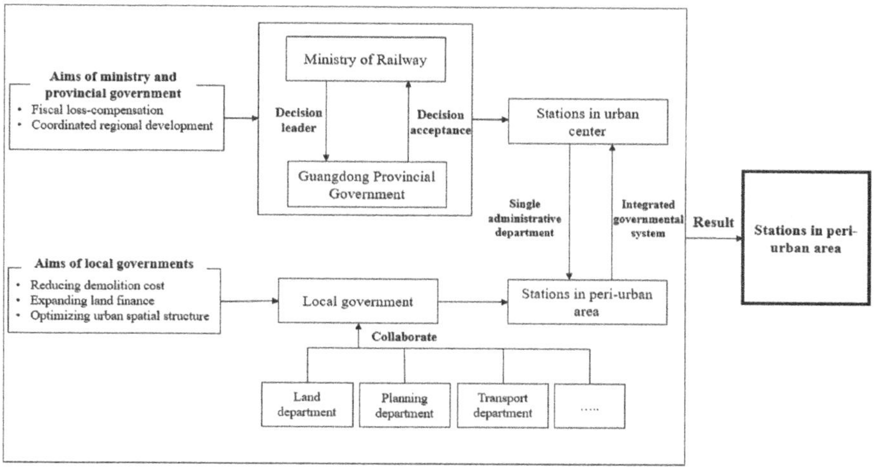

Fig. 7.4 Second phase of inter-scalar bargaining (2005–2008). *Source* Author

areas rather than in the urban center and in a highly populated area. This aim was largely rooted in the low cost of land development and housing demolition in those preferred areas, as well as the higher land profits that local governments could obtain from these areas because of the high development potential created by the railway. The inter-city railway was a crucial infrastructure project to improve urban accessibility and local socioeconomic vitality in the suburban areas, new town areas, and undeveloped areas. A local official in Guangzhou expressed this view:

> The local government rarely considers site selection for railway stations from the regional perspective. As for the municipal government, the inter-city railway is like a pie falling from the sky (天上掉下的馅饼), because they don't need to bear the operating loss and the investment for construction, operation and management. They want to maximize this opportunity for expanding urban land stock through putting sites in the peri-urban area. Such a regional infrastructure project provides a precious opportunity for them to create land finance and also to promote the new town's development (Interview with a local official, 11/2017).

The divergence in site selection between the Ministry of Railways and local governments was the primary tension in the second phase of inter-scalar bargaining, and the Guangdong Provincial Government stayed in a subordinate role in this stage. However, the local governments played an active role in this phase of bargaining. Within the negotiation process, the local governments integrated the different related governmental departments (e.g., the land department, planning department, and transport department) to share information and make strategies to negotiate with the Ministry of Railways, operating as an integrated governmental system (Interview with a local official, 11/2017). In contrast, the Ministry of Railways acted merely as a special administrative department, lacking the capacity to integrate across different national departments and information on the local socioeconomic developments. Given such an institutional difference between the integrated government system among the local governments and the single administrative department of the railway ministry, the capability to obtain information within the bargaining process was different, which greatly affected the results of inter-scalar negotiation (Interview with a local official, 12/2017).

Within the negotiation process, to boycott the claim of the Ministry of Railways on site selection, the local governments argued that the division of property rights, the high cost of housing demolition, and the related social resistance from local residences would be difficult to address if the railway site was located in a highly populated area. Given the lack of information for the Ministry of Railways on the local socioeconomic environment, the local governments were in advantageous positions during the negotiation. Specifically, the primary aim of the Ministry of Railways was to strengthen administrative regulation and control of inter-city railway development rather than compensation for operation losses through putting the railway stations in urban centers or highly populated areas as national finances provided strong financial support for the Ministry of Railways. However, social resistance from local residents was a major concern for the Ministry under the national policy of creating a Socialist Harmonious Society (*shehuizhuyi hexie shehui,* 社会主义和谐社会) (Interview with a local official, 12/2017). As a result, in the second phase of inter-scalar bargaining, the Ministry made some concessions in site selection. The stations of the inter-city railway, especially for the Guangzhou–Zhuhai inter-city railway, were located primarily in suburban areas, new town areas, and undeveloped areas.

Overall, the bargaining result of moving stations away from the urban centers to peripheral areas was triggered by the divergent relative advantages between an integrated government system and a single administrative department in the negotiation process, although the latter was higher in hierarchical position.

7.4.2 Inter-scalar Politics for Site Selection of SGSIR Under Provincial–Local Cooperation

7.4.2.1 Inter-scalar Bargaining for Site Selection of the SGSIR Stations

Apart from ministerial–provincial–local cooperation, another level of the operating model of provincial–local cooperation also took place in the process of inter-scalar bargaining for site selection. To closely examine the dynamics, the SGSIR station selection was investigated as a case study. Initially, the *Investment Group* negotiated with the local governments, on behalf of the Guangdong Provincial Government. However, the political logic of inter-scalar bargaining at this stage from 2008 to 2010 was similar to that in the previous stage, and the *Investment Group* also confronted the acute issue of lacking information or information asymmetry in negotiations with the local government's on-site selection of the SGSIR stations.

The initial agreement on selecting sites for the SGSIR stations was initially reached before 2008 under the operating model of ministerial–provincial–local cooperation. However, with the shifting to the collaboration mode, the site selection of the SGSIR stations was re-negotiated between the Guangdong Provincial Government and the local governments when the SGSIR project moved into the implementation stage. According to the initial draft, the locations of the SGSIR stations in Guangzhou, Dongguan, and Shenzhen passed through highly populated areas or urban centers in order to attract a sizeable passenger flow. After the provincial–local cooperation mode was established in 2008, negotiations on the site selection of the SGSIR stations were continually conducted between the Guangdong Provincial Government and the different local governments.

(1) Guangzhou

According to the initial version of SGSIR, the starting station in GSIR in Guangzhou was Pazhou, which was an urban center and closed to the commercial centers of Guangzhou, e.g., the Guangzhou Baiyun International Convention Center and the Poly International Plaza. However, during the subsequent provincial–local negotiations, the Guangzhou Municipal Government opposed the selection of Pazhou. As told by a local official in Guangzhou:

> The government department argues that the urban planning and urban design of Pazhou have been confirmed and the set-up of SGSIR in Pazhou will largely change the previous urban planning and design scheme and incur a huge institutional cost. Besides, most of the land in Pazhou has been leased, and the cost of readjustment of land use is too high (Interview with a local official, 12/2017).

During the provincial–local negotiations, the *Investment Group* failed to obtain sufficient local information on Pazhou and persuaded the Guangzhou Municipal Government to establish a station in Pazhou. Finally, the Guangdong Provincial Government made an agreement with the Guangzhou Municipal Government to re-allocate Xintang as the SGSIR station in Guangzhou. Xintang was in the Zengcheng District, which was a suburban area of Guangzhou (Fig. 7.5).

(2) Shenzhen

In terms of the initial version, the SGSIR line was designed to pass through the Shajing, Fuyong area and finally connected to the station at the Shenzhen Airport East (Fig. 7.6). In addition, SGSIR acquired the right to use the transport corridor of Line Eleven of the Shenzhen Subway to connect with the central area of Futian. However, during the provincial–local negotiation, the Shenzhen Municipal Government argued that SGSIR had a strong functional overlap with Line Eleven, which would have an adverse impact on the operation of the said subway line. Hence, the Shenzhen Municipal Government proposed a new solution, which was to connect SGSIR to the station at Shenzhen Airport East. Moreover, it should concatenate Luohu and Futian by merging with Line One. Considering that Line One was a slow line, the *Investment Group* opposed this solution (Interview with an urban planner, 12/2017). The negotiation lasted until 2009 and tentatively located the station in the Songgang

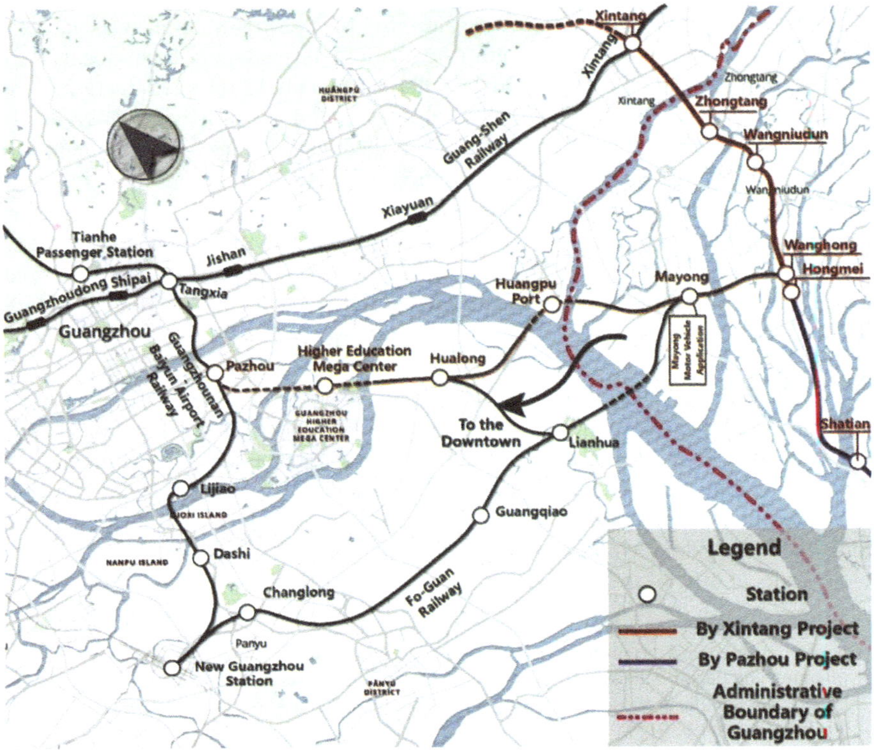

Fig. 7.5 Station peripheralization of SGSIR in Guangzhou. *Source* Author

area, on the outskirts of Shenzhen (Fig. 7.6). It was expected that the site selection would be further re-negotiated in the future.

(3) Dongguan

According to the initial version, the stations of Dongguan far outnumbered those of Shenzhen and Guangzhou. Therefore, the cost of land acquisition and housing demolition was far higher in Dongguan than in Shenzhen and Guangzhou as well. SGSIR in Dongguan started from the station of Mayong Town, connected with a bus station, and then merged with the Convention Center station of Line R1 of an existing subway. Then, it also merged with Line R2 of the subway and connected with the Humen and Chang-an stations. The initial site selection had also undergone complex provincial–local negotiations to rearrange the specific site location of the related station.

The Dongguan Municipal Government argued that a functional overlap occurred between SGSIR, R1, and R2. Such an arrangement would largely weaken the function of R1 and R2 with their similar speeds and station settings. The R1 and R2 lines had been confirmed by the "Dongguan Railway Transit Planning" and would

Fig. 7.6 Station peripherization of SGSIR in Shenzhen. *Source* Author

be difficult to reroute (Interview with a local official, 12/2017). In addition, because Shenzhen and Guangzhou had relocated the stations from urban centers to outskirt areas, Dongguan could not receive any economic benefits that radiated from their booming urban economies. Hence, the Dongguan Municipal Government also negotiated with the *Investment Group* to relocate the station to a peripheral area. Consequently, SGSIR was routed through the Shuixiang District in the western and outskirt areas of Dongguan (Fig. 7.7).

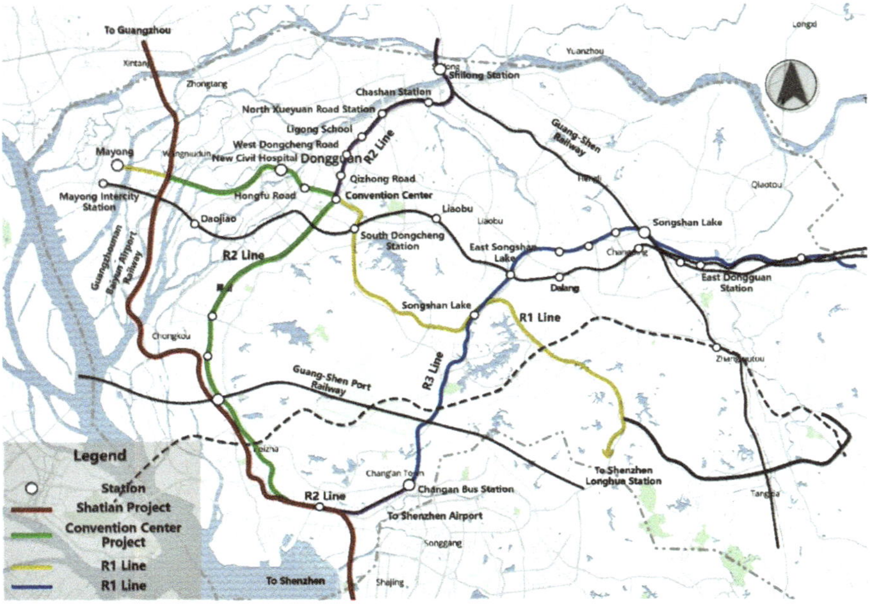

Fig. 7.7 Station peripheralization of SGSIR in Dongguan. *Source* Author

7.4.2.2 Conceptualizing the Inter-scalar Politics of Site Selection of the SGSIR Stations

With the aforementioned analysis, the inter-scalar politics of site selection for SGSIR stations emerged under the provincial–local mode of cooperation. The Guangzhou, Shenzhen, and Dongguan Municipal Governments resisted the site selection of the SGSIR stations in the initial version, triggering the reshuffling of the power configuration of the inter-city railway. The inter-scalar politics on the site selection of SGSIR stations in Guangzhou, Shenzhen, and Dongguan led to the spread of stations to peri-urban areas. As shown in Table 7.3, local governments resisted the Guangdong Provincial Government on the initially agreed-upon locations of the SGSIR stations based on the unsupportive local socioeconomic situations.

Through investigating the relative advantages, aims in negotiation, participation levels, and interest orientation of different actors in this bargaining process of site selection, the aims of the negotiation for the local governments were to optimize urban land income and boost the new town development through relocating the SGSIR stations to suburban areas. These aims were driven by the pro-growth and economy-first ideologies based on the GDP-based political evaluation system (Zhou, 2007). In order to acquire higher promotion benefits, local governments attempted not only to compete against each other to attract mobile capital, but also to maximize land revenue through the land-based finance system (Zhou, 2007). Specifically, not only was it easier for the local government to levy land and remove housing from

Table 7.3 Station peripheralization of SGSIR in Guangzhou, Shenzhen, and Dongguan

City	Site selection (Guangdong provincial government)	Site selection (Municipal government)	Local pretense	Result
Guangzhou	Pass through Pazhou (urban center)	Pass through Xintang (suburban area)	• Urban planning and design are difficult to readjust • Land around Pazhou has been leased	Passed through Xintang (suburban area)
Shenzhen	Merge with Line Eleven of the subway	Merge with line one of the subway	• Functional overlapping between SGSIR and Line Eleven	Temporally located in Songgang (suburban area) The second half was pending
Dongguan	Merge with R1 and R2 subway lines (urban center)	Pass through Shuixiang District (suburban area)	• Functional overlapping between SGSIR and R1 and R2 subway lines • Weak economic radiation from Guangzhou and Shenzhen	Passed through Shuixiang District (suburban area)

Source Author

the suburban and outskirt areas, but such locations in suburban areas also provided more land income *vis-a-vis* with the secondary development land in an urban center.

Likewise, the Guangdong Provincial Government, also viewed the PRD inter-city railway network as a crucial regional project to promote the economic performance of its jurisdiction. Hence, the Guangdong Provincial Government's aim in the negotiations was the optimization of the provincial economy through using the PRD inter-city railway network to optimize the regional spatial layout. Thus, the Guangdong Provincial Government claimed that the SGSIR stations should be located in an urban center or a highly populated area.

However, in terms of the relative advantages of these actors, information on local socioeconomic development in the provincial–local interaction was asymmetric and was the key reason for moving the stations to the peri-urban areas. Similar to the inter-scalar politics under the operation model of ministerial–provincial–local cooperation, the *Investment Group* was a single administrative entity and faced the issue of missing local information. When the local governmental departments combined as an integrated government system to negotiate with the *Investment Group*, its negotiation capability was higher and the local pretense was difficult to refute. In addition, when SGSIR was implemented at the local level, the Guangdong Provincial Government still needed collaboration from local governments in terms of land acquisition and housing demolition. The inter-scalar politics under the operation model of provincial–local cooperation are conceptualized in Fig. 7.8. As shown, the local governments operated as an integrated governmental system that could effectively

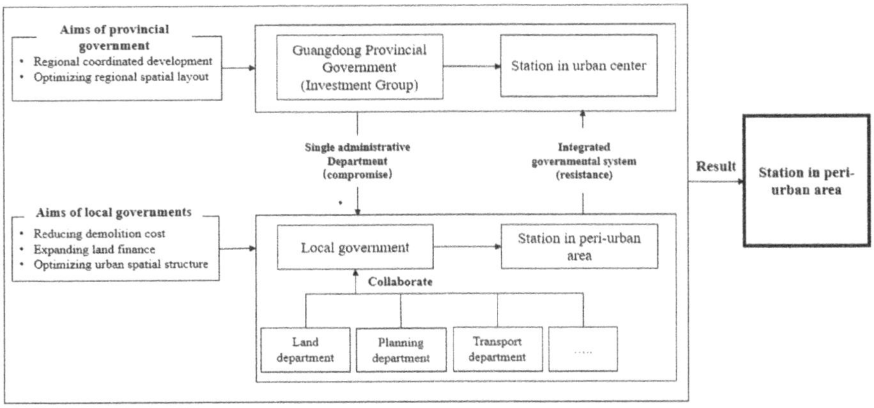

Fig. 7.8 Inter-scalar politics under provincial–local cooperation. *Source* Author

integrate the information about local socioeconomic development across the different departments. When they negotiated with the *Investment Group*, which operated as a single administrative department, they won because of their advantageous position and succeeded in moving the stations to peri-urban areas.

7.4.3 Summary

In the site selection for the inter-city railway station such as the Guangzhou–Zhuhai inter-city railway and SGSIR, the inter-scalar politics reshuffled from collaboration-based politics to a multi-scalar game. The negotiations on site selection across the Ministry of Railways, the Guangdong Provincial Government, and local governments were also deeply affected by the changes in the stakeholders' aims in negotiations and the different operating models (see Sect. 7.3.2). Based on the investigation above, three major findings resulted.

First, generally, due to the existing responsibility allocation in the operation model (see Table 7.2), the Ministry of Railways and the Guangdong Provincial Government maintained that the inter-city railway station should be located in the urban center or highly populated area, but the local governments hoped that the station could be located in a suburban area or the outskirts. Inter-scalar negotiations on the site selection of the PRD inter-city railway station led to the peripherization of the station from urban centers or highly populated areas to suburban areas or outskirts. The key intuitional factor to illustrate this result was the asymmetric information-obtaining capacity on local socioeconomic development for the Ministry of Railways, the Guangdong Provincial Government, and local governments. The Ministry of Railways and the *Investment Group* constituted a single administrative department that encountered difficulty in negotiating with local governments as an integrated governmental system. Thus, some local pretenses stemming from the local socioeconomic environment made the Ministry of Railways and the Guangdong Provincial Government compromised the local governments on site selection. The local resistance was supported by the relative advantages of local governments, in terms of their strong integrative capability to gather local socioeconomic information across governmental departments. Thus, higher-level governments or authorities could not dominate the inter-scalar negotiations.

Second, under the operating model of ministerial–provincial–local cooperation, the Ministry of Railways and local governments as the major actors for site selection negotiated with each other according to different aims and interests. To cover the operation losses, the Ministry of Railways maintained that the selected station should be in an urban center or highly populated area. Conversely, the local governments, aiming to reduce related costs, expand land finance, and optimize urban spatial development, argued that the station should be in a suburban area or outskirts to avoid planning modification and social resistance. Due to the information missing on local development, the strong finance base from national finance, and the strategy to build a Socialist Harmonious Society, the local governments obtained advantageous positions in the negotiation and used the stations to catalyze peri-urban development.

Third, regarding the construction of SGSIR, under the operating model of provincial–local cooperation, the *Investment Group*, on behalf of the Guangdong Provincial Government as the major actor, upheld that the inter-city railway station should be located in the urban center or a highly populated area with consideration for regional coordinated development and the optimization of the regional spatial layout. Similar to the Ministry of Railways, the *Investment Group*, as the single administrative department with weak ability to obtain information, also failed to negotiate equally with local governments in site selection. The phenomenon of locating stations in the peri-urban areas therefore occurred widely in Shenzhen, Dongguan, and Guangdong.

To summarize, for the site selection of the inter-city railway station, a multi-scalar game occurred among different major actors motivated by their own aims in the railway project. The conflicts and diverse aims among urban governments, regional governments, and the railway authority of the central government and the unequal advantages in terms of influence in obtaining information and drawing support within the inter-scalar system, were the key institutional factors that drove the multi-scalar game and caused the spread of stations in the peri-urban areas. Different from the inter-scalar relation in the planning approval process and negotiations of the operating model (see Sect. 7.3), the inter-scalar relations in the station selection process shifted from collaboration to local government-led, forcing the higher-level Ministry of Railways and the Guangdong Provincial Government to compromise and accede to the wishes of the lower-level local governments.

7.5 Financial Crisis Under the Peri-Urban Location of SGSIR

7.5.1 *Financial Crisis on the Operation Deficit of SGSIR*

The inter-scalar negotiation on the site selection of the SGSIR stations triggered station peripheralization and developed beyond the expectation of the Guangdong Provincial Government. With the implementation of the PRD inter-city railway network, the Guangdong Provincial Government faced a heavy financial burden from both the bank loan for the construction funds and the unexpected operation deficit, especially for SGSIR (Interview with a local official, 12/2017).

According to the calculation from the *Investment Group*, the SGSIR project had an operation deficit of 0.71 billion Yuan in the first year of its completion, 0.6 billion Yuan in the second year, and 0.05 billion Yuan in the tenth year. SGSIR is predicted to have a total operating loss of 3.57 billion Yuan in the ten years after its completion (Interview with a local official, 12/2017). This figure means that it would require ten years for SGSIR to reach a balance of payment. Moreover, the Guangdong Provincial Government had to bear the construction costs of SGSIR independently, according to the provincial-local cooperation mode.

In addition to SGSIR, the Guangdong Provincial Government bore a large financial burden for other inter-city railways, such as the Dongguan–Shenzhen line and the Foshan–Zhaoqing line. The operating losses of the Dongguan–Shenzhen line and the Foshan–Zhaoqing line in ten years after their completion will reach 1.5 billion Yuan and 2.83 billion Yuan, respectively. Considering the bank loan for construction funds, the Guangdong Provincial Government should pay more than 15 billion Yuan for SGSIR construction and operation, the Dongguan–Shenzhen line, and the Foshan–Zhaoqing line, for ten years in total (Interview with a local official, 12/2017). As a result, station peripheralization led to an unexpected operating loss due to the low amount of passenger flow in peri-urban areas, which further enlarged the financial crisis of the Guangdong Provincial Government, making the construction and operation of the PRD inter-city railway network more difficult.

7.5.2 *Hollowing-Out of Financial Power for the Guangdong Provincial Government*

Compared with the national finances, in terms of the tax base, the financial capacity of the Guangdong Provincial Government was weak, because it had no right to conduct land financing, which is an important source of revenue for urban governments. Moreover, given the unexpected heavy operating losses incurred by station peripheralization, supporting the construction and operation of SGSIR would be difficult for the Guangdong Provincial Government. The public financial expenditure of the

Guangdong Provincial Government in 2005 reached 1.28 trillion Yuan, with 201.7 billion Yuan in the field of transportation.[1] Regarding SGSIR, the Dongguan–Shenzhen line, and the Foshan–Zhaoqing line alone, the Guangdong Provincial Government needed to pay 4 billion Yuan annually (Interview with a local official, 12/2017). Hence, a heavy burden was placed on the Guangdong Provincial Government.

From the existing power configuration across the governmental system and to tackle the issue of a financial crisis, the hollowing-out of financial power became the primary obstacle for the Guangdong Provincial Government to implement the PRD inter-city railway network under station peripheralization.

Regarding the local governments, they could have taken on the infrastructure construction through land financing. For example, although the local governments were responsible for the construction of the urban subway, they could raise construction funds from the land financial system. Land financing provided substantial funds for local governments to construct and operate the urban subway and compensate for its operating losses. In addition, the local governments tended to integrate the construction and operation of a subway with the land development around a subway station to yield profits (Interview with an urban planner, 12/2017). However, the Guangdong Provincial Government had no right to operate land financing and no related governmental system to raise money through the land financing system. Thus, the operating losses of SGSIR placed a heavy financial burden on the Guangdong Provincial Government due to the hollowing-out of its financial power.

To summarize, under the financial crisis, the limited provincial financing failed to provide funding to the Guangdong Provincial Government to implement SGSIR. In addition, the lack of the right-to-operate land financing triggered the hollowing-out of financial power for the Guangdong Provincial Government, given its duty to construct and operate the PRD inter-city railway network. In the next stage, the Guangdong Provincial Government attempted to re-configure its power relations with the local governments to engage in local land development and cover the large operating deficit.

7.6 Local Land Development: Reconfiguring the Inter-scalar Relations

Before the construction of SGSIR, the Guangdong Provincial Government attempted to negotiate with the local governments to participate in local land development, aiming to cover the large operating losses. However, the negotiation between the *Investment Group* and the local governments failed to form a provincial government-involved land development model due to resistance from the local governments. As a local official stated:

---

[1] Guangdong Provincial Financial Department, (2016). The implementation of the public payout budget of the Guangdong Provincial Government in 2005 and the drafted budget of the 2016 report (*Guangdongsheng 2015nian yusuan zhixingqingkuang he 2016nian yusuancaoan de baogao*).

> The Investment Group wants to participate in land development around the inter-city railway station and this decision totally goes against our interests. We have reached an agreement across different governmental departments that we will resist firmly the involvement of the provincial government in land development. Because we have already born the cost of land acquisition and demolition of housing, why should we have to pay for the operating losses? In addition, land development around the stations will bring a large cost for the infrastructure building and institutional cost for the construction land quota (Interview with a local official, 12/2017).

By the same token, the weak negotiating capacity of the *Investment Group* as a single governmental department led to the failure in negotiation on land development with the integrated local government, comprised of a range of departments. Therefore, in the subsequent process for the PRD inter-city construction, especially for SGSIR, the *Investment Group* needed to seek institutional support from the Guangdong Provincial Government. The following section explains how the Guangdong Provincial Government reconfigured powers to facilitate its involvement in land development.

7.6.1 Power Integration Across Provincial Departments

During the negotiations, the *Investment Group*, as a single administrative department, showed a weak capacity to grasp the local political-economic environment. This weakness was the primary reason why the *Investment Group* failed to obtain information on local socioeconomic development.

To repair the fragmentation among provincial departments, the Guangdong Provincial Government conducted two activities to integrate the information across departments, especially in the case of SGSIR. The first was the set-up of the "Task Group for Comprehensive Land Development of the Inter-city Railway of Guangdong Province" (hereafter the *Task Group*). The second was the carrying-out of a general survey across provincial departments to obtain information on local land development.

In January 2012, the Guangdong Provincial Government issued the important document "Suggestions on the Optimization of Comprehensive Land Development Mechanism around PRD Inter-city Railway Station," which stipulated the following:

> The Guangdong Provincial Government will set up a Task Group that takes responsibility for the arrangement of comprehensive land development and coordination of crucial issues in the land development process. The group leadership is held by a vice-governor; members are composed of local officials from related local government, the Provincial Development and Reform Commission, Provincial Finance Department, Provincial Resource Department, Provincial Housing and urban–Rural Development Department, State-owned Assets Supervision and Administration Commission, Guangzhou Railway Group Company, the GPRCIR and The Pearl River Delta Inter-city Railway Co. Ltd and Provincial Development

and Reform Commission is in charge of the daily operation of Task Group[2] *(Guangdong Provincial Government, 2012).*

On the basis of this governmental document, the *Task Group* was created in July 2012, aiming to build a new institution to integrate the provincial departments on information exchange, notably including the Guangdong Provincial Development and Reform Commission, Department of Housing and Urban–Rural Development of Guangdong Province, and Department of Land Resource of Guangdong Province. Within the *Task Group*, the Guangdong Provincial Development and Reform Commission focused primarily on how to allocate the duty of covering operating losses between the provincial government and the local governments. The Department of Housing and Urban–Rural Development of Guangdong Province and the Department of Land Resource of Guangdong Province oversaw the formulation of land and planning regulatory policies for the implementation of cooperative land development (Interview with a local official, 12/2017).

As for the general survey, it aimed to collect information on land use around the station of SGSIR and was primarily undertaken through the collaboration between the Department of Housing and Urban–Rural Development of Guangdong Province and the Department of Land Resources of Guangdong Province. To acquire thorough information on local land development, the Guangdong Provincial Government embarked on two rounds of the general survey. The first round was undertaken in mid-2010 and was led by the Department of Housing and Urban–Rural Development of Guangdong Province. In addition, they organized a Consultant Group to handle technological issues. On the one hand, the master plan, regulatory plan, and land use plan of the related cities were reported to the Department of Housing and Urban–Rural Development of Guangdong Province, and the *Investment Group* summarized the alignment and engineering standard for the entire PRD inter-city railway. On the other hand, the Consultant Group combined all the materials above and the PRD regional plan for field trips. On the basis of the field trips, the Guangdong Provincial Government systematically evaluated the development potential of the land around the inter-city railway station within a range of 800 m and issued a governmental report. In light of this report, most of the land was ascertained to have strong development potential largely due to the proximity to the railway station (Interview with a local official, 12/2017).

However, the first round of the general survey failed to identify the land ownership, which further triggered local resistance on the pretext of mismatched land ownership. Accordingly, the Guangdong Provincial Government conducted the second round of general surveys in November 2010. At this stage, the Consultant Group integrated the two types of land use plans in the Land Department and the Housing and Urban–Rural Development Department and the land ownerships and created new standards for evaluating land development potential. According to the second round of the general survey, merely 7 out of all 53 stations had no development potential

[2] Guangdong Provincial Government. (2012, January 19). Suggestions on the Optimization of Comprehensive Land Development Mechanism around PRD Inter-city Railway Station. Retrieved from http://zwgk.gd.gov.cn/006939748/201201/t20120130_302347.html#

(Interview with a local official, 12/2017). Consequently, the provincial government gained a clear understanding of the local land development through the two rounds of the general survey. In addition, the *Investment Group* obtained abundant materials and information on local land development and secured a favorable position in the negotiations with the local governments.

7.6.2 Power Recentralization from Local Government to Provincial Governments

The narrative above shows that in addition to the power integration across governmental departments, power recentralization from the local government upwards to the provincial government was another crucial vehicle for achieving the restructuring of inter-scalar relations. The Guangdong Provincial Government recentralized power through three activities: (1) putting a freeze on the land development around the SGSIR stations; (2) withdrawing the plan-making and assessment powers of the land around SGSIR from local governments; and (3) rearranging the responsibility for covering the operating losses between the provincial and local governments.

Putting a freeze on the land development around the SGSIR stations was largely triggered by the fact that when local governments negotiated with the provincial government on the land development mode, they accelerated land acquisition from the farmers to maximize the land revenue before jointly developing the land with the provincial governments (Interview with an urban planner, 12/2017). To freeze land development, the provincial government promulgated a series of policies to suspend local governments from granting land use rights and approving related plans and the new land projects around the SGSIR stations. In addition, to enforce these policies, the Guangdong Provincial Government required the Department of Housing and Urban–Rural Development of Guangdong Province and the Department of Land Resources of Guangdong Province to strengthen supervision of local planning and land departments. The vertical linkage across governmental departments appeared to have been fortified through strengthening the regulatory power of the provincial department. On one hand, the policies slowed down the land acquisition and banking actions of the local government. On the other hand, it forced the local government to negotiate with the Guangdong Provincial Government and the *Investment Group* (Interview with a local official, 12/2017).

Withdrawing the plan-making and examination powers of the land around SGSIR from the local governments was the key strategy to achieve power recentralization from the local governments to the provincial government. First, withdrawing the planning power was manifested primarily as the "PRD inter-city railway station TOD comprehensive development planning" formulated by the Department of Housing and Urban–Rural Development of Guangdong Province. In the outline of this plan, the Department of Housing and Urban–Rural Development of Guangdong Province divided the land around the inter-city railway stations into three levels: regional, urban, and district levels, on the basis of their positions in the regional strategy. The Department also provided specific guidelines for the different stations in terms of land use (Interview with a local official, 12/2017). The remainder of the plan provided detailed guidelines for every inter-city railway station. Specifically, the plan proffered integrative planning for the land use around the stations, involving the TOD development strategy, land-use planning, urban design, transportation planning, concept design, and architectural design. In doing so, the "PRD inter-city railway station TOD comprehensive development planning" reflected the strong regulatory capacity of the Guangdong Provincial Government for local land development through incorporating the main contents of the master plan, regulatory plan, and site plan (Interview with a local official, 12/2017). Second, the Guangdong Provincial Government established a system of record-keeping and examination for the regulatory plan of the land around the inter-city railway station. This system required the Department of Housing and Urban–Rural Development of Guangdong Province to be the unique principal body to make the regulatory plan for the land around the inter-city railway station, and this regulatory plan had to be approved by the Guangdong Provincial Government.[3] The former regulatory plan and land use plan of the local government had to adjust in line with the "PRD inter-city railway station TOD comprehensive development planning" within half a year. In particular, the revised regulatory plan needed to be recorded and examined by the Department of Housing and Urban–Rural Development of Guangdong Province. Essentially, withdrawing its planning and examination powers weakened the regulatory capacity of local planning and land departments and increased the institutional costs of local land development.

Furthermore, under the heavy burden of covering losses, the Guangdong Provincial Government rearranged the responsibility for losses between the provincial and local governments. According to "Suggestions on the Optimization of Comprehensive Land Development Mechanism around PRD Inter-city Railway Station," the responsibility for covering losses was directly linked to the construction volume and investment. Specifically, for the local governments, if the percentage of the cost of land acquisition and housing demolition in the total investment of the inter-city railway was higher, then the local government took more responsibility for covering losses. If the local governments paid more money for housing demolition and land acquisition, then it bore more responsibility for covering the operating losses. In addition, the Guangdong Provincial Government proclaimed that if the local governments conducted the land development independently, then it also needed to bear all the operating losses.

In view of the power reconfiguration, the inter-scalar relation between provincial and local governments led to two divergent institutional forms in land development. First, some advanced cities, such as Shenzhen and Guangzhou, took the initiative to resist the power embeddedness of the Guangdong Provincial Government in

[3] Guangdong Provincial Government. (2012, January 19). Suggestions on the Optimization of Comprehensive Land Development Mechanism around PRD Inter-city Railway Station. Retrieved from http://zwgk.gd.gov.cn/006939748/201201/t20120130_302347.html#

land development. Their aims in the negotiations included obtaining and controlling the liberty of urban development, which was a precondition for its prosperous urban economy. To boycott power infusion, the Shenzhen and Guangzhou Municipal Governments bore all the responsibility for covering the operating losses based on their urban finances. These cities provided funds from their urban revenues for the *Investment Group* to cover the operating losses. Second, some ordinary cities, such as Dongguan and Qingyuan, compromised with the Guangdong Provincial Government to collaborate with it on land development around the inter-city railway stations. However, for SGSIR, the land development in Dongguan encountered dramatic resistance from the lower levels of authority. Due to the two-tier system (i.e., a municipality–town system) of Dongguan, its administrative tradition of a strong town and village and a weak municipal government, and considering that certain stations were located in the territories of the town and the village, the town and village collective held land ownership, which presented a tremendous obstacle for land acquisition. Although Dongguan selected a collaborative model with the Guangdong Provincial Government to undertake land development, the intricate land ownership between different levels of authority arguably resulted in slow progress for land development in Dongguan (Interview with a local official, 12/2017).

7.6.3 Summary

To tackle the financial crisis triggered by the station peripheralization, the Guangdong Provincial Government attempted to participate in local land development actively through power reconfiguration. In this issue, the aim of the Guangdong Provincial Government was to co-develop the land around the SGSIR stations with local governments, while local governments continued to resist the initiative from the Guangdong Provincial Government. Thus, the Guangdong Provincial Government took advantage of its administrative power over local governments through integrating powers across provincial departments and recentralizing power from localities to Guangdong Province.

Two major findings result from the investigation above.

First, in this issue, the Guangdong Provincial Government primarily focused on the issues of information asymmetry between the *Investment Group* and the local governments. On one hand, it integrated provincial departments by creating a cooperative model and agency among different departments and provided information on local land-use and socioeconomic development for the *Investment Group*. The aim of such power integration was to assist the *Investment Group* to win an advantageous position in negotiating with local governments. On the other hand, the Guangdong Provincial Government recentralized some administrative powers, such as planning and examination powers of the land around SGSIR, from local governments, compelling them to accept collaboration with the provincial government in the land development around the SGSIR stations. These two approaches helped the *Investment Group* obtain an advantageous position in the negotiating process.

Second, although the Guangdong Provincial Government reconfigured the inter-scalar relations based on its administrative and regulatory power over localities, the local governments also attempted to resist the regulation from the Guangdong Provincial Government based on their relative advantages. Some advanced cities, such as Guangzhou and Shenzhen, directly bore all the operating losses of SGSIR, based on their prosperous urban financial incomes. However, laggard cities, such as Dongguan, had to adopt the collaboration model and worked with the Guangdong Provincial Government to embark on land development around the SGSIR stations.

7.7 Discussion

To systemically examine how governance structure reshuffles to enable infrastructure projects, this chapter identifies the relative advantages, participation levels, aims in negotiation, and value orientations of different stakeholders in SGSIR, revealing the dynamic evolution of regional governance structures.

7.7.1 Changes in Regional Governance: Evolution of Inter-scalar Relations of SGSIR

Through investigating the case of SGSIR, this book finds that the government-led integrative process in China has reshuffled towards changing regional governance across government hierarchies to enable regional infrastructure projects. This book further identifies four underlying mechanisms in the governmental hierarchy of China that trigger the dynamic changes in regional governance.

First, according to the central-planned administrative system in China, the administrative powers and resources are allocated top-down (Ma, 2005). Such a formal administrative setting enables high-level governments to share the administrative controlling and regulating powers over low-level governments. Because of this, the provincial government shares the absolute administrative controlling and regulating powers over municipal governments. Essentially, this top-down institutional alignment is the legacy of the state authoritarianism of a socialist system that attempts to retain the dominant role of the central government in the governmental hierarchy even in the post-reform era (Ma, 2005). Such top-down administrative controlling is the key driving force, enabling the recentralization of power from local to provincial levels.

Second, a provincial government is able to overcome its lower position in the administrative configuration within China's government hierarchy. For example, compared with the national ministries and commissions, although the provincial government shared equal status with them in terms of administrative power, it was far from the central government geographically and therefore operated as a local authority rather than a central authority (Zhou, 2007). Such a deficiency gives rise to difficulties for regional development, even it has the power of administrative approval. These deficiencies enabled the provincial government to select the collaborative model with other levels of authority to integrate administrative powers and resources (Zhou, 2007).

Specifically, given that the PRD inter-city railway network was a provincial government-led trans-jurisdictional infrastructure project, the Guangdong Provincial Government proactively sought to enable it through establishing a collaborative model with different levels of authorities, such as the Ministry of Railways in the planning approval process and operation model from 2000 to 2007, collaboration with local governments in the operation model from 2000 to 2010, and land development after its financial crisis. Essentially, such collaborative initiatives of the Guangdong Provincial Government, on one hand, resulted from the relatively weak capacities in some public fields, such as planning approval, and financing; on the other hand, they were driven by the growth-oriented ideology.

In other words, to acquire regional development and political promotions for officials, the provincial government still sought collaboration across different levels of authority to overcome its deficiencies in the national state apparatus and compete against other governments and jurisdictions at the same levels (Interview with an urban planner, 12/2017). Thus, regarding the provincial government-led infrastructure project for regional development, the provincial government served as a prominent driving force in changing regional governance, initiating collaboration across different levels of authorities. In the case of SGSIR, based on its relative advantages, the Guangdong Provincial Government initially collaborated with the Ministry of Railways through delegating the regulatory power over the PRD inter-city railway network to the ministry. In addition, it attempted to collaborate with local governments through recentralizing some land-based power from the local to the provincial scale.

Third, the rise of entrepreneurial government in China indicates that economic growth in urban development has gradually become the target of local government (Li & Chiu, 2018) and the growth-first ideology within the urban scale has become the primary driving force for them to participate in regionalization. The provincial government-led regional project is like a pie falling from the sky in that it provides great opportunities for local governments in breaking economic deadlocks. They need to pay only a minor cost for this type of regional project, mainly land acquisition and related housing demolition costs (see Table 7.2). Driven by urban-based interests, local governments actively collaborated with the provincial government, aiming to facilitate the implementing of the regional project and ensure its involvement in this project, even they are only subordinate actors.

However, when the urban-based interest is challenged by the regional or the Ministry's interests, the local government shows a strong integration capacity across different departments to resist the regulation and collaboration initiatives from upper-level governments and authorities. Driven by the asymmetric information controlling capacities between integrated governmental systems and single administrative departments, local governments won the advantageous position in the inter-scalar politics, and upper-level government compromised with local government and further gave rise to the collapse of the agreed-upon collaborative models. This case reveals that the different capacities among different levels of government in terms of their capability to draw support from across governmental units horizontally and vertically are inherent characteristics of socialist China, and the ability to manipulate a reconfiguration of power is an important means for governments to overcome their institutional weakness.

Fourth, the provincial government can mobilize resources across the different departments of the provincial government to establish information systems about the local socio-economic development of affected local areas, diminishing the relative advantages of local governments in the negotiation process (Luo & Shen, 2008). The work of this mechanism is based on an institutional alignment so that high-level governmental departments share administrative controlling and regulating powers over the corresponding departments at the lower level. In the case of SGSIR, the large scale of general surveys conducted was an example to show the mobilization capacity of the provincial government. Putting a freeze on land development in local areas showcased the passing of top-down regulation from high-level departments to low-level ones.

In general, this case details the changing regional governance and reveals the different forms of inter-scalar relations, manifesting as collaboration, compromise, and resistance. Generally speaking, provincial governments in China should serve as the most influential actors in the inter-scalar relation of regional governance. The lack of land-based revenue enables the provincial government to reconfigure the power relations with local governments to create a collaborative model in local land development. The evolution of inter-scalar relations in SGSIR leads to four conclusions in the provincial government-led regional project. First, provincial governments always seek collaboration models with different levels of authority to overcome their deficiencies so that they can compete against other governments and jurisdictions at the local level, especially to overcome the lack of land-based revenue. This situation is the primary driving force for the changing regional governance. Second, such provincial government-led regional projects, such as SGSIR, provide great opportunities for local governments to break economic deadlocks as they need to pay merely for minor costs but benefit hugely from the projects. Local governments thus actively respond to collaborations with the provincial government. In addition, when their interests in urban development were challenged, local governments showed a strong institutional integrative capacity to resist the provincial government; their power reconfigurations can serve as an approach for coping with inter-scalar conflicts. Third, confronted with resistance from local governments, the provincial government showed a strong capacity to mobilize politically across different provincial departments, diminishing the relative advantages of local governments on the information about local socio-economic development. Fourth, the absolute administrative controlling and regulating power of higher-level authorities over lower-level authorities was the key and determining factor for the success of power reconfiguration. It needs to be noted that overtly uneven power relations existed among Shenzhen, Guangzhou, and Dongguan in the case of SGSIR, although the local governments were viewed as unrelated and separated actors in the negotiations with the Guangdong Provincial Government and the Ministry of Railways. Such uneven power relations were generally determined by the uneven levels of socioeconomic development rather than the unequal administrative statuses. For example, adequate fiscal income and strong urban finances enabled Shenzhen and Guangzhou to resist the implementation of the collaborative governance model in land development from the Guangdong Provincial Government. In contrast, due to its weak financial situation and the crisis in industrial upgrading, Dongguan was unable to bargain with upper-level authorities about the mode of land development.

7.7.2 Positioning the Changing Regional Governance: Beyond State Rescaling

This book investigates the reshuffling of regional governance in the provincial government-led infrastructure project under the PRD city-regionalism. The intricate inter-scalar politics problematize the two-track state rescaling paradigm in illustrating the formation of regional governance in China.

The conventional literature describes city-regionalism and regional governance through a paradigm of state rescaling (see Sect. 2.4.3). Li and Wu (2012) identify two scale-building processes for China's regionalization, manifesting as a bottom-up model initiated by the local government and a top-down model driven by the central government; further, this scaling-building process is not smooth but ridden with struggles and tensions. Li and Wu (2018) systemically identify city-regionalism in post-reform China and describe the regional scale of building mechanism as "entrenched political intervention embedded within administrative structure" (p. 321). These state-rescaling studies outline the rationale of institutional configuration and reconfiguration in response to city-regionalism. However, simplifying the politico-economic evolution of post-reform China and treating scale as merely material and not discursive has been criticized by some scholars (Li et al., 2015; Lim, 2017; Xu, 2017). This book restores the complicated inter-scalar politics to connect with the criticism of the theory of state-rescaling. Investigating the inter-scalar politics in the case of SGSIR has identified two theoretical implications that have been to illuminate the deficiencies of the theory of state rescaling.

First, in the actual context of city-regionalism in post-reform China, scaling up or down is generally deployed as a state-designed strategy to overcome a local crisis, not the static, mechanistic, and two-track "upward" and "downward" institutional reconfiguration process. In the case of SGSIR, the Guangdong Provincial Government up-scaled the issue of planning approval of the PRD inter-city railway network from the provincial level to the national level by inviting the Ministry of Railways to cooperate in its operation. This up-scalar strategy was essentially driven by the local crisis of inefficient administrative examination and approval of development planning. After overcoming the local crisis in planning approval, to tackle the slow-moving implementation of the PRD inter-city railway network and strict regulation from the Ministry of Railways, the Guangdong Provincial Government selected a down-scalar strategy by excluding the Ministry of Railways and delimiting the inter-city railway project as provincial issues. Thus the theoretical implications of scaling within the hierarchy in this case are not just the movement and reconfiguration of power, but its use as a strategic tool to overcome crises flexibly.

Second, the two cases of SSSCZ and SGSIR pertain to the two-track movement and (re)configuration of power, either upward or downward, decentralization or recentralization, in exemplifying the regional governance or city-regionalism of post-reform China. In the case of SSSCZ, driven by the initiatives of regional coordinated development, the Guangdong Provincial Government decentralized power to localities through political mobilization and deregulation. Along with the decentralization of the Guangdong Provincial Government, local governments aspired to establish the inter-jurisdictional cooperation model to dismantle potential hurdles through political decentralization. However, in the case of SGSIR, to compel local governments to accept the collaboration model in local land development, the Guangdong Provincial Government recentralized some plan-making and administrative examination and approval powers from the local to the provincial scale. This recentralization shows that the movement of power and institutional reconfiguration between provincial and local scales follows the case-by-case paradigm; the theoretical dialogue across cases will contribute to the understanding of actual city-regional politics in China. In this book, the two cases were mobilized by the provincial government, but the latter tended to decentralize power to enable industrial cooperation projects and recentralize power to enable trans-jurisdictional infrastructure projects. These actions verify the concept of the "Institutional Power of Reach," which argues for the horizontal or vertical imagery of the geography of state spatiality and power (Allen & Cochrane, 2010). The changing geography of state power can be referred to as "the ability of authorities to reach into the politics of the region" (Allen & Cochrane, 2010, p. 1087). The different power reaches of the Guangdong Provincial Government in these two cases reveal the complicated state spaces in city-regions with varied power reaches in different directions, rather than stretching just horizontally or vertically in conventional state spaces.

7.8 Conclusions

This chapter discusses the reshuffling process of governance structures to enable effectual inter-scalar infrastructure integration. In this book, the case of SGSIR was selected to probe inter-scalar politics, with a particular focus on the relations between the Ministry of Railways, the Guangdong Provincial Government, and the local governments.

This case reveals that the inter-scalar relations in China cannot be simplified merely through the state-rescaling theory as argued by Xu and Wang (2012), who provide a typological view of socio-economic evolution in post-reform China as a synergistic effect between decentralization and recentralization. To facilitate a trans-territorial infrastructure project, relations across different authority levels have been dynamically restructured in line with the interest demands of different actors. The SGSIR case reveals that inter-scalar relations occur in three forms: collaboration, compromise, and resistance, plus their interactions. By tracing the complicated inter-scalar politics in the different stages of SGSIR, this book provides three conclusions.

First, under the prevailing city-regionalism in China, the inter-scalar relations manifested as changing institutional forms in accordance with specific local crises for socioeconomic development. According to the conventional theory of state rescaling, city-regionalism can be categorized as two reversed scale-building processes: a bottom-up model initiated by local governments and a top-down model driven by the central government (Li & Wu, 2018). In addition, within the depiction of the socioeconomic evolution and institutional (re)configuration, the state power and spatiality have been viewed as horizontal and vertical mechanistic "upscaling and downscaling" processes (Allen & Cochrane, 2010; Lim, 2017). However, this book argues that inter-scalar relations change dynamically in diversified institutional forms at different stages. Specifically, in the planning process, inter-scalar relations formed the collaboration model between the Ministry of Railways and the Guangdong Provincial Government. The operation models shifted from ministerial-provincial-local cooperation to provincial-local cooperation; the local governance resisted the initiatives of the Ministry of Railways and the Guangdong Provincial Government in the stage of site selection. Due to information asymmetry, the Ministry of Railways and the Guangdong Provincial Government compromised with the local governments and decided on the peri-urban location of the stations; to compensate for the operating losses, the Guangdong Provincial Government integrated power across its branch departments and recentralized power from local to provincial scales. These changing inter-scalar relations indicate that movement along the hierarchical scale was deployed as a strategic tool rather than as material and socioeconomic trends, and the reshuffling processes of regional governance in China were complicated by city-regional politics ridden with collaboration, compromise, and resistance.

Second, the provincial government not only serves as regulator, mediator, collaborator and animator (see conclusions in Sect. 6.8), but also acts as a competitor with local governments within the changing regional governance in post-reform China. In the case of SGSIR, the Guangdong Provincial Government attempted to compete with local governments for the right to develop local land. Regarding its changing role, due to the economy-first and pro-growth ideologies across provinces and intrinsic deficiencies of the provincial government in state apparatus, the provincial government always seek to collaborate with different levels of government/authority to overcome these deficiencies and compete against other jurisdictions and governments at the same level. This competition is also the primary driving force for the changing regional governance in post-reform China.

Third, within the intricate inter-scalar politics in changing regional governance, the starting point of local governments is to maximize the urban-based interests which are driven by the pro-growth ideology. Regarding the provincial government-led infrastructure project, local governments generally actively participated in the project to acquire urban socioeconomic development. However, if the strategy formulated by the provincial government challenges the urban interest, local governments have a strong capacity to integrate within the inter-scalar politics to formulate some solutions and resist the collaboration model entrenched by the provincial government. This mechanism is another driving force that gives rise to the changing regional governance rather than the mechanistic "upscaling and downscaling" processes.

Fourth, to maintain its predominant role in regional governance, the provincial government could mobilize and integrate administrative powers and resources to create a new integrated institutional entity to negotiate with local governments. Such a reconfiguration of power could reduce the relative advantages of local governments. Aside from the equal status when negotiated with local governments, the top-down imposition of administrative powers and resources in the governmental hierarchy was the primary factor for the provincial government's success in reconfiguring power, moving from the local to the provincial levels.

In sum, the case study of SGSIR reveals that the government-led integration process in China has reshuffled to change regional governance centered on several rounds of inter-scalar collaboration, compromise, and resistance to enable effectual infrastructure integration. This changing typology of regional governance still hinges on power reconfiguration and scalar relations within the state sector. The case study of SGSIR not only problematizes the two-track state rescaling paradigm in elaborating the city-regionalism of China. It also identifies that the collaborative initiatives of the provincial government in overcoming its deficiencies within the state apparatus are unsuccessful, so it must resort to its administrative power to control and counteract the urban-based interests of local governments. Thus, while inter-scalar collaboration does occur in regional governance and regional projects, the provincial government plays a leading and controlling role by exerting its administrative supremacy as it must overcome its financial deficiencies incurred by the lack of land-based financial sources.

References

Allen, J., & Cochrane, A. (2010). Assemblages of state power: Topological shifts in the organization of government and politics. *Antipode, 42*(5), 1071–1089. https://doi.org/10.1111/j.1467-8330.2010.00794.x

Li, J., & Chiu, L. R. (2018). Urban investment and development corporations, new town development and China's local state restructuring–the case of Songjiang new town, Shanghai. *Urban Geography, 39*(5), 687–705. https://doi.org/10.1080/02723638.2017.1382308

Li, Y., & Wu, F. (2012). The transformation of regional governance in China: The rescaling of statehood. *Progress in Planning, 78*(2), 55–99. https://doi.org/10.1016/j.progress.2012.03.001

Li, Y., & Wu, F. (2018). Understanding city regionalism in China: Regional cooperation in the Yangtze River Delta. *Regional Studies, 52*(3), 313–324. https://doi.org/10.1080/00343404.2017.1307953

Li, Y., Wu, F., & Hay, I. (2015). City region integration policies and their incongruous outcomes: The case of Shantou-Chaozhou-Jieyang city region in east Guangdong Province, China. *Habitat International, 46*, 214–222. https://doi.org/10.1016/j.habitatint.2014.12.006

Lim, K. F. (2017). State rescaling, policy experimentation and path dependency in Post-Mao China: A dynamic analytical framework. *Regional Studies, 51*(10), 1580–1593. https://doi.org/10.1080/00343404.2017.1330539

Luo, X., & Shen, J. (2008). Why city region planning does not work well in China: The case of Suzhou–Wuxi–Changzhou. *Cities, 25*(4), 207–217. https://doi.org/10.1016/j.cities.2008.04.003

Ma, L. J. (2005). Urban administrative restructuring, changing scale relations and local economic development in China. *Political Geography, 24*(4), 477–497. https://doi.org/10.1016/j.polgeo.2004.10.005

Xu, J. (2017). Contentious space and scale politics: Planning for intercity railway in China's mega-city regions. *Asia Pacific Viewpoint, 58*(1), 57–73. https://doi.org/10.1111/apv.12142

Xu, J., & Wang, J. J. (2012). Reassembling the state in urban China. *Asia Pacific Viewpoint, 53*(1), 7–20. https://doi.org/10.1111/j.1467-8373.2012.01472.x

Zhou, L. (2007). Governing China's local officials: An analysis of promotion tournament model. *Economic Research Journal, 42*(7), 36–50. (in Chinese).

Chapter 8
Conclusions

8.1 Introduction

Over the past three decades, PRD has experienced tremendous reform and restructuring economically, socially, and institutionally. The repositioning from city-based development to the broader city-regional scale triggered a series of struggles between institutional alignment and economic restructuring—an important theme in the literature of China's city-regionalism. SDH, as the most important sub-region of PRD, has often been confronted with impromptu and complicated situations in building an integrative economy and institutions. To decode the underlying economic-institutional interplay in China's city-regionalism, investigating the economic and institutional integrations as well as the underlying mechanism of how the state sector was structurally reconfigured in response to the economic restructuring has a great significance for regional studies and territorial practices of city-region making. Thus, the case of SDH has been examined to identify key concerns of China's city-regionalism. This book began with a comprehensive assessment of the extent of economic and institutional integrations of SDH to provide an overall landscape of the city-regionalism of one of the three sub-regions in PRD. After an investigation of the macro-economic and institutional integrations of SDH, SSSCZ and SGSIR were selected as empirical cases to demonstrate how the state was assembled across different levels, jurisdictions, and public and private domains in different regional projects to accelerate city-regionalism.

This chapter summarizes the major findings on SDH's city-regionalism and provides further insights and overall conclusions on the investigation, followed by a review of the contributions of this book to literature, while acknowledging its limitations. The next section summarizes the findings of the first seven chapters in terms of the three research questions that are addressed. To tackle the fourth research question, the third section presents the arguments on policy implications and provides recommendations to facilitate China's city-regionalism and ways to build competitive city-regions; the fourth section outlines how this book contributes to the existing

X. Zhang, *Regional Integration and Governance Reshuffling in the Making of China's City-Regions*, Urban Sustainability, https://doi.org/10.1007/978-981-99-2792-0_8

literature empirically and conceptually. The last section systematically presents the limitations of this book and suggestions for further research.

8.2 Summary of Major Findings

This book provides an in-depth case analysis of China's city-regionalism with reference to its economic and institutional extent as well as the underlying reshuffling process of governance structures. This book proposes four inter-connected research questions: (1) To what extent does SDH operate as an economically integrated entity as reflected in price similarity, industrial structure similarity, and infrastructure development? (2) To what extent does SDH operate as an institutionally integrated entity by redefining the governance structure across government hierarchies, between cities, and between the state and non-state domains? (3) How does the governance structure reshuffle itself to enable or encumber the functioning of SDH as an integrated economic entity against the backdrop of China's reorientation strategy to develop city-regions as loci of economic development? (4) How does the territorial praxis of SDH contribute to the theoretical debates on China's city-regionalism? This section summarizes the findings of the book to answer the first three research questions; the last question will be addressed in the next two sections.

8.2.1 Extent of Economic Integration of SDH

The extent of the economic integration of SDH was explored from three perspectives, those of market, industrial, and infrastructure integration, as well as on two scales: the overall SDH and inter-city group scales. Relative price variance was employed to assess the extent of market integration on the overall SDH and inter-city group scales. Second, to explore the phenomenon of industrial isomorphism in SDH, the degree of industrial structure similarity was determined in identifying the extent of industrial integration on the overall SDH scale. Moreover, the modified Krugman's index was used to assess the degree of industrial integration on the inter-city group scale. Third, the infrastructure integration of SDH was demonstrated through assessing the equality and connectivity in the infrastructure system. The infrastructure equality on the overall SDH scale was measured by a coefficient of variation, and the degree of infrastructure development similarity was used to assess the infrastructure equality on the inter-city group scale. The infrastructure connectivity index was measured to assess the infrastructure connectivity on the overall SDH and inter-city group scales.

Overall, the market integration of SDH has experienced a steadily escalating trend, especially after 2009. The extent of industrial integration and infrastructure equality increasingly weakened from 2004 to 2015, whereas the extent of infrastructure connectivity of SDH gradually rose from 2004 to 2015. Specifically, Shenzhen was integrated well with Dongguan's and Huizhou's consumer markets and infrastructure, whereas industrial integration between them was poor, manifested by strong industrial isomorphism. The market and infrastructure integration between Dongguan and Huizhou remained at a medium degree, and their industrial integration continued to weaken. Considering Shenzhen's integration with outlying cities, market integration between Shenzhen and Shanwei remained high, whereas market integration between Shenzhen and Heyuan remained at a medium degree. Shanwei and Heyuan shared high degrees of industrial integration with Shenzhen, but their infrastructure integration stayed at low and medium levels, respectively.

Based on the theoretical argument that the degree of integration of a regional market is indicative of the final state of economic integration (see Chap. 2), the above findings affirm that the actual situation of the economic integration of SDH was improving and became steadier after 2009. In keeping with the conventional literature, the industrial isomorphism of SDH worsened and led to a latent crisis potentially hampering regional integration, as a high level of isomorphism may induce fierce inter-city industrial competition. While the infrastructure equality of SDH deteriorated from 2004 to 2015, the infrastructure connectivity of SDH improved through gradually forming the regional infrastructure network. This result indicates that the construction of infrastructure served as a significant driving force for the economic integration of SDH by accelerating the flow of goods and enabling efficient utilization of factors of production.

The regional core city, Shenzhen, shared a high degree of market integration with Dongguan and Huizhou, which was facilitated by a high extent of infrastructure integration. Such a highly integrated extent of infrastructure not only accelerated the building of a regional market, but also further enhanced the social integration between cities. The escalating industrial isomorphism between Shenzhen and Dongguan and Shenzhen and Huizhou indicated that there was an underlying industrial competition among them, further restraining economic integration. The best example of that was the construction of "Pin-Qing-Xin Industrial Cooperation Demonstration Park" in the border region among Shenzhen, Dongguan, and Huizhou. The conflicting demand for industrial development was the primary reason why it had become stagnant in recent years. However, Shenzhen had a high degree of industrial integration with Shanwei and Heyuan, but the infrastructure conditions among them needed to be enhanced in the future. The spatial expansion of SDH provided a valuable opportunity for these cities to break a development deadlock through creating new inter-city linkages.

To summarize, SDH was moving towards the goal of becoming a regional market unit, and its infrastructure has increasingly evolved towards being a network of a regionally integrative system. However, serious industrial isomorphism worsened over the entire SDH sub-region, especially among Shenzhen, Dongguan, and Huizhou compared with the isomorphism between Shenzhen and outlying cities Shanwei and Heyuan.

8.2.2 Extent of Institutional Integration of SDH

This book investigates the extent of institutional integration of SDH along three relations derived from the organization-embraced approach and the concept of MLG grounded in the EU regionalism, namely, inter-scalar, inter-city, and state-market-civil society relations. First, the analysis of inter-scalar relations focused primarily on the existing administrative setting as well as the adjustment of administrative divisions during the integration process. Second, inter-city relations manifested as the spatial and temporal orders, cooperation area, and cooperation mode. Third, state-market-civil society relations were systematically assessed in terms of the degree of involvement, cooperation area, organizational form, and development status.

In terms of inter-scalar relations, first, the intensified administrative control that the central government imposed on local governments in SDH was essentially a legacy of state authoritarianism in the Mao era. Through such a central government-monopolized administrative setup, the central government retained its dominant role in local socioeconomic development. Second, informal institutional innovations emerging as a result of the economic autonomy of Shenzhen were actively constituted as an institutional strategy to alleviate the conflict between rapid socioeconomic development and rigid administrative constraint. Flourishing local institutional innovations resulted from political decentralization, including two types of reconfiguration of power. The first one was the decentralization of power from the central government to local governments, empowering the economic autonomy of the Shenzhen Municipal Government. The second one was the deregulation and devolution of the Guangdong Provincial Government through assigning development opportunities and resources to localities. Third, the lack of formal regional authority in inter-scalar relations indicates that the tiers of jurisdictional arrangement in SDH cannot be literally called an MLG paradigm. The provincial government served as a sub-regional fabric to actually shore up the initiatives as they enhanced regional economic development and reduced uneven regional development.

Regarding inter-city relations, first, SDH integration manifested as a voluntary process of city grouping, centered on Shenzhen. Second, SDH integration was a growth-oriented and economy-first inter-city cooperation process instead of the multi-objective integrating process in the EU (see Chap. 2). Shenzhen was the sole regional core and privileged aid activity with peripheral cities as the main spatial-fixed strategy. Shenzhen also interacted with Dongguan and Huizhou for economic development and institutional design. Third, the inter-city cooperation in SDH was settled by abundant informal institutional arrangements, such as unofficial communication and investigation, joint action, and organizing activities; these informal institutional alignments were subtly guided by formal institutional arrangements, such as joint conferences and cooperation agreements. In other words, SDH integration was spearheaded by informal institutional arrangements, whereas the formal ones steadily established an action framework for informal institutional alignments.

In the state-market-civil society relations, first, the institutional integration of SDH remained a government-led process that positioned the quasi- and non-state actors on the margins of influence. Second, compared with the dominant role that governments played in all subject areas, market and social actors focused primarily on certain regional issues that were relatively marginal. Market actors specifically in Shenzhen engaged mainly in economic development and assisting low-level cities to develop local economies as required by the provincial government. Social actors centered in Shenzhen participated mostly in activities with social development and aid activities. Third, the organizational forms of markets and social actors in the regional integration of SDH were limited, including primarily the chambers of commerce, public institutions, NGOs, and communities.

In summary, the institutional integration of SDH was a top-level government-led integration process and departed from the MLG paradigm in the EU context. The inter-scalar relations of SDH manifested as interwoven processes dominated ultimately by the highest authority in the region. The central government's control was entangled with flourishing local institutional innovations stemming from political decentralization. The provincial government set the sub-regional context for economic growth under an even regional development strategy. The inter-city relations of SDH were growth-oriented and economy-first voluntary processes of city grouping, centered on Shenzhen and settled by informal institutional alignments under the guidelines of the formal institutional arrangements. The state-market-civil society relations of SDH in the government-led region-building process emerged in the form of the market and social actors keeping marginal positions and embarking on marginal regional issues with limited organizational forms.

8.2.3 Reshuffling Process of Governance Structure in Response to the Industrial Integration of SDH

In this book, the case of SSSCZ has been examined to illustrate the underlying reshuffling of governance structures to enable the effective industrial integration of SDH. The book reveals that the government-led integration process of SDH (see Chap. 5) involved reorganization across inter-scalar and inter-city relations in SSSCZ in order to develop a steady collaborative framework. Specifically, the common aims in the negotiation process, the relative advantages of cities, political mobilization by the Guangdong Provincial Government, and stable and mutually beneficial mechanisms of allocating profits sharing were important for building a steady collaborative base. Such a collaboration governance model gave rise to the highly efficient construction of SSSCZ. However, considering the geographic locational disadvantages and weak local socioeconomic bases, as well as poor public service standards, the market response to governmental efforts to construct SSSCZ was hesitant. The key stakeholders in constructing SSSCZ were identified as the Guangdong Provincial Government, the Shenzhen and Shanwei Municipal Governments, and enterprises. Probing the four critical issues of negotiating the construction of SSSCZ, legitimizing

SSSCZ, daily operations, and the industrial development of SSSCZ produced three major findings.

First, to enable productive industrial collaboration, the inter-scalar and inter-city relations underwent radical restructuring to create a collaboration-oriented mode of governance. The Guangdong Provincial Government embarked on political decentralization and deregulation to empower great economic and institutional autonomy at local levels to facilitate cross-boundary alliances (i.e., Shenzhen-Shanwei cooperation). It did so through macro-level political mobilization, a series of concessionary policies, and providing administrative resources. In other words, compared with a conventional provincial government, the Guangdong Provincial Government acted as not only a coordinator and regulator but a proactive animator and collaborator through decentralizing power to local government and injecting development and administrative resources.

Second, the inter-city relations between Shenzhen and Shanwei showed a high degree of integration based on the complementary nature of their relative advantages and aims in negotiation, political mobilization, concessionary policy, and administrative resources from the Guangdong Provincial Government. According to Luo and Shen (2009), under city-regionalism, "cities with the same political, economic and social status tend to compete with each other" (p. 60). This argument indicates that the complementary nature of the relative advantages of different cities remains the precondition and driving force for successful inter-city cooperation. In addition, there was an explicit division of responsibility and shared supervision between Shenzhen and Shanwei in the making of SSSCZ. For example, Shenzhen took responsibility for the economic development and formulated re-territorization strategies, such as industrial concessionary policies, and for building inter-city infrastructure projects, based on its capital, technology, and industrial advantages. Shanwei took charge of social development, owing to its regulatory power for SSSCZ. Additionally, the two cities shared supervision through alternate personnel appointments in the administrative system of SSSCZ.

Third, this government-led collaborative governance was a reshuffling process within the state sector where the market actors were excluded from the making of SSSCZ. Therefore, while the rise of collaborative governance distinctly differed from the conventional regional governance in inter-scalar and inter-city relations, it remained a form of the government-led integration process in SDH. The restriction of market actors in the collaborative framework largely limited the effectiveness of the SSSCZ in industrial integration.

In summary, the case study of SSSCZ reveals that the government-led process of industrial integration in SDH produced the reshuffling of inter-scalar and inter-city relations in the collaborative governance mode within the state sector. This collaborative mode of governance among different levels of government and between cities has become an institutionalized prescription for a government to break economic development deadlocks in the regionalization process. Specifically, the political decentralization and deregulation of the Guangdong Provincial Government and the high extent of integration between Shenzhen and Shanwei enabled the highly efficient construction of SSSCZ. This collaborative framework was crystallized and stabilized by the setup of a mutually agreed-upon mechanism for allocating profit-sharing.

8.2.4 Reshuffling Process of Governance Structure in Response to the Infrastructure Integration of SDH

Using the case of SGSIR, this book systematically demonstrates how the governance structure reshuffled itself across different levels of government to enable the infrastructure integration of SDH. Key stakeholders in SGSIR have been identified, primarily the Ministry of Railways, the Guangdong Provincial Government, and local governments. This case reveals that the government-led integration process of SDH reshuffled the regional governance to enable collaboration and compromise, driven by the divergent aims in negotiation and the relative advantages among different stakeholders. Three significant findings are produced by analyzing the changing inter-scalar relations in the issues of plan-making and approval, negotiation of operating models, site selection of SGSIR stations, and negotiation of local land development.

First, this book reveals that regional governance in China was a dynamically changing process in keeping with the changing aims in negotiation and the changing relative advantages of different actors, rather than a mechanistic "upscaling and downscaling" process maintained by state rescaling theory. In the case of SGSIR, this changing regional governance was manifested in four features: (1) the collaboration model between the Ministry of Railways and the Guangdong Provincial Government in the planning approval process; (2) the operating model shifting from ministerial-provincial-local cooperation to provincial-local cooperation; (3) the resistance of the local government and compromise by the Ministry of Railways and the Guangdong Provincial Government in selecting sites of railway stations, driven by an information asymmetry; and (4) reconfiguration of power driven by the Guangdong Provincial Government to resist local governments in local land development. The changing forms of regional governance indicate that regional governance manifests as a dynamic state-orchestrated process rather than a static and one-fold scale-building process (Lim, 2017).

Second, within the changing regional governance ridden with inter-scalar collaboration, compromise, and resistance in this case, to enable a productive city-regionalism, the Guangdong Provincial Government performed both as a coordinator, regulator, animator, and participator and as a competitor against local governments. This role of competitor was chiefly driven by the pro-growth nature of provincial governments. In addition, within the inter-scalar relations, the role shifting of the provincial government was essentially driven by the aim of overcoming the functional deficiencies in the national state apparatus. In the case of SGSIR, the functional deficiencies of the provincial government manifested in three dimensions: (1) weak administrative power in facilitating the planning approval of the PRD inter-city railway network by the central government; (2) limited capacity to acquire information about local socioeconomic development and land use; and (3) weak fiscal revenues and capacity to increase its revenues. To overcome these deficiencies, the provincial government sought to establish collaborations with different levels of authority, such as collaboration with the Ministry of Railways for planning approval and collaboration with local governments for land development. This power (re)configuration driven by the provincial government was the primary driving force of dynamically changing regional governance under the prevalent city-regionalism of China.

Finally, the role of local governments was relatively passive in the planning approval process and in the operating models because the PRD inter-city railway project was a provincial government-led regional project driven by a top-down imperative administrative order. However, the role of local government shifted from a passive participant to a proactive actor when the project hurt the interests of local governments in urban development, especially regarding the tax basis of land revenues. The information and power asymmetry between a single administrative department and an integrated governmental system was the key factor that enabled local governments to win advantageous positions in negotiating with the Ministry of Railways and the Guangdong Provincial Government in selecting the sites of railway stations. Wu (2016) argued that, theoretically, "Chinese state spatiality has experienced a shift from national state space to local city-based state space and has now been rescaled toward the city-regions" (p. 1148). Importantly, this book signifies that the city-based accumulation regime still played a subdued, but irreplaceable role in emerging city-regionalism. In addition, due to the top-down administrative powers arranged by the central-induced administrative setting, the provincial government could regain control and leadership in the final stage through horizontal power integration across different provincial departments and vertical power recentralization from local to provincial levels.

In summary, the case study of SGSIR reveals that the government-led integration process of SDH manifested as dynamically changing regional governance within the state sector beyond a mechanistic "upscaling and downscaling" process in response to the infrastructure integration of SDH.

8.2.5 Overall Regional Integration and Governance Reshuffling of SDH

This book systematically proposes a context-sensitive analytical framework to assess the economic integration of SDH and an MLG-based analytical framework to assess the institutional integration of SDH. In addition, this book illustrates the multifaceted nature of regional governance under the city-regionalism of China by probing the governance reshuffling process in industrial collaboration and infrastructure integrations.

First, SDH is increasingly forming a regional economic unit and a steady regional market between cities. This well-integrated market is driven by the high extent of infrastructure connectivity. However, serious industrial isomorphism and inequality left the underlying crises of the economic integration of SDH. Shenzhen has a relatively strong market and infrastructure integrations with Dongguan and Huizhou. However, the industrial integration between them is weak. Regarding the outlying cities (i.e., Shanwei and Heyuan), Shenzhen also has relatively good market and industrial integrations with them, while the infrastructure integration between them remains weak.

Second, the institutional integration of SDH remained a government-led integration process. In terms of inter-scalar relations, the vigorous administrative control of central government was entangled with flourishing local institutional innovation to reconcile the conflicts between fast-tracked socioeconomic growth and ossified administrative constraints. Under the lack of regional authority, the provincial government performed as an actual sub-regional context in the regional governance of China. In terms of inter-city relations, institutional integration of SDH manifested as a voluntary process of city grouping centered on Shenzhen and remained an economy-first and growth-oriented process that developed via flourishing informal institutional alignments and was settled by formal institutional alignments. In terms of state-market-civil society relations, SDH kept a government-led administrative process where the market and social actors remained marginal with limited institutional forms.

Third, the two cases of SSSCZ and SGSIR signify that the city-regionalism in China is a complicated and multifaceted process, orchestrated and ultimately propelled by governments. In the project of industrial collaboration, the government-led integration process reshuffled towards a collaborative model that enabled the provincial government to perform as a proactive collaborator and animator to shore up both pro-growth and coordinated development initiatives. In the project of developing regional infrastructure, the government-led integration process produced a dynamically changing regional governance ridden with inter-scalar collaboration, compromise, and resistance. Different from the collaborator and animator roles of the provincial government through political mobilization, the provision of developmental resources and policy incentives, and political deregulation, the provincial government acted as a competitor of local governments in the issue of local land development. The former eventually got the upper hand through integrating power across departments in the provincial government and recentralizing power from the local to provincial levels. Regarding inter-city cooperation, the case of SSSCZ indicates that the complementary aims and relative advantages between cities in coping with their respective crises of economic development deadlock served as a precondition and primary driving force for successful inter-city cooperation. The city-based economic development model and GDP-based evaluation mechanism for local officials still impeded the city-regionalism in China through reinforcing the city rather than the scale of the city-region as the overarching arena for capital accumulation (Wu, 2016). The case of SGSIR signals that when the urban-based land revenue was challenged by the provincial government, local governments showed a strong integration capacity to resist the intervention of the provincial government. However, they were no competitors for the provincial government when it determined to take the lead.

8.3 Policy and Institutional Implications and Recommendations

Based on the major findings of this book, certain policy and institutional implications and recommendations can facilitate China's city-regionalism as seen through an economic and institutional lens.

To enable productive economic integration of city-regions, two integration-oriented policy implications should be implemented at the overall city-regional scale. First, policymakers who formulate policies for city-regions should make industry specialization and differentiation between cities critical elements of integration. Creating a complementary industrial structure on the city-regional scale and defining the leading industries of cities based on their roles in the urban system of city-regions are the significant standards by which to assess the effectiveness of regional industrial policies. Moreover, maintaining a certain degree of industrial isomorphism can promote benign competition among cities. Hence, to pitch the right extent of industrial isomorphism in city-regions is important, though not easy to gauge for policymakers. This extent should be considered in accordance with the developmental stage of city-regions and related socioeconomic environments of localities.

Second, establishing an infrastructure linkage between regional core cities and surrounding and outlying cities, especially the transportation linkage, is important for constructing an infrastructure network to facilitate the functional integration of city-regions. This book signals that Shenzhen is the sole regional core in SDH and has a strong overflow effect in terms of its economic development and inter-city cooperation. The positive condition of regional infrastructure is conducive to accelerating the flow of essential productive factors, such as labor, capital, and goods. Therefore, building a regional infrastructure network centered on a core city is an effective way to enable productive economic integration, especially between the core city and outlying cities.

Revisiting the institutional integration of the SDH, three policy implications can be proposed for building an integrated institutional entity. First, setting up a region-based government and task-specific authority can effectively contribute to city-regional integration if the regional government can acquire certain powers from upper-level government. The primary objectives for these authorities are to provide public services to localities and to serve as a regulatory body in coping with regional affairs, such as environmental protection and industrial collaboration. Municipal governments in the Chinese administrative system have been primarily assessed by GDP-ism with the chief consideration of political performance in economic growth for a jurisdiction. The inter-jurisdictional competition under the GDP-based performance appraisal system serves as the underlying driving force for the actions of local governments. Likewise, provincial governments can operate as an entrepreneurial governments. Aside from the socioeconomic regulation and coordination functions, the provincial government must be thoroughly involved in economic development and operate as an entrepreneurial government in close conjunction with securing city-regionalism. The issue of the provision of public services across jurisdictions becomes a serious obstruction for regional integration under the hollowing-out of the financial power of the provincial government. According to this book, the provincial government serves as a regulator, coordinator, animator, collaborator, and competitor for regional development; it also confronts some potential functional deficiencies in fulfilling these roles under China's bureaucratic system and state apparatus. It is urgent to establish region-based authorities to extract the functions of the provincial government. Thus, building region-based institutions to serve as a regional fabric is actually needed to create integrative institutions on the city-regional scale. The main tasks for these region-based institutions are providing public services at the entire city-regional scale and resolving conflicts between different jurisdictions to enable large-scale projects and agreements for regional affairs.

Second, inter-city cooperation in China should be conducted further in the field of social development. Existing inter-city cooperation is economy- and institution-oriented, which puts social affairs behind economic and institutional development. Under such circumstances, inter-city cooperation for social development should be mobilized and forged by the upper-level government to enable a multifaceted city-regionalism. Governmental agendas on inter-jurisdictional cooperation for social issues should be formulated, and specialized regional authorities should be established to regulate and coordinate inter-city cooperation and to facilitate the social development of city-regions.

Third, related concessionary policies to enhance market and social involvement in regional integration should be formulated to forge network governance on the city-regional scale. Diversified organizational forms of market and social actors should be supported and forged by governments, especially to foster the independent socioeconomic status of such actors in participating in the socioeconomic development of a city-region. Professional standards for different forms of market and social actors, such as NGOs and chambers of commerce, must be formulated and optimized under the encouragement of the urban government. Urban governments should lower the threshold for market and social actors' involvement in regional affairs by creating and allowing multiple public–private partnerships in the decision-making process for regional affairs.

Analysis of the two empirical cases demonstrating the governance reshuffling process in the infrastructure and industrial integrations reveals four policy and institutional implications. First, the setup of task-specific regional authority can contribute to rational regional governance to enable different regional affairs. These task-specific regional authorities should be given further administrative powers in certain special fields to promote inter-jurisdictional coordination for large-scale infrastructure projects. In the case of SGSIR, the *Investment Group*, a government-affiliated agency, is incapable of negotiating with local governments due to power and information asymmetry. Independent or semi-independent agencies can efficaciously collect information on local socioeconomic development and provide professional consulting for governments. These task-specific agencies are also crucial components of the local network of regional governance across different jurisdictions (Luo & Shen, 2009; Shen, 2018).

Second, capturing complementary interest demands and relative advantages between cities is important for regional policymakers when they formulate integration-oriented policies. The crises of economic fluctuations and local socioeconomic environments of different cities should be given full consideration when inter-city cooperation is designated in regional policies. In other words, the initiatives and political mobilization from the provincial government should assist in breaking development deadlocks in localities.

Third, the provincial government should leave additional power space for localities to enable effective inter-city cooperation (Luo & Shen, 2009; Shen, 2018). Concessionary policies and the injection of development resources from provincial governments are needed in regional integration to address the institutional and financial barriers in inter-city cooperation.

Finally, before making regional policies, government and market actors must negotiate, especially on site selection of large-scale regional projects such as industrial transfer parks. The coordination between governments and market players should be institutionalized and routinized as a regional policy-making process to achieve feedback from market actors on their consideration for policies. Further institutional design and cost–benefit analysis for regional policies, such as the public service and transportation improvement for industrial transfer parks, should be based on the rational choice of and respect for market forces.

8.4 Contribution to the Literature

By demonstrating the economic and institutional integration of the SDH, this book provides three primary theoretical contributions to the existing literature on China's city-regionalism.

First, this book develops a well-rounded and context-sensitive analytical framework to investigate the economic integration of city-regions through examining the integration within market, industry, and infrastructure sectors. The investigation sheds light on economic integration in the city-regionalization process. First, the extents of market, industrial, and infrastructure integrations are invariably inconsistent on inter-city group scales. For example, the inter-city group of Shenzhen and Dongguan had high market and infrastructure integrations, whereas its industrial integration remained low. In addition, on the overall city-regional scale, the regional market was increasingly integrated, and the regional connectivity of the infrastructure network was strengthened. However, industrial integration gradually worsened because of industrial isomorphism. These findings and such a context-sensitive analytical framework contribute to the existing literature on the economic integration of a city-region beyond the network-based analysis paradigm (Hall & Pain, 2006; Yeh et al., 2015) by developing a problem-solving and case-by-case paradigm.

Second, this book expands the concept of institutional integration as an empirical slogan and an inclusive theoretical term through building an MLG-based conceptual framework. This conceptual framework employs the concept of organization-embraced institutions rather than "institution-as-rule" and integrates the EU's integrative approaches to formulate a framework to examine inter-scalar, inter-city, and state-market-civil society relations. An operation-oriented conceptual framework is also provided to assess and illustrate institutional integration and enrich the scope of territorial organization from state and city to a hierarchical governmental system, inter-jurisdictional network across local authorities and localized public private partnership. Using this MLG-based conceptual framework, this book not only sheds light on the de facto extent of institutional integration under China's city-regionalism by building a bridge between an empirical slogan and an inclusive theoretical term for institutional integration but also examines MLG and its integrative approaches in China.

Finally, the two empirical cases in this book contribute to the existing theoretical paradigm in illustrating China's city-regionalism based on state rescaling theory. By demonstrating how the state is assembled across different levels of governments, inter-jurisdictions, and public and private domains to enable productive economic integration, this book problematizes state rescaling theory in deciphering China's city-regionalism as periodizing "regulatory reconfiguration as static shifts 'upwards' or 'downwards' from the national scale" (Lim, 2017, p. 1582). This research finds that the state is assembled as different power and institutional (re)configurations varied across cases, and these varied state assemblages in city-regionalism challenge the two-track (i.e., decentralization and recentralization) and static conceptual framework of state rescaling. Within the making of city-regions in PRD, the political logic of the provincial government is not understood as a power decentralization for creating a city as an overarching scale for capital accumulation, an agent of wealth creation (Li & Wu, 2012, 2018), and a power recentralization to perform as a "watchful eye" of the central government in urban development (Xu et al., 2009), but as an institutional power-reaching process that differed vastly from case to case (Allen & Cochrane, 2010). In addition, these distinct city-regional politics ridden with multi-site struggles and tensions (Li & Wu, 2018) signify that the concept of scale in deciphering the city-regionalism of China can not only be treated as a material theoretical tool that agglomerates certain politic-economic activities and relations (Chung & Xu, 2016; Xu, 2017) but also be best understood as scale politics, which is a strategic tool deployed by different actors in city-regional politics (Xu, 2017).

8.5 Limitations and Future Research Directions

Despite the stated major findings and three theoretical contributions to conventional literature, this book also has certain limitations in relation to the assessments of economic and institutional integrations.

Industrial integration in the assessment of economic integration is assessed merely from the perspective of industrial isomorphism. The measurement of industrial isomorphism is the comparison of internal industrial structures between cities, and industrial linkage and differentiation have not been considered and articulated in this book due to a lack of available data. This linkage reveals the industrial division and cooperation on the city-regional scale as well as the internal intensity of industrial linkage, which is important in economic integration. In addition to the research limitation in measuring industrial linkage, the underlying interactions and correlations among markets, industries, and infrastructure remain less than transparent. As for the assessment of institutional integration, news items about inter-city cooperation, which are collected from websites, also have limitations in this book. First, certain inter-city cooperation news was cancelled due to age and refreshed websites. Second, the implementation of inter-city cooperation is difficult to probe in inter-city cooperation news items. Third, certain inter-city cooperation news is not reported on websites and is overlooked in this book.

Hence, in future studies, industrial linkages should be incorporated into the assessment of economic integration. Data, such as details of advanced producer services, can also be deployed to provide a comprehensive assessment of industrial integration. In addition, mathematical models should be further employed to investigate the correlations among the markets, industries, and infrastructure of a city-region. Moreover, given the intrinsic imperfection of news about inter-city cooperation, a qualitative analysis focusing on the implementations of regional projects and their actual results, such as semi- or in-depth interviews and focus group discussions, can be methodologically integrated into the analysis of news about inter-city cooperation to remove this data imperfection. In addition, considering the different arrangements of political power and their reaches in different cases, the biggest challenge in future studies is to assemble a series of inter-connected empirical materials to be examined under a coherent and thought-provoking theoretical framework to re-ascertain and give a fuller portray of the city-regionalism of China (Lim, 2018).

8.6 Conclusions

This book proffers a systematical examination of China's city-regionalism by using SDH as an empirical case. Economic and institutional integrations have been quantitatively deciphered through formulating the "market-industry-infrastructure" framework and an MLG-based framework, respectively. Two empirical cases demonstrate the nuanced state assemblages across different levels of government, inter-jurisdictional networks, and public and private domains. Based on the major findings, policy implications, theoretical contributions, and related limitations, this well-rounded work can be instrumental in deepening the understanding of the internal mechanism of China's city-regionalism, as well as providing an analytical framework to investigate the intricacies of MLG underlying regional integration across cities in urban hierarchies.

Under China's city-regionalism, MLG has been strategically launched by different levels of government to create competitive city-regions. However, this book finds noticeable differences between the MLG in the EU context and its creation in the Chinese context. First, the regional authorities in the EU devote themselves to the provision of public services and coordinate the struggles and tensions among member cities that take responsibility for urban development. However, the provincial government in Guangdong plays multiple roles in regional development, such as being a competitor in leading regional growth, animator for regional integration, and collaborator in the provision of public services. Explicit responsibility and divisions of power between regional and local governments in regional development have not been formed in China, and regional integration is ridden with inter-scalar tensions and conflicts. Second, the city-based economic mode still plays an essential role in China's inter-city cooperation. The economic growth of urban development remains the key factor that determines the success or failure of inter-city cooperation. This situation is distinctively different from that in the EU, where regional integration, regional community, new regionalism, and fuzzing boundaries are prevalent over economic growth. Finally, concerning the role of the market and social sectors, compared with territorial practices of MLG in the EU that place the private sector as the principal body in regional governance, China's regional governance retains the legacy of government-dominance and marginalizes market and social actors.

References

Allen, J., & Cochrane, A. (2010). Assemblages of state power: Topological shifts in the organization of government and politics. *Antipode, 42*(5), 1071–1089. https://doi.org/10.1111/j.1467-8330.2010.00794.x

Chung, C. K. L., & Xu, J. (2016). Scale as both material and discursive: A view through China's rescaling of urban planning system for environmental governance. *Environment and Planning C: Government and Policy, 34*(8), 1404–1424. https://doi.org/10.1177/0263774X15619630

Hall, P., & Pain, K. (Eds.). (2006). *The polycentric metropolitan region: Learning from mega-city regions in Europe*. Earthscan.

Li, Y., & Wu, F. (2012). The transformation of regional governance in China: The rescaling of statehood. *Progress in Planning, 78*(2), 55–99. https://doi.org/10.1016/j.progress.2012.03.001
Li, Y., & Wu, F. (2018). Understanding city regionalism in China: Regional cooperation in the Yangtze River Delta. *Regional Studies, 52*(3), 313–324. https://doi.org/10.1080/00343404.2017.1307953
Lim, K. F. (2017). State rescaling, policy experimentation and path dependency in post-Mao China: A dynamic analytical framework. *Regional Studies, 51*(10), 1580–1593. https://doi.org/10.1080/00343404.2017.1330539
Lim, K. F. (2018). Researching state rescaling in China: Methodological reflections. *Area Development and Policy, 3*(2), 170–184. https://doi.org/10.1080/23792949.2017.1382380
Luo, X., & Shen, J. (2009). A study on inter-city cooperation in the Yangtze river delta region, China. *Habitat International, 33*(1), 52–62. https://doi.org/10.1016/j.habitatint.2008.04.002
Shen, J. (2018). *Urbanization, regional development and governance in China.* Taylor & Francis.
Wu, F. (2016). China's emergent city-region governance: A new form of state spatial selectivity through state-orchestrated rescaling. *International Journal of Urban and Regional Research, 40*(6), 1134–1151. https://doi.org/10.1111/1468-2427.12437
Xu, J. (2017). Contentious space and scale politics: Planning for intercity railway in China's mega-city regions. *Asia Pacific Viewpoint, 58*(1), 57–73. https://doi.org/10.1111/apv.12142
Xu, J., Yeh, A., & Wu, F. (2009). Land commodification: New land development and politics in China since the late 1990s. *International Journal of Urban and Regional Research, 33*(4), 890–913. https://doi.org/10.1111/j.1468-2427.2009.00892.x
Yeh, A. G., Yang, F. F., & Wang, J. (2015). Producer service linkages and city connectivity in the mega-city region of China: A case study of the Pearl River Delta. *Urban Studies, 52*(13), 2458–2482. https://doi.org/10.1177/0042098014544762

Appendix

Interview questions

Shenzhen-Shanwei Special Cooperation Zone (SSSCZ)

- **Interview with Urban planners**

1. What kinds of concessionary policy did the Guangdong Provincial Government offer to the construction of SSSCZ? How did these concessionary policies impact SSSCZ?
2. During the construction process of SSSCZ, what kinds of resource did SSSCZ get from the Guangdong Provincial Government?
3. What were the primary challenges for SSSCZ after its site selection process?
4. In this stage, how did the land quotas of SSSCZ constitute and how did the different levels of government negotiate to determine the land quotas?
5. In the SSSCZ land development and redevelopment processes, how did different levels of government, specifically referred to municipal government, county government, and town government, negotiate and address the conflicts?

- **Interview with local officials**

6. What was the evolution of policy in PRD in terms of industrial transfer after 2000, and what was the actual effectiveness of these polices?
7. What was the primary challenge for the land and industrial development of Shenzhen and how did the Shenzhen Municipal Government cope with these challenges?
8. How did Shenzhen negotiate with its surrounding cities, such as Dongguan, and Huizhou, for the industrial collaboration and what were the actual effectiveness?
9. What was the land compensation policy the local government made with regard to land acquisition issue and what was the basis for the compensation policy?

X. Zhang, *Regional Integration and Governance Reshuffling in the Making of China's City-Regions*, Urban Sustainability, https://doi.org/10.1007/978-981-99-2792-0

10. What were the attitudes of related enterprises for the land redevelopment of SSSCZ? How did the local government negotiate with them with regard to land-based issues and concessionary policies?
11. What kinds of concessionary policy SSSCZ got from the Shenzhen Municipal Government in terms of industrial development and what kinds of concessionary policy SSSCZ got from the Guangdong Provincial Government in terms of provision of quota of construction land?
12. How many enterprises in SSSCZ in this stage and what types of them?
13. During SSSCZ planning and construction processes, how did the Shenzhen Municipal Government and the Shanwei Municipal Government negotiate with regard to the division in the future? How did this division reflect in the administrative system of SSSCZ?
14. In terms of industry selection, what were the purposes for the Shenzhen Municipal Government and the Shanwei Municipal Government respectively? How did they negotiate with each other with regard to industry orientation and development?
15. What were the concessionary policies within industry introduction policies driven by the Shenzhen Municipal Government? How did the Shenzhen Municipal Government negotiate with industrial guild with regard to industry introduction?
16. Did the Shenzhen Municipal Government have some formal or informal institutional arrangements to achieve industrial introduction and cooperation?

- **Interview with Enterprises**

17. To encourage related enterprises to relocate to SSSCZ, what kinds of concessionary policy the Shenzhen Municipal Government were offered? Were these concessionary polices in line with the interests of enterprise?
18. Did the relocating enterprise have the branches or industrial basis in Shanwei previously?
19. For the related enterprises that relocated to SSSCZ, what were the attractions of SSSCZ?
20. What were the primary reasons for the enterprises that were encouraged by local government but did not relocate to SSSCZ?
21. How about the enterprise revenues after they relocated to SSSCZ? What were the primary reasons?
22. After relocated to SSSCZ, did they establish some linkages or the upstream

and downstream relations of the production chain with local enterprises? Did the local government treat the enterprises, which relocates to SSSCZ, and local enterprises differently? What aspects did it performance and what are the effects?

Sui-Guan-Shen Intercity Railway (SGSIR)

- **Interview with Local Officials**

23. During the SGSIR construction process, how did project funds collect? How much did the national ministry, provincial government and local government pay respectively?
24. During the SGSIR project fund collection process, how did different levels of government negotiate? What were their attitudes for the project funds collection and how did they reach a consensus?
25. Can you introduce the whole SGSIR planning making process? Who initiated this planning and which institution approved it? How did they reach a consensus on this planning?
26. How did the different actors arrange their responsibility in the daily operation of PRD inter-city railway network, especially for SGSIR? How did local governments respond to the operation model and why?
27. Within site selection process of SGSIR, how did the Guangdong Provincial Government, the Shenzhen Municipal Government and the Dongguan Municipal Government negotiate with regard to the specific site selection?
28. Within site selection process of SGSIR, did the site selection experience a significant revision and what caused it?
29. What were the main responsibilities of the *Investment Group*? How did the *Investment Group* negotiate with local government with regard to its responsibilities and site selection of the SGSIR stations?
30. What was the role of the Guangdong Provincial Government in the operation model of SGSIR and how did the Guangdong Provincial Government fulfilled its role and what were the outcomes?
31. What were the profits and costs of PRD inter-city railway network? Did these lines have operation loss? If they had, how did the government cope with the issue of operation loss?
32. How did local government plan the land use around the SGSIR stations? Did they have some special plans with regard to land use? If any, which institution made them? If not, when the land use contradicted with urban master planning, how to negotiate them?
33. How did the *Investment Group* negotiate with local government with regard to land development around the SGSIR stations?
34. What kinds of solutions did the provincial government adopt to help the *Investment Group* in negotiating with local government? What were results of these solutions?

- **Interview with Urban Planners**

35. How did the Guangdong Provincial Government arrange the investment and responsibility sharing of SGSIR and why?
36. What was the primary purpose of the local government in site selection process of SGSIR?
37. What was the primary purpose of local government in the SGSIR relocation of route process and what was the outcome of their negotiation?
38. What aspects were taken into consideration when local government arranged the land use around the SGSIR stations? How did the local government select the related corporation?
39. From your point of view, how did the different levels of governments negotiate the site selection and what were the results?
40. For the investment projects around the SGSIR stations, how did local government negotiate with upper-level government? What types of agreement were signed?

Uncited References

Cai, L., Ma, X., Chen, W., Luan, X., & Li, G. (2013). Characteristics of functional polycentricity of PRD urban region based on passenger traffic flow. *Economic Geography, 33*(11),52–57 (*in Chinese*). DOI:https://doi.org/10.15957/j.cnki.jjdl.2013.11.009

Findlay, A., Mason, C., Harrison, R., Houston, D., & McCollum, D. (2008). Getting off the escalator? A study of scots out-migration from a global city region. *Environment and Planning A, 40*(9), 2169. https://doi.org/10.1068/a39256

Harrison, J., Hoyler, M., & Kanai, J. M. (2014). Capital of the amazon rainforest: Constructing a global city-region for entrepreneurial Manaus. *Urban Studies, 51*(11), 2387–2405. https://doi.org/10.1177/0042098013493478

Heeg, S., Klagge, B., & Ossenbruügge, J. (2003). Metropolitan cooperation in Europe: Theoretical affairs and perspectives for urban networking 1. *European Planning Studies, 11*(2), 139–153. https://doi.org/10.1080/0965431032000072846

Heyuan City Bureau of Statistics. (2015, December 4). *Heyuan economy and society developed statistical bulletin 2014.* (Heyuanshi 2014 guominjingji yu shehuifazhan tongji gongbao).

Leck, H., & Simon, D. (2013). Fostering multiscalar collaboration and co-operation for effective governance of climate change adaptation. *Urban Studies, 50*(6), 1221–1238. https://doi.org/10.1177/0042098012461675

Logan, J. R. (2018). People and plans in urbanising China: Challenging the top-down orthodoxy. *Urban Studies, 55*(7), 1375–1382. https://doi.org/10.1177/0042098018763552

Lu, C. (2004). Regional integration between Zhujiang River Delta and its surrounding region. *Scientia Geographica Sinica, 24*(5), 522–527. (in Chinese).

Pike, A., Pose, A. R., & Tomaney, J. (2006). *Local and regional development.* Taylor & Francis.

Scott, A. J. (2002). *Global city regions: Trends, theory, policy.* Oxford University Press.

Shanwei City Bureau of Statistics. (2015, March 4). *Shanwei economy and society developed statistical bulletin 2014.* (Shanweishi 2014 guominjingji yu shehuifazhan tongji gongbao).

Shenzhen City Bureau of Statistics. (2009, March 24). *Shenzhen economy and society developed statistical bulletin 2008.* (Shenzhenshi 2008 guominjingji yu shehuifazhan tongji gongbao).

Shenzhen City Bureau of Statistics. (2015, April 22). *Shenzhen economy and society developed statistical bulletin 2014.* (Shenzhenshi 2014 guominjingji yu shehuifazhan tongji gongbao).

Vogel, R. (Ed.). (2009). Governing global city regions in China and the West. *Progress in Planning, 73*, 1–75. https://doi.org/10.1016/j.progress.2009.12.001

Yu, T., & Li, N. (2005). Study on the regional integration in the Yangtze River Delta area. *Planners, 21*(4), 17–24. (in Chinese).

Zhang, J. (2008). On regional cooperation from view of new regionalism in the Pan-Pearl River Delta. *Wuhan University Journal (philosophy & Social Sciences), 61*(3), 351–357. (in Chinese).

X. Zhang, *Regional Integration and Governance Reshuffling in the Making of China's City-Regions*, Urban Sustainability, https://doi.org/10.1007/978-981-99-2792-0

Zhang, J., & Tang, Y. (2007). Research on the mechanism of environmental coordination among governments to harness the pollution of valleys-taking harnessing the Xiao Dong river as an example. *Journal of Public Management, 4*(3), 50–56. (in Chinese).
Zhang, J., Zou, J., Wu, Q., & Chen, X. (2001). Research on territorial spatial form of megacity region. *City Planning Review, 25*(5), 19–23. (in Chinese).

The manufacturer's authorised representative in the EU is Springer Nature Customer Service Centre GmbH, Europaplatz 3, 69115 Heidelberg, Germany. If you have any concerns regarding our products, please contact ProductSafety@springernature.com

Printed and bound by CPI Group (UK) Ltd, Croydon, CR0 4YY
30/09/2026
02225318-0009